CAMBRIDGE LIBRARY COLLECTION

Books of enduring scholarly value

Mathematical Sciences

From its pre-historic roots in simple counting to the algorithms powering modern desktop computers, from the genius of Archimedes to the genius of Einstein, advances in mathematical understanding and numerical techniques have been directly responsible for creating the modern world as we know it. This series will provide a library of the most influential publications and writers on mathematics in its broadest sense. As such, it will show not only the deep roots from which modern science and technology have grown, but also the astonishing breadth of application of mathematical techniques in the humanities and social sciences, and in everyday life.

Principles of Geometry

Henry Frederick Baker (1866–1956) was a renowned British mathematician specialising in algebraic geometry. He was elected a Fellow of the Royal Society in 1898 and appointed the Lowndean Professor of Astronomy and Geometry in the University of Cambridge in 1914. First published between 1922 and 1925, the six-volume *Principles of Geometry* was a synthesis of Baker's lecture series on geometry and was the first British work on geometry to use axiomatic methods without the use of co-ordinates. The first four volumes describe the projective geometry of space of between two and five dimensions, with the last two volumes reflecting Baker's later research interests in the birational theory of surfaces. The work as a whole provides a detailed insight into the geometry which was developing at the time of publication. This, the sixth and final volume, describes the birational geometric theory of surfaces.

Cambridge University Press has long been a pioneer in the reissuing of out-of-print titles from its own backlist, producing digital reprints of books that are still sought after by scholars and students but could not be reprinted economically using traditional technology. The Cambridge Library Collection extends this activity to a wider range of books which are still of importance to researchers and professionals, either for the source material they contain, or as landmarks in the history of their academic discipline.

Drawing from the world-renowned collections in the Cambridge University Library, and guided by the advice of experts in each subject area, Cambridge University Press is using state-of-the-art scanning machines in its own Printing House to capture the content of each book selected for inclusion. The files are processed to give a consistently clear, crisp image, and the books finished to the high quality standard for which the Press is recognised around the world. The latest print-on-demand technology ensures that the books will remain available indefinitely, and that orders for single or multiple copies can quickly be supplied.

The Cambridge Library Collection will bring back to life books of enduring scholarly value (including out-of-copyright works originally issued by other publishers) across a wide range of disciplines in the humanities and social sciences and in science and technology.

Principles
of Geometry

VOLUME 6:
INTRODUCTION TO THE THEORY OF
ALGEBRAIC SURFACES AND HIGHER LOCI

H.F. BAKER

CAMBRIDGE
UNIVERSITY PRESS

CAMBRIDGE UNIVERSITY PRESS

Cambridge, New York, Melbourne, Madrid, Cape Town, Singapore,
São Paolo, Delhi, Dubai, Tokyo, Mexico City

Published in the United States of America by Cambridge University Press, New York

www.cambridge.org
Information on this title: www.cambridge.org/9781108017824

© in this compilation Cambridge University Press 2010

This edition first published 1933
This digitally printed version 2010

ISBN 978-1-108-01782-4 Paperback

PRINCIPLES OF GEOMETRY

LONDON
Cambridge University Press
FETTER LANE

NEW YORK · TORONTO
BOMBAY · CALCUTTA · MADRAS
Macmillan

TOKYO
Maruzen Company Ltd

PRINCIPLES OF GEOMETRY

BY

H. F. BAKER, Sc.D., LL.D., F.R.S.

LOWNDEAN PROFESSOR, AND FELLOW OF ST JOHN'S COLLEGE,
IN THE UNIVERSITY

VOLUME VI

INTRODUCTION TO THE THEORY OF

ALGEBRAIC SURFACES

AND HIGHER LOCI

CAMBRIDGE

AT THE UNIVERSITY PRESS

1933

PREFACE

THE origin and final purpose of this volume, and the preceding, have been stated in the preface to the latter. It may be useful to describe in outline the contents of the present volume. The first chapter deals with the theory of correspondence, mainly of points on one or two curves, with inclusion of the treatment by transcendental methods, and the connection with the theory of defective integrals. The second chapter attempts an exposition of Schubert's remarkable ideas, which are as interesting logically as geometrically, and of the extension of the theory of correspondence to aggregates of any dimension. The third chapter is in part a reminder of theorems which belong to plane geometry, and in part a sketch of general theorems for rational surfaces. In the fourth chapter the elementary preliminary properties of surfaces in ordinary space, and in space of four dimensions, are dealt with. Chapter V is that which is concerned with the most interesting and the most novel ideas of the volume. For this reason it is written in a tentative introductory manner, and will best have served its purpose if it leaves the reader convinced of the importance of the theory involved, and with a desire to follow it further. The next chapter develops in detail the theory of the intersections of manifolds in space of four dimensions. The last chapter collects together various particular theorems, and some easy applications of foregoing theory. Only want of space has led to the exclusion from the volume of many other results which are of interest.

I should like to give expression to my sense of how much this, and preceding volumes, owe to those who have been students with me during their composition; my experience has been of a remarkable and unremitting keenness in the prosecution of the matters treated. Without this encouragement and co-operation, I might not have persevered in the formulation of the ideas, especially in these last two volumes. But besides this personal reference, I would add that it is clear that the purely geometrical and descriptive aspects of the subject are felt by many of our students to offer a discipline which is a welcome complement to others which are open to them.

And, I would repeat, readers of these volumes are under much obligation to the staff of the University Press for the trouble and attention with which the printing has been executed.

H. F. B.

5 *October* 1933

TABLE OF CONTENTS

CHAPTER I. ALGEBRAIC CORRESPONDENCE

PAGES

Part I. Elementary methods . . . . . 1—46

Function expressing a correspondence with valency . . 5
Set of coincidences in the correspondence; number of co-
incidences 8, 9
Coincidences of $r+1$ points in a linear system of freedom r . 10
Valency of direct lateral correspondence 11
Bitangents of a curve from theory of correspondence . . 13, 14
Ruled surface by joins of corresponding points of two curves 15
Transformation of canonical series; Zeuthen's formula . 19
Analytic treatment of correspondence between two curves . 20—24
Formula for a curve on a ruled surface 25
Torsal lines, genus of double curve and number of triple
points, on a ruled surface . . . . . . 26—28
Common chords of two curves, trisecants and quadrisecants
of a single curve . . . . 28—34
Sets common to an involution and a linear series on a curve 35—37
Condition that sets of an involution belong to a linear series 37, 38
de Jonquières' formulae for contacts of curves . . . 39—43
Linear spaces incident with a curve in higher space . . 44—46

Part II. Transcendental methods . . . . . 46—59

Relations for valency matrix in general . . . . 46—53
Valency matrix is numerical when the curve has general
moduli 53
Coincidences in a general correspondence . . . . 54—56
Representation of correspondence by a curve, on the product
of two curves . . . . 57—59

Part III. Correspondence and defective integrals . 59—68

Least number of period columns for a defective system of
integrals 62
Theorem of complementary systems of defective integrals . 62—65
Application to theory of correspondence . . . . 65—68

CHAPTER II. SCHUBERT'S CALCULUS. MULTIPLE CORRESPONDENCE

Part I. Schubert's methods. Preliminary, as to
notations. Characters of a manifold . . . 69—86

Fundamental conditions for linear spaces. The characters of
a manifold . . . . 69—77
The calculus of conditions applied to a line. Examples of
more general results . . . . 77—86

Contents

PAGES

Part II. The problem of multiple tangents of a
manifold . . . . 86—92

Part III. Correspondence of points of two manifolds 92—105
Schubert's notation; application to two and three dimensions 96—98
General formula, and examples . . . . . . 99—105

Part IV. Pairs of corresponding linear spaces . . 105—108
Appendix. Some enumerative formulae . . . . 108—111

CHAPTER III. TRANSFORMATIONS AND INVOLUTIONS
FOR THE MOST PART IN A PLANE

Cremona transformations in a plane . . . . . 112—120
Bertini's four types of involution in a plane . . . 121—130
The four involutions and rational double planes . . . 130, 131
Involutions of sets of more than two points . . . 131—137
Surface representing a plane involution cannot contain an
irrational pencil 138, 139
The rationality of the representative surface of a plane
involution . . . . 139—141
Number of cyclical sets in a correspondence . . . 142—145
Note. Surfaces with a pencil of rational curves . . . 145—147

CHAPTER IV. PRELIMINARY PROPERTIES OF SURFACES
IN THREE AND FOUR DIMENSIONS

Elementary properties for general surface. Jacobian of a net
of curves thereon . . . . 148—156
The surface with a double curve. Salmon's formulae . . 157—169
Surfaces in space of four dimensions . . . . . 169—175
Note I. In regard to pinch points . . . . . 176—180
Note II. Some formulae given by Noether 180, 181

CHAPTER V. INTRODUCTION TO THE THEORY OF THE
INVARIANTS OF BIRATIONAL TRANSFORMATION
OF A SURFACE, PARTICULARLY IN SPACE
OF THREE DIMENSIONS

Preliminary; a general survey . . . . . . 182—185
Definition and illustration of the invariant I . . . 185—189
Definition of I in terms of an irrational pencil of curves . 189—191
Introductory definition of the invariant ω . . . . 191—198
Effect of isolated nodal points of the surface upon the
definition of the invariants . . . . . . 198—200

Contents

PAGES

Applications to easy cases . . . . 201—206
Sketch of the proof of the invariance of I . . . . 206—214
The invariance of ω in a birational transformation . . 214, 215
The canonical system of curves of a surface . . . 215—217
The modification of the canonical system in a birational transformation . . . . 217—221
Return to the definition of the invariant ω . . . 221—224
The class of immersion and the canonical number of a curve on a surface 225
The computed characters of an exceptional curve . . 225, 226
The modification of the invariant ω in a birational transformation 226
Note I. Examination of fifteen examples cited by Noether 226—232
Note II. The adjoint surfaces of a given surface; the exceptional curves . . . . 232—237
Note III. The birational transformation of the Kummer and Weddle surfaces . . . . 237—240
Note IV. Miscellaneous examples 240, 241

CHAPTER VI. SURFACES AND PRIMALS IN FOUR DIMENSIONS. FORMULAE FOR INTERSECTIONS

The chord curve and trisecant curve for a surface in four dimensions. Rational surfaces . . . . . 242—247
Intersections of loci in space of four dimensions, introductory 247, 248
Residual intersection of three primals with a common curve 248—251
Residual intersection of two surfaces with a common curve 251—255
Residual intersection of a primal and a surface having a common curve . . . . 255—257
Residual intersection of two primals having a surface in common . . . . 257—263
The postulation of a surface for primals containing it . . 263—266
Residual intersection of three primals having a surface in common . . . . 266—268
Various examples . . . . 268—270

CHAPTER VII. ILLUSTRATIVE EXAMPLES AND PARTICULAR THEOREMS

Particular examples of intersections . . . . . 271—282
The surface representing pairs of points of one or two curves 282—294
Note on the multiple correspondence of two surfaces . . 294, 295
On complete sections of a non-singular primal . . . 295—298
Miscellaneous theorems and examples . . . . 298—300

INDEX . . . . 301—308

CHAPTER I

ALGEBRAIC CORRESPONDENCE

Part I. Elementary methods. There is said to be an (s, r) correspondence between points P, Q of a single given curve when, if P is assigned, anywhere on the curve, there is determined a set, $Q_1, Q_2, \ldots, Q_r$, of r points, these being the r positions of Q corresponding to P, and, when Q is assigned, there is similarly determined a set $P_1, \ldots, P_s$, of s points, these being the s positions of P corresponding to Q. In particular, the set of points P so determined where Q is at any one of $Q_1, \ldots, Q_r$, say Q_i, is a set $P_1, \ldots, P_s$ of which P is one; and, when Q is at another of $Q_1, \ldots, Q_r$, say Q_j, the set so determined may be a different set, $P_1', \ldots, P_s'$, but, among these, P is still one. The operation by which we pass from P to $Q_1, \ldots, Q_r$ is called the *forward*, or *direct*, operation; it may be denoted by a symbol, say T, so that we may write $Q = TP$, but it must be borne in mind that TP has r significations. The operation by which we pass from Q to $P_1, \ldots, P_s$ is called the *reverse* operation, and may be denoted by T_1 (and, later, by T^{-1}), and we may write $P = T_1 Q$. It is supposed, in what follows, unless the contrary be stated, that the set $Q_1, \ldots, Q_r$ is determined from P by algebraical processes, so that any rational symmetrical function of the coordinates of $Q_1, \ldots, Q_r$, say $(z_1, t_1), \ldots, (z_r, t_r)$, is a rational function (in virtue of the equation of the curve) of the coordinates (x, y) of P. In particular there are two equations, satisfied respectively by $z_1, \ldots, z_r$ and by $t_1, \ldots, t_r$, say

$$u_0 z^r + u_1 z^{r-1} + \ldots + u_r = 0, \quad v_0 t^r + v_1 t^{r-1} + \ldots + v_r = 0,$$

in which $u_0, \ldots, u_r, v_0, \ldots, v_r$ are rational polynomials in x and y. These equations must be consistent with one another in virtue of the equation of the curve, $f(z, t) = 0$; and, if the coordinates be chosen with generality, it should be possible, by means of $f(z, t) = 0$, to obtain from the second equation a rational expression for t in terms of x, y, z. The relations are then expressed by two equations, say $u (x, y, z) = 0$, $t = R (x, y, z)$, where $u = 0$ denotes the first equation, and R is rational in x, y, z, the equation $f(z, t) = 0$, or

$$f[z, R (x, y, z)] = 0,$$

being satisfied in virtue of $f(x, y) = 0$ and $u = 0$. Thus the correspondence is given by $u = 0$, $t = R$, $f (x, y) = 0$, which involve $f(z, t) = 0$. And from these similar equations are determinable,

$$w (z, t, x) = 0, \quad y = S (z, t, x), \quad f (z, t) = 0.$$

A particularly simple case is that for a rational curve. Then any two points (x), (z), of the curve, can be characterised by parameters, say θ and ϕ. The expression of the correspondence is then by a single equation, which we suppose irreducible, say $R(\theta, \phi) = 0$, where R is a rational polynomial both in θ and ϕ, of respective orders s and r in these. A special example is given, for the correspondence of the points $(\theta^2, \theta, 1)$, $(\phi^2, \phi, 1)$ of the conic $x - y^2 = 0$, by the equation

$$\theta^2 (a\phi^2 + h\phi + g) + \theta (h\phi^2 + b\phi + f) + g\phi^2 + f\phi + c = 0;$$

this expresses that the two points (z_1, t_1), (z_2, t_2) of the conic, which correspond to the point (x, y) thereon, are the intersections of the conic with the polar of (x, y) taken in regard to another fixed conic.

More generally, if (x, y), (z, t) be points of a curve $f(x, y) = 0$, or $f(z, t) = 0$, connected by a relation $R(x, y, z, t) = 0$, where R is a rational (non-homogeneous) polynomial in x, y, z, t, then, to a point (x, y) of the curve, there correspond the intersections with $f(z, t) = 0$ of the curve $R(x, y, z, t) = 0$, in which (z, t) are current coordinates, and to a point (z, t) the intersections with $f(x, y) = 0$ of the curve $R(x, y, z, t) = 0$, in which (x, y) are current coordinates. Such a relation $R(x, y, z, t) = 0$ is necessarily of the form

$$u_0(x, y)v_0(z, t) + \ldots + u_k(x, y)v_k(z, t) = 0,$$

where u_i, v_j are rational polynomials; this equation expresses that the points (z, t) which correspond to a given (x, y), are points of a set belonging to the linear series, on the curve $f(z, t) = 0$, obtained by its intersections with the curves $\lambda_0 v_0(z, t) + \ldots + \lambda_k v_k(z, t) = 0$, namely, that set for which $\lambda_0, \ldots, \lambda_k$ are in the ratios of

$$u_0(x, y), \ldots, u_k(x, y).$$

Similarly, the points (x, y) which correspond to a given (z, t) are points of a set of a certain linear series. We have given a special example, of a $(2, 2)$ correspondence on a conic. Another very obvious example is that between the point of contact (x, y), of a tangent at any point of a curve $f(x, y) = 0$, and the points (z, t), other than the point of contact, in which the tangent meets the curve. The correspondence is then expressed by an equation of the form $f_1(x, y).z + f_2(x, y).t + f_3(x, y) = 0$, in which f_1, f_2, f_3 are the partial derivatives of f. In this case, if $f = 0$ be of order n, the $(n-2)$ points (z, t) which correspond to the point (x, y), when taken with the point (x, y) itself, counted twice, are all the points of a particular set of the linear series determined by $\lambda z + \mu t + \nu = 0$, where λ, μ, ν are parameters. And so in general, for the curve in (z, t) expressed (with given x, y) by the equation,

$$u_0(x, y)v_0(z, t) + \ldots + u_k(x, y)v_k(z, t) = 0,$$

it may happen that, beside fixed zeros common to all of

$$v_0(z, t) = 0, \ldots, v_k(z, t) = 0,$$

(which we ignore), there is a certain number of intersections, with the fundamental curve $f(z, t) = 0$, which coincide with (x, y). In such case, if this number be denoted by γ, and the set of points (z, t), or Q, which correspond to (x, y), or P, be denoted by (Q), the composite set $\gamma P + (Q)$ forms a set of a certain linear series; and then, if (Q') be the points Q which correspond to another point (P'), the two sets $\gamma P + (Q)$ and $\gamma P' + (Q')$ belong to the same linear series, or, in the phraseology of Vol. v, Chap. IV, are *equivalent*, which we write as $\gamma P + (Q) \equiv \gamma P' + (Q')$. It is not to be assumed, however, though the set (Q) is determined from γ coincident points P, when the correspondence is established, that the complete linear series to which the composite set $\gamma P + (Q)$ belongs, is of freedom γ. We have indeed had above a simple example on a conic, in which the sets belong to a series of freedom 2, though γ is zero. And it will appear below that there are correspondences for which γ is negative, the equivalence above written being then understood to mean $-\gamma P' + (Q) \equiv -\gamma P + (Q')$. The number γ, in correspondences of the kind now considered, is called the *valency*.

A correspondence which is entirely defined by a single equation of the form $u_0(x, y)v_0(z, t) + \ldots + u_k(x, y)v_k(z, t) = 0$, has, clearly, the same valency as its reverse. For, if the left side, regarded as a function of (z, t), vanishes to order γ when (z, t) approaches to (x, y), regarded as a fixed point, then it is equally true that, regarded as a function of (x, y), it vanishes to order γ when the point (x, y) approaches to the point (z, t), regarded as fixed.

Conversely, we may define a correspondence, of indices (s, r), with (positive or zero) valency γ, between the places (x), (z) of the curve, by the properties: (1), that, when (x) is given, there is determined, algebraically, a set of r places $(z_1), \ldots, (z_r)$, any rational symmetrical function of these latter being expressible rationally by the coordinates x, y, of the point (x); and (2), that, when (z), one of $(z_1), \ldots, (z_r)$, is given, there is a similar determination of the set, $(x_1), \ldots, (x_s)$, of places, which correspond to (z) in the reverse correspondence; with (3), the further property, that, in the direct correspondence, the composite set consisting of $(z_1), \ldots, (z_r)$, and the place (x) taken γ times, are the points of a set of a linear series, which is the same for all positions of (x). For, with these hypotheses, we can construct a function, rational in the coordinates of both the places (x) and (z), which, as a function of (z), vanishes at $(z_1), \ldots, (z_r)$ as well as γ times at (x), and, as a function of (x), vanishes at $(x_1), \ldots, (x_s)$ as well as γ times at (z). This function, by its vanishing, expresses then both the direct and the reverse correspondence.

To prove this, we argue as follows: the meaning of the hypothesis (3)—if $(c_1), \ldots, (c_r)$ be the positions of $(z_1), \ldots, (z_r)$ corresponding to a position (a) of (x)—is, that there exists a rational function of (z), on the curve, having, for its zeros, the place (x), taken γ times, and the places $(z_1), \ldots, (z_r)$, which correspond to (x), taken each once, and having, for its poles, the place (a), taken γ times, and the places $(c_1), \ldots, (c_r)$, which correspond to (a), each taken once. Such a function is definite, save for a constant multiplier; dividing the function by its value at an arbitrary place (c) of the curve, we may then agree that it reduces to unity at (c). So determined, let the function be denoted by

$$R_c\left(z; \frac{x^\gamma, z_1, \ldots, z_r}{a^\gamma, c_1, \ldots, c_r}\right).$$

The actual expression of such a function is to be found, by, first, forming the most general rational function of (z) which has $(a)^\gamma, (c_1), \ldots, (c_r)$ as poles, and, then, limiting the arbitrary constants, which enter linearly in such a function, by the condition that $(x)^\gamma, (z_1), \ldots, (z_r)$ are its zeros. This limitation is by linear equations; some, relating to the zeros $(x)^\gamma$, being explicitly rational in (x), the others, relating to the zeros $(z_1), \ldots, (z_r)$, being symmetrical in these, and, therefore, by the initial definition, also ultimately rational in (x). The function of (z) thus formed, depending on (x), is thus rational in (x).

Let us now consider this as a function of (x), depending on (z). As such, it vanishes to order γ when (x) approaches to (z). To the place (z) there correspond, by hypothesis, in the reverse correspondence, places $(x_1), \ldots, (x_s)$. When (z) is at any one of the places $(z_1), \ldots, (z_r)$, which correspond to (x) in the direct correspondence, then (x) will be at one of the places $(x_1), \ldots, (x_s)$, which correspond to this position of (z) in the reverse correspondence. Thus, conversely, when (x) is at one of $(x_1), \ldots, (x_s)$, then (z) will be at one of $(z_1), \ldots, (z_r)$; and the function of (x), now under consideration, will vanish to the first order. The function will, we assume, have no other zeros than those mentioned. A like statement is true for the poles: the function of (z), considered above, has a pole of order γ when (z) is at (a), and has also a pole, of the first order, when (z) is at any one of the places $(c_1), \ldots, (c_r)$, which correspond to (a) in the direct correspondence; and this function is of the form

$$R\left(z; \frac{x^\gamma, z_1, \ldots, z_r}{a^\gamma, c_1, \ldots, c_r}\right) \div R\left(c; \frac{x^\gamma, z_1, \ldots, z_r}{a^\gamma, c_1, \ldots, c_r}\right),$$

and is infinite to the first order when (c) is at any one of the places $(z_1), \ldots, (z_r)$, which correspond to (x) in the direct transformation; this arises when, and only when, (x) is at any one of the places

Correspondence; elementary methods 5

$(a_1), ..., (a_s)$, which correspond to (c) in the reverse transformation; the quotient is also infinite, to order γ, when (x) is at (c); and is not otherwise infinite; and it reduces to unity when (x) is at (a), and, therefore, $(z_1), ..., (z_r)$ are, respectively, at $(c_1), ..., (c_r)$; also, the *infinities* of the numerator and denominator, of the quotient just written do not depend on the position of (x). Thus the function, regarded as depending on (x), may equally be represented by

$$R_a\left(x; \frac{z^\gamma, x_1, ..., x_s}{c^\gamma, a_1, ..., a_s}\right).$$

Thus we have the result: *For an existing (s, r) correspondence, between the places (x), (z) of the given curve, assumed to be of valency γ (zero or positive) in the direct correspondence from (x) to $(z_1), ..., (z_r)$, and such that any rational symmetric function of $(z_1), ..., (z_r)$ is rational in (x), while to (z) there correspond $(x_1), ..., (x_s)$ in the reverse correspondence, if (a), (c) be two arbitrary places, we can construct a function $\phi(x, z; a, c)$, rational in regard to both (x) and (z), and in regard to both (a) and (c); this function, regarded as a function of (z), is zero to order γ at the place (x), and, to the first order, at each of the places $(z_1), ..., (z_r)$ which correspond to (x) in the direct correspondence; it is also infinite to order γ at the place (a), and, to the first order, at each of the places $(c_1), ..., (c_r)$, which correspond to (a) in the direct correspondence. This function of (z) expresses, by its zeros, the direct correspondence. But, regarded as a function of (x), it expresses the reverse correspondence also, being zero to order γ at the place (z), and, to the first order, at the places $(x_1), ..., (x_s)$, which correspond to (z) in the reverse correspondence, beside having a pole of order γ at the place (c), and poles, of the first order, at the places $(a_1), ..., (a_s)$, which correspond to (c) in the reverse correspondence. Thus the reverse correspondence is equally of valency γ.*

A simple illustration is that remarked above, where the direct correspondence leads from a place (x) to the $(n-2)$ places $(z_1), ..., (z_{n-2})$, in which the curve is met by the tangent of the curve at the place (x). The reverse correspondence then leads from a place (z), of the curve, to the $(n'-2)$ places $(x_1), ..., (x_{n'-2})$, which are the points of contact of the tangents to the curve drawn from (z). The equation of the tangent at (x), in which (z) is used as current coordinate, is equally the equation, with (x) as current coordinate, of the first polar of a point (z). The fact that the valency of the reverse correspondence, like that of the direct correspondence, is 2, is the fact that the first polar, of a point (z) of the curve, touches the curve at this point.

There are two general definitions in regard to correspondence which it is convenient to employ: (a), If we have two correspondences, in which there correspond to P, in the direct correspondences, respectively the sets (Q) and (R), then we shall regard the composite set, consisting of the sets (Q), (R) together, say the set $(Q)+(R)$, as being in correspondence with P, directly, in a correspondence which we call the *sum* of the two given correspondences. It can then be seen that the reverse of this sum correspondence is likewise the sum of the reverses of the two original correspondences. For the correspondence reverse to T, which we denote by T_1, is defined by the fact that T_1, applied to any one of the points TP, gives rise to a set of which P is one point. Now consider $(T+U)P$, which is defined as the aggregate of the sets $(Q)=TP$ and $(R)=UP$. The operation $(T+U)_1$, reverse to $(T+U)$, is that which, applied to any one of the set TP, or to any one of the set UP, gives a set including P. This is satisfied only if $(T+U)_1=T_1+U_1$. In symbols,

$$(T_1+U_1)(T+U)P=(T_1+U_1)(TP+UP)$$
$$=T_1TP+T_1UP+U_1TP+U_1UP,$$

and P is among the latter aggregate. Beside this definition of the sum of two correspondences, is, (b), that of the *product* of two correspondences: If to P, in a direct correspondence T, corresponds the set (Q), of which Q_i denotes every element in turn; and to Q_i, in another direct correspondence U (independent of i), corresponds a set $(R)_i$, the set (Q) consisting of q points, and each set $(R)_i$ consisting of r points, then we regard the aggregate of rq points forming all the sets $(R)_i$, as corresponding to P in a direct correspondence, which we call the product of T by U, and denote by UT, the operation T being that first applied. It is easy to see that the reverse correspondence $(UT)_1$, is T_1U_1.

For correspondences with valency, as considered above, the valency of $T+U$ is the sum of the valencies of T and U. For, if, in a notation employed already,

$$(Q)+\gamma P\equiv(Q')+\gamma P', \quad (R)+\delta P\equiv(R')+\delta P',$$

then also

$$(Q)+(R)+(\gamma+\delta)\,P\equiv(Q')+(R')+(\gamma+\delta)P'.$$

For the product of two correspondences, however, the valency of UT is the negative product of the valencies of T and U. For, from $\Sigma Q_i+\gamma P\equiv\Sigma Q_i'+\gamma P'$, $(R)_i+\delta Q_i\equiv(R')_i+\delta Q_i'$, of which the latter gives $\Sigma(R)_i+\delta\Sigma Q_i\equiv\Sigma(R')_i+\delta\Sigma Q_i'$, we have

$$\Sigma(R)_i-\delta\gamma P\equiv\Sigma(R')_i-\delta\gamma P'.$$

Thus, also, the (generally different) correspondences UT and TU have the same valency, which is also the valency of U_1T, TU_1, UT_1, T_1T_1, U_1T_1, T_1U_1.

A simple correspondence is obtained by considering, on the fundamental curve, any linear series of sets of points which has freedom unity. Any point P of the curve then determines the other points (Q), of the set to which P belongs; and there is thus a correspondence of valency unity, in which the points corresponding to P are these other points (Q), of the set of the linear series determined by P. In this case, since a set of the series is equally determined by any one of its points, the reverse correspondence is the same as the original. Further, by what has been said, the product of any two such simple correspondences, determined by any two linear series of freedom unity, is a correspondence whose valency is negative unity. By the sum and product of such simple correspondences we can thus set up a correspondence of any arbitrary valency, positive, zero or negative.

The chief utility of the theory of correspondence, in geometry, arises by the consideration of the *united* points, or *coincidences*; these are the positions of P in which one of the points Q, corresponding thereto, coincides with P.

Consider first the easy case of a rational curve: let a correspondence, between two points represented by parameters θ, ϕ, be given by an equation $(\theta, \phi) = 0$, of order s in θ and order r in ϕ. The coincidences arise then corresponding to the roots of the equation $(\theta, \theta) = 0$, and are $s + r$ in number, provided the original relation contains the term $\theta^s \phi^r$, that is, provided the parameters are so chosen that $\theta = \infty$ is not a coincidence. Further, the set of points forming the coincidences is equivalent with the aggregate of the two sets (Q), (Q)$_1$, which correspond to a point P in the direct and reverse correspondences; indeed, on a rational curve, any two sets of the same number of points are equivalent with one another. The same equivalence, of the set of coincidences with the aggregate of the sets corresponding to any point in the direct and reverse correspondences, holds on any curve, for the coincidences in a correspondence which is of valency zero.

The proof of this remark, as to the coincidences for a correspondence, on any curve, which is of valency zero, arises from the expression of the correspondence by the vanishing of a polynomial $\Sigma u_i(x, y) v_i(z, t)$, the numerator of the function $\phi(x, z; a, c)$ investigated above for any correspondence. This, as a polynomial in (z, t), has, when the valency is zero, no zeros coinciding with (x, y). The remark made is the same as the statement that the zeros of $\Sigma u_i(x, y) v_i(x, y)$ are a set equivalent to the aggregate of the set (Q), given by $\Sigma u_i(x_0, y_0) v_i(z, t) = 0$ with $f(z, t) = 0$, and the set (Q)$_1$, given by $\Sigma u_i(x, y) v_i(x_0, y_0) = 0$ with $f(x, y) = 0$. The theorem is contained in what follows, but is necessary as a lemma.

In general, however, the set of coincidences (U), of a correspondence of valency γ, in which the direct and reverse sets corresponding to a point P are (Q) and (Q)$_1$, upon a curve on which a canonical set, of $2p-2$ points, is denoted by (K), satisfies the equivalence $(U) \equiv (Q) + (Q)_1 + \gamma [2P + (K)]$. To prove this important result, remark, first, that, if true for each of two correspondences, it is true for the correspondence which is their sum; for the coincidences in the sum are evidently the aggregate of those in the separate correspondences. Then, next, the formula is clearly true for the simple correspondence, of valency unity, in which P and (Q) together, as also P and $(Q)_1$ together, are a set of a linear series of freedom unity, on the curve. For, in this case, it becomes the fundamental result $(J) \equiv (L) + (M) + (K)$, in which (L) and (M) are any two sets of the linear series, and (J) is the Jacobian set of the series (Vol. v, p. 84). From these facts it follows that the formula is true for any correspondence, of positive valency γ, which can be obtained as the sum of γ such simple correspondences. Being assumed to be true, in virtue of the lemma preceding, for any correspondence of zero valency, it is therefore true, for any *general* correspondence, of negative valency $-\gamma$, not necessarily assumed to be obtained by composition of the simple correspondences referred to. Hence, also by the lemma, it is true for any general correspondence of positive valency γ.

We may also deduce this result for the coincidences directly, by use of the function $\phi(x, z; a, c)$ obtained above. Suppose the valency to be positive. Consider the function of (z) which is expressible, in the notation employed (p. 4, above), in either of the forms

$$(z-x)^{-\gamma} R_c \left(z; \begin{array}{c} x^\gamma, z_1, \ldots, z_r \\ a^\gamma, c_1, \ldots, c_r \end{array} \right), \quad (z-x)^{-\gamma} R_a \left(x; \begin{array}{c} z^\gamma, x_1, \ldots, x_s \\ c^\gamma, a_1, \ldots, a_s \end{array} \right),$$

(x) being a general place; and therein suppose (z) to approach to coincidence with (x). For simplicity of statement, suppose that the curve, $f(x, y) = 0$, of order n, has n distinct places at $x = \infty$. Putting, in terms of the representative parameter θ for the neighbourhood of (x), $z - x = \theta(dz/dx) + \frac{1}{2}\theta^2(d^2z/dx^2) + \ldots$, where (dz/dx), etc., are the values at (x), a factor θ^γ, in R_c, will cancel a factor $\theta^{-\gamma}$ in $(z-x)^{-\gamma}$. Further, as this function R_c vanishes to the first order when (z) is at any of the places, $(z_1), \ldots, (z_r)$, which correspond to (x) in the direct correspondence, the function of (x) obtained by putting (z) at (x) will vanish when (x) is at any of the places $(z_1), \ldots, (z_r)$; also, as we see with the help of the second form of the function, this function of (x) will have poles of order γ both at (a) and (c), and poles of the first order at each of $(c_1), \ldots, (c_r), (a_1), \ldots, (a_s)$. The function (dz/dx), regarded as a function of (x), vanishes to order $k-1$ when (x) approaches a branch place of index k (where there is a cycle of k values); in all, then, (dz/dx) has $\Sigma(k-1)$, or, say, w zeros, at such places—whose aggregate we may denote by (w); and it is known that $w = 2n + 2p - 2$; the function of x, (dz/dx), is also infinite to the second order at each of the n places at $x = \infty$. We may denote the aggregate of these n places by (∞).

On the whole, then, when (z) coincides with (x) we obtain a function of (x) whose zeros and poles are those represented by the symbols, in the numerator and denominator, respectively, in the fractional scheme

$$\frac{(U)(\infty)^{2\gamma}}{(w)^\gamma (a^\gamma)(c^\gamma)(c_1) \dots (c_r)(a_1) \dots (a_s)}.$$

In making this deduction, it is assumed that the function of (x) which is the factor of θ^γ in R_c, when (z) approaches to (x), gives only those zeros and poles which we have stated. From this scheme, if ν be the number of coincidences of the correspondence, equating the numbers of zeros and poles of this rational function, we hence deduce $\nu + 2\gamma n = (2n + 2p - 2)\gamma + 2\gamma + r + s$, or $\nu = r + s + 2\gamma p$.

Further, if u denote any everywhere finite integral of the curve, the rational function of (x), given by du/dx, has zeros and poles which are indicated by the scheme

$$\frac{(K)(\infty)^2}{(w)},$$

where (K) denotes a canonical set. Dividing then the previous function of (x) by $(du/dx)^\gamma$, we obtain a rational function whose zeros and poles are those denoted by the scheme

$$\frac{(U)}{(K)^\gamma (a^\gamma)(c^\gamma)(c_1) \dots (c_r)(a_1) \dots (a_s)},$$

so that we have the equivalence $(U) \equiv (Q) + (Q)_1 + \gamma [2P + (K)]$ originally stated. A function of the character of $\phi(x, z; a, c)$ can be constructed when the valency γ is negative, and then a similar argument leads to the same equivalence for this case.

Remark. A correspondence, such as the simple correspondences we have constructed from a linear series of freedom unity, in which the distinction between the direct and the reverse correspondence is lost, or say $T_1 = T$, is called *symmetrical*. The number of distinct points of coincidence of such a correspondence is $\frac{1}{2}\nu$, namely is $r + p\gamma$.

Ex. 1. To illustrate the formation of the function $\phi(x, z; a, c)$, and the limit of $(z - x)^{-\gamma}\phi(x, z; a, c)$ when (z) approaches to (x), we may consider the simple example of the correspondence, on a plane cubic curve $f(x, y) = 0$, in which, to (x) corresponds the point (z), in which the tangent at (x) meets the curve again, while, to (z) correspond the four points $(x_1), \dots, (x_4)$, in which the tangents drawn from (z) touch the curve. Then $\gamma = 2$. Denoting the function $(z - x)\partial f/\partial x + (t - y)\partial f/\partial y$, or, say, $(z - x)f_1 + (t - y)f_2$, by $T_{x,z}$, the function $\phi(x, z; a, c)$ is

$$T_{x,z} T_{a,c}/T_{a,z} T_{x,c}.$$

If, further, H denote the Hessian form $|\,\partial^2 F/\partial \xi_i \partial \xi_j\,|$, where

$$F = \xi_3{}^3 f(\xi_1 \xi_3{}^{-1},\ \xi_2 \xi_3{}^{-1}),$$

we easily compute that $H = -4\xi_3{}^7 (f_1{}^2 f_{22} - 2f_1 f_2 f_{12} + f_2{}^2 f_{11})$, and, when (z) approaches to (x),

$$(z-x)^{-2} T_{x,z} = -\tfrac{1}{2} f_2{}^{-2}(f_1{}^2 f_{22} - 2f_1 f_2 f_{12} + f_2{}^2 f_{11}).$$

Thus, putting $\xi_3 = 1$, $\xi_1 = x$, $\xi_2 = y$, $(z-x)^{-2}\phi(x, z;\, a, c)$ becomes

$$[\tfrac{1}{3} H/(\partial f/\partial y)^2][T_{a,c}/T_{a,x} T_{x,c}].$$

Here, the Hessian H vanishes in the $1 + 4 + 2.1.2$, or 9, inflexions of the curve, which are the coincidences of the correspondence, while $T_{a,x}$ vanishes twice at (a), and once at its tangential (c_1), and $T_{x,c}$ vanishes twice at (c), and once at the four points of contact of tangents from (c); while $\partial f/\partial y$ vanishes once at each of the six points of contact, (w), of tangents parallel to $x = 0$, and, being of the second order, is infinite to the second order at the three places (∞). The function which expresses the equivalence of the inflexions with the aggregate $(a^2)(c^2)(c_1)(a_1)...(a_4)$ is then $H/T_{a,x} T_{x,c}$.

Ex. 2. Considering, upon a curve of genus p, a linear series $g_r{}^n$, of sets of n points, of freedom r, and the sets of this series in which r points are taken coincident, there is, between each such point of coincidence, P, and the $n - r$ remaining points (Q) of the set determined by P, a correspondence of valency r; in this, there correspond to P the $n - r$ points (Q), and to Q correspond, reversely, a certain number, say ξ, of points P. Let the number of coincidences of a point Q with P be denoted by $x_{n,r}$; this is the number of sets of the series $g_r{}^n$ for which $(r+1)$ points of the set are coincident. The sets from $g_r{}^n$ of which a given point Q is one point, form a linear series g_{r-1}^{n-1}; in this are $x_{n-1,\,r-1}$ sets which contain r coincident points; this is then the number, ξ, required in the correspondence (P, Q), of points P which correspond reversely to Q. Thus, using the formula found above for the number of coincidences in a correspondence, we see that $x_{n,r} = x_{n-1,r-1} + n - r + 2pr$; if we put

$$y_{n,r} = x_{n,r} - (r+1)[n + r\,(p-1)],$$

this equation is the same as $y_{n,r} = y_{n-1,r-1}$. But

$$y_{n,1} = 2n + 2p - 2 - 2\,[n + p - 1] = 0;$$

thus, in general, $y_{n,r} = 0$, and the required number $x_{n,\,r}$ is given by

$$x_{n,r} = (r+1)\,[n + r\,(p-1)].$$

This formula gives the total number of cases; in any particular application it is necessary to consider whether (when r is not a prime) the r-fold coincidences can arise by repetitions of r_1-fold coincidences, in which r_1 is a factor of r. Further, it is to be remarked that it is not assumed that the series $g_r{}^n$ is complete; if, for example, we had $r = n - p$, then $x_{n,r}$ would reduce to $(r+1)^2 p$.

Ex. 3. If we have, upon the curve, two correspondences, (s, r, γ) and (s', r', γ'), of respective valencies γ and γ', in which there correspond to P the respective sets (Q) and (Q'), of r and r' points, then the number of times it happens that, for the same P, a point Q coincides with a point Q', is $rs' + r's - 2p\gamma\gamma'$.

Ex. 4. Two linear series, $g_1{}^h$, $g_1{}^k$, have a pair of points in a set of one series coinciding with a pair of points in a set of the other series, in $(h-1)(k-1)-p$ ways.

Ex. 5. On a general plane quartic curve, a direct correspondence from P to three points Q, is given by conics, drawn through three fixed points A, B, C of the quartic curve, which touch the curve at P. Shew that the reverse correspondence, from Q to (P), is of index 12, and of valency 2, being given by quintic curves, with double points at A, B, C, which touch the quartic curve at Q. If $f(x, y, z) = 0$ represent the quartic curve, referred to A, B, C as triangle of coordinates, the coordinates of P, Q being respectively (x, y, z) and (ξ, η, ζ), prove that the correspondence is given by

$$\eta\zeta x^2 \frac{\partial f}{\partial x} + \zeta\xi y^2 \frac{\partial f}{\partial y} + \xi\eta z^2 \frac{\partial f}{\partial z} = 0.$$

Generalise this to the case of an $(r^2 p, p)$ correspondence, on a plane curve of order m, with multiple points (i), when (Q) is determined from P by adjoint curves of order $m-2$, with a number of fixed points on the curve given by $\frac{1}{2}(m-2)(m+1) - \frac{1}{2}\Sigma i(i-1) - r$, which have r-pointic contact at P with the fundamental curve. Obtain the curves, with r-ple intersection at Q, which determine the corresponding $r^2 p$ points (P).

There is a particular theorem, sometimes of use, which it is convenient to explain here, while the notation is familiar.

The direct correspondence, from (x) to (z_1), ..., (z_r), enables us to establish a correspondence between, say, (z_1), and the other points of the set (z_1), ..., (z_r). For this, when (z_1) is given, we should pass, reversely, from (z_1) to (x_1), ..., (x_s), and then, directly, from any one of these, say (x_i), to a set consisting of (z_1) and $(r-1)$ places, say, $(z_{i,1})$, ..., $(z_{i,r-1})$. There exists, therefore, a direct correspondence of (z_1) with $s(r-1)$ places $(z_{i,j})$. Evidently this is a symmetrical correspondence, the reverse set consisting also of $s(r-1)$ places. We proceed to find the valency of this correspondence.

As $(x_i)^\gamma$, $(z_1), (z_{i,1})$, ..., $(z_{i,r-1})$ are the zeros of a rational function, so are the $s(r+\gamma)$ places

$$(x_1)^\gamma, ..., (x_s)^\gamma, (z_1)^s, (z_{1,1}), ..., (z_{1,r-1}), ..., (z_{s,1}), ..., (z_{s,r-1});$$

but so also are $(z_1)^\gamma$, (x_1), ..., (x_s), and hence also $(z_1)^{\gamma^2}$, $(x_1)^\gamma$, ..., $(x_s)^\gamma$. Thus the places, obtained by taking the latter set from the former, namely $(z_1)^{s-\gamma^2}$, $(z_{1,1})$, ..., $(z_{1,r-1})$, ..., $(z_{s,1})$, ..., $(z_{s,r-1})$, are the zeros of a rational function; this shews that, in the symmetrical correspondence spoken of, of indices $s(r-1)$, $s(r-1)$, between (z_1) and $(z_{i,j})$, the valency is $s-\gamma^2$. The number of distinct coincidences, of (z_1) with a $(z_{i,j})$ is therefore $s(r-1) + p(s-\gamma^2)$.

We may call this correspondence the *direct lateral correspondence* derived from the original correspondence T. If it be denoted by X, and the correspondence on the curve which leaves any point P unaltered, say the *identical* correspondence, be denoted by I, the symbolical representation of the process we have followed (if T_1 be the reverse of T), is $TT_1 = sI + X$. The valency of X is thus obtainable by subtracting from the valency of TT_1 (which is $-\gamma^2$), s times the valency of I. This identical correspondence, expressing that the point P, and its corresponding point Q, are identical, or $P - Q \equiv 0$, is, however, of valency -1. Thus the valency of X is $-\gamma^2 - (-s)$, or $s - \gamma^2$, as we have found.

More generally, denoting by α_i, β_i the indices, by γ_i the valency, and by u_i the number of coincidences of a correspondence, suppose we have several correspondences, from P to the set (Q), from P to the set (R), and so on, which are such that there exists the equivalence

$$kP + k_1(Q) + k_2(R) + \ldots \equiv 0;$$

as we have $(Q) + \gamma_1 P \equiv 0$, $(R) + \gamma_2 P \equiv 0$, etc., this is merely the same as $k = k_1\gamma_1 + k_2\gamma_2 + \ldots$. Then we have

$$k_1(u_1 - \alpha_1 - \beta_1) + k_2(u_2 - \alpha_2 - \beta_2) + \ldots = 2kp,$$

as was remarked by Cayley (*Papers*, VI, p. 12).

An application of the formula for the valency of the direct lateral correspondence is to find the number, due to Brill, given in Ex. 3 preceding, for the number of pairs of points which correspond to one another in *two* correspondences. For this number is the number of coincidences in the direct lateral correspondence of the sum of the two given correspondences, less the numbers of coincidences in the direct lateral correspondences of both the given correspondences—the difference obtained to be halved, on account of the symmetrical character of the resulting correspondence.

We now proceed to give applications of the theory of correspondence of increasing difficulty, which for clearness we tabulate as Exx. 1–15 (pp. 12–46).

Ex. 1. Apply the formula to obtain the *general* number, $2(n'-3)(n-4) + 2(n'-12)p$, of osculating planes of a curve of order n, and class n', and genus p, in ordinary space, which touch the curve again. (Here $s = n'-3$, $r = n-3$, $\gamma = 3$.)

Ex. 2. A natural application of the formula is to the bitangents of a plane curve, of order n. Between the point of contact of a tangent of the curve, and the remaining intersections of this line with the curve, is a direct correspondence (s, r) for which $r = n-2$; if n' be the class of the curve, the number of proper tangents from an arbitrary point, we may take, in the reverse correspondence, $s = n'-2$, the points (x), reversely corresponding to (z), not being supposed to include the cusps of the curve, if such exist; denote the number of cusps by κ. Then the formula, $2s(r-1) + 2p(s-\gamma^2)$, for the number of coincidences in the direct lateral correspondence, each counted twice, gives, for *twice* the number of bitangents of the curve, $2(n'-2)(n-3)+2p(n'-6)-\kappa(n'-3)$, the correction $\kappa(n'-3)$ being for the $n'-3$ proper tangents to the curve from each of the cusps. If however we allow the cusps, as points of contact of tangents to the curve from (z), we must take $s=n'-2+\kappa$; then twice the number of bitangents is given by

$$2(n'-2+\kappa)(n-3)+2p(n'-6+\kappa)-2\kappa(n'-3)-(\kappa^2-\kappa),$$

where there is correction, (1), for tangents drawn from a cusp, (2), for lines from a cusp to other cusps.

The second result agrees with the first, and with the number $2\delta+(n'-n)(n'+n-9)$, where δ is the number of double points of the curve.

Ex. 3. For the particular case of a plane quartic curve, if $(z_1),(z_2)$ be the two further intersections of the tangent at a point (x), the correspondence, from (x) to (z_1) and (z_2), having $r=2$, $s=10$, $\gamma=2$, there are $s(r-1)+p(s-\gamma^2)$, or $10+3.6$, or 28 coincidences, which is the number of bitangents of the curve. If the equation of the curve be written, in the usual symbolical way, as $a_x^4=0$, the correspondence between (z_1) and (z_2) is at once found to be expressible by $9(a_{z_1}^2 a_{z_2}^2)^2-16a_{z_1}^3 a_{z_2}.a_{z_1}a_{z_2}^3=0$; this is the condition that the chord from (z_1) to (z_2) touches the curve. The points of contact, (x), of the bitangents of the quartic curve, thus lie on a curve obtainable by finding the limit, when (z) approaches to (x), of

$$[9(a_z^2 a_x^2)^2-16a_z^3 a_x.a_z a_x^3](z-x)^{-6}.$$

From this, or directly, it follows, that, if (ζ) be any point of a bitangent of which one point of contact is (x), then

$$[2(a_x a_\zeta^3)^2-3a_x^2 a_\zeta^2.a_\zeta^4](\zeta-x)^{-6}=0.$$

Using (x, y) and (ζ, σ) for the non-homogeneous coordinates of (x) and (ζ), and eliminating σ by means of the equation of the tangent of the curve at (x) (which is of the form $u\zeta+v\sigma+w=0$, where u, v, w are definite functions of (x)), the factor $(\zeta-x)^{-6}$ will identically disappear. In other words, if the expression here multiplying $(\zeta-x)^{-6}$ be denoted by $\phi(x, y, \zeta, \sigma)$, the function $v^6\phi(x, y, \zeta, \sigma)$ divides by $(\zeta-x)^6$ in virtue of $u\zeta+v\sigma+w=0$. The quotient will then be of order, in x, y, equal to $6.3+2-6$, or 14. This gives a curve of order 14, containing all the points of contact of the 28 bitangents. Cf. also Salmon's *Higher Plane Curves*; and Dersch, *Math. Ann.* VII. (Below, p. 298.)

In particular shew, for the quartic curve $x^4+y^4+z^4=0$, that the curve of order 14 may be taken to be $x^2y^2z^2(x^8+y^8+z^8)=0$. The 28 bitangents, in this case, consist of 16, each with an equation of the form $\theta x+\phi y+z=0$, where $\theta^4=1$, $\phi^4=1$, together with 12 others each meeting the curve in four coincident points, these 12 bitangents being $(y^4+z^4)(z^4+x^4)(x^4+y^4)=0$.

Ex. 4. It is known that there are sets of six pairs, of the bitangents of a plane quartic curve, such that the eight points of contact, of any two of these pairs, lie upon a conic. The points of contact of the 28 bitangents of the curve lie on the degenerate curve of order 14 which is composed of seven such conics, properly taken (and in various ways).

This we may most easily see by regarding the quartic curve as the projection of the profile of a cubic surface, as seen from any point of this surface. It is well known that, if a, b' be two intersecting lines of a cubic surface, meeting another line, c, of the surface, and a', b be two other intersecting lines, also intersecting c, the four bitangents of the quartic curve which arise by projection of a, b', a', b have their points of contact on a conic (*Proc. Lond. Math. Soc.* IX, 1911, p. 163). Hence, if f, g, h be three coplanar lines of the surface, and f_1, f_1' be intersecting lines which meet f, other than g or h, the other such pairs of intersecting lines which meet f being f_2, f_2'; f_3, f_3' and f_4, f_4', then we obtain one such conic as contemplated by projection of f_1, f_1' and f_2, f_2', and another such conic by projection of f_3, f_3' and f_4, f_4'. Proceeding similarly with the lines g and h, we thus obtain six conics in the plane, from f, g and h. Another conic contains the points of contact of the bitangents which arise by projection of f, g, h themselves; and this conic contains the points of contact of the bitangent arising from the neighbourhood of the point of the surface from which the projection is made. The whole 27 lines of the surface are given by f, g, h and the couples f_i, f_i'; g_j, g_j'; h_k, h_k'. Thus the 56 points of contact of the 28 bitangents lie on the seven conics described.

The suggestion that the 56 points of contact lie on 7 conics was made to the writer by Dr P. Du Val, who obtained another solution than that just given. We may for instance take six conics containing the points of contact of the bitangents projecting the 6 quadruples of lines (in Schläfli's notation)

$$(a_1, b_4; c_{23}, c_{56}), (a_2, b_5; c_{31}, c_{64}), (a_3, b_6; c_{12}, c_{45}),$$
$$(a_4, b_2; c_{15}, c_{36}), (a_5, b_3; c_{14}, c_{26}), (a_6, b_1; c_{24}, c_{35});$$

here, each quadruple consists of two pairs of intersecting lines, having a common transversal line (these lines being, respectively, c_{14}, c_{25}, c_{36}, c_{24}, c_{35}, c_{16}). With c_{34}, c_{16}, c_{25}, which, on projection, lead to a seventh conic, these 6 quadruples exhaust the 27 lines of the cubic surface.

It may be interesting to remark, in this connexion, on the fact that the 27 lines of the cubic surface are its complete intersection with a surface of order 9 (noted in Salmon's *Solid Geometry*, 1882, p. 510). One particular surface of order 9, with this property, consists of the two quadric surfaces which contain respectively the six lines $a_1, a_2, a_3, b_4, b_5, b_6$ and the six lines $a_4, a_5, a_6, b_1, b_2, b_3$ together with 5 planes each containing 3 of the fifteen lines c_{ij}; such a set of 5 planes can be chosen in 6 ways. (In a phraseology due to Sylvester, we can arrange the 15 duads (i, j) in 5 synthemes in 6 ways; see Vol. II, p. 220.) But in fact (as follows from a result remarked by Dr H. S. M. Coxeter, *Phil. Trans.* CCXXIX (1931), p. 419) the 27 lines lie in 9 tritangent planes, as, for instance, those given by the columns in the scheme

$$c_{12}, \ a_1, \ b_6, \ c_{36}, \ c_{15}, c_{26}, c_{45}, \ b_5, \ a_2,$$
$$c_{34}, c_{13}, a_4, \ c_{25}, \ b_1, \ b_2, \ c_{16}, a_3, \ c_{24},$$
$$c_{56}, \ b_3, \ c_{46}, c_{14}, a_5, \ a_6, \ c_{23}, c_{35}, \ b_4.$$

Ex. 5. Suppose that a correspondence exists between one algebraic curve in space, C, of order n, and another curve C', of order n', whereby, to each point P, of the curve C, there correspond α' points P', of the curve C' (so that the rational symmetric functions of the coordinates of the points P' are expressible as rational functions of the coordinates of P), and to each point, P', of C', there correspond α points, P, of C (of which the rational symmetric functions are rational in P'). It may be required to find the order of the ruled surface generated by the line PP', joining a pair of corresponding points of the two curves.

First, suppose the curves to be so situated that corresponding points of the two curves never coincide. Take an arbitrary fixed line, l, which we may suppose to be given by $x = 0$, $y = 0$. To any point P, of the curve C, correspond α' points of C'; the α' planes, joining the line l to these points P', meet the curve C in $n\alpha'$ points, (Q), which are thus determined when P is assigned. Reversely, a point Q, of the curve C, joined to l, determines n' points of the curve C', and hence $n'\alpha$ points (P) of the curve C. We thus have a correspondence (P, Q), on the curve C, of indices $(n'\alpha, n\alpha')$. And this correspondence is of valency zero; for the product of the α' planes joining l to the points P', of the curve C', which correspond to P, is of the form $(x - ym_1)\ldots(x - ym_{\alpha'}) = 0$, and hence of the form $u_0 x^{\alpha'} + u_1 x^{\alpha'-1}y + \ldots + u_{\alpha'} y^{\alpha'} = 0$, where $u_0, \ldots, u_{\alpha'}$ are rational polynomials in the coordinates of P. The points (Q) arise from P by the intersections of these planes with the curve C, and hence, for all positions of P, are sets of the same linear series on C; and P is not a member of the general set of this series. Thus the valency of the correspondence (P, Q) is zero.

Hence the number of coincidences is $n'\alpha + n\alpha'$. And such a coincidence, of Q with P, arises when, and only when, the line PP' meets l. As this is an arbitrary line, the number $n'\alpha + n\alpha'$ is the order of the ruled surface generated by the line PP'.

The problem thus illustrates the theory of correspondence of points on a curve which is not necessarily rational. But it can be otherwise treated, as involving a correspondence in the rational pencil of all planes through the arbitrary fixed line l. For we may regard two planes, (ξ) and (ξ'), of this pencil, as corresponding when they contain, respectively, points P, P', of the two curves C, C', which correspond to one another. Then, any plane (ξ) meets C in n points, to which there correspond $n\alpha'$ points of C', each of which gives rise to a plane ξ'; similarly, to any plane ξ' correspond $n'\alpha$ planes ξ. There are thus $n\alpha' + n'\alpha$ coincidences of corresponding planes; and, if no coincidence of corresponding points P, P' occurs, such can only arise when the line PP' meets the line l. But we see also that if there be i coincidences of corresponding points P, P', the number $n\alpha' + n'\alpha$ includes the i planes joining l to these; and the order of the ruled surface is then only $n\alpha' + n'\alpha - i$.

A corollary from the result is that, if two curves in the same plane are in (α, α') correspondence, and there be no coincidence of corresponding points, then the lines joining corresponding points envelop a curve of class $n\alpha' + n'\alpha$.

Ex. 6. If there be a correspondence, of indices α, α', and valency γ, between two points, P and P', of a *single* curve C, in space, of order n and genus p, we may similarly seek the order of the ruled surface formed by the lines PP' which join corresponding points.

In this case, as in the second solution in Ex. 5, we may similarly consider a correspondence between pairs of planes, ξ and ξ', drawn through an arbitrary fixed line l—two planes corresponding when they respectively contain corresponding points, P and P', of the curve. Then, a plane ξ gives rise to n points P, and hence to $n\alpha'$ points P', and hence to $n\alpha'$ planes ξ'; and, reversely, a plane ξ' gives $n\alpha$ planes ξ. The coincidences of corresponding planes of the pencil arise, both when the line PP' meets the line l, and also when P' coincides with P. The number of the latter coincidences is $\alpha + \alpha' + 2\gamma p$; the number of lines PP' which meet l, or the desired order of the ruled surface, is thus

$$n(\alpha + \alpha') - (\alpha + \alpha' + 2\gamma p), \text{ or } (n-1)\alpha + (n-1)\alpha' - 2\gamma p.$$

This result is also obtainable by the theory of correspondence on the curve C. To any point P correspond a set (Q), of $(n-1)\alpha'$ points, on this curve, obtained as the residual intersections of the α' planes which join the line l to the α' points P' which correspond to P. Reversely, a point Q, joined to l, gives $n-1$ points P', and hence gives a set (P) of $(n-1)\alpha$ points. We thus have a correspondence (P, Q), of indices $(n-1)\alpha, (n-1)\alpha'$, derived from the original. This derived correspondence is clearly what we have described as the product of the two correspondences, (1), the original correspondence between P and P'; (2), the correspondence between P' and Q, established by means of the line l. The latter is of valency unity; the product correspondence is thus of valency $-\gamma$. We therefore obtain the same number as before for the number of coincidences.

As a particular application, calling *torsal* a chord of a curve in ordinary space when the tangents of the curve at its extremities meet one another, prove that the torsal chords of a curve of order n and rank r, without cusps, form a ruled surface whose order, in general, is $(n-3)r$.

Ex. 7. We have used the notion of an (α', α) correspondence between two curves (C', C). The existence of such a correspondence, however, implies relations between the curves. Consider the particular hypothesis of a $(1, \alpha)$ correspondence, each point P' of C' corresponding to α points P of C, any one of which corresponds to the same point P'. For instance, between the circle whose equation is $x'^2+y'^2=1$, and the quartic curve expressed by $x^4+y^4=1$, there exists such a correspondence, given by $x^2=x'$, $y^2=y'$, whereby, to any point of the circle correspond 4 points of the quartic curve, or $\alpha=4$. In general, with such a $(1, \alpha)$ correspondence between P' on the curve C', and P on the curve C, as the point P' moves on the curve C', we obtain sets of α points (P) on C; and any point P of such a set, through its determination of the point P' of the curve C', determines all the other points (P) of that set. We say then that there exists, on the curve C, an involution $I_1{}^\alpha$, of sets of α points, of freedom 1; and we say that this involution is of genus p', this being the genus of the curve C' described by the point P' which represents a set of the involution on C. We may also say that there is a symmetrical correspondence, of which both indices are of value $\alpha-1$, between P and the other points of the set (P); but these sets (P) do not all belong to the same linear series on C, unless $p'=0$; so that this symmetrical correspondence is not one of those of the kind hitherto considered, which have a valency. When the sets of the involution $I_1{}^\alpha$ do belong to a linear series, it is a familiar fact that the number of coincidences of two points of a set (P) is $2\alpha+2p-2$, these coincidences forming the Jacobian set of the linear series. We proceed now to shew that, in general, for an involution of sets of α points, on a curve of genus p, of which the sets correspond to the points of another curve of genus p', there are d coincidences of two points of a set, where $d=2\alpha+2p-2-2\alpha p'$.

The proof we give assumes the result for the simpler case of a linear series. Consider a linear series, of sets of k points, of freedom 1, on the (supposed plane) curve C'. The points of any set of this series will be given by the intersections of C' with a curve of equation $\lambda\phi(x', y')+\psi(x', y')=0$, where the parameter λ varies from set to set. By hypothesis, there are α points on C corresponding to (x', y'), and, if (x, y) be any one of these, then x' and y' are both rationally expressible in terms of x and y, by use of the equations of the two curves; conversely, any rational symmetric function of the α points (x, y) is rational in (x', y'). To the points of the set, on C', given by $\lambda\phi\,(x', y') + \psi\,(x', y') = 0$, correspond therefore, on C, the points of a set given by an equation $\lambda\Phi\,(x, y) + \Psi\,(x, y) = 0$, namely, also a linear series, of freedom 1, of sets of $k\alpha$ points. Now let G' denote any set of the series on C', and J' the Jacobian set of this linear series, the points of coincidence of pairs of points in sets of this series, while K' denotes a canonical set on C', when $p' > 0$. Then we have the equivalence, on the curve C', expressed by $K' \equiv J' - 2G'$, which remains conventionally true when $p' = 0$ (or $p' = 1$), if we omit K'. Also let tK', tJ', tG' denote the sets on the curve C which arise by taking the α points which correspond to every point of the sets K', J', G', respectively. As *any* linear series on C' gives rise, in this way, by the argument above, to a linear series on C, we infer that, on C, $tK' \equiv tJ' - 2tG'$.

Take now the linear series on C given by $\lambda\Phi\,(x, y) + \Psi\,(x, y) = 0$, of which a set consists of k subsets, each of α points, each such subset arising from one point of a set of the original linear series on C'. A coincidence in a set of this series on C may arise, (1), from coincidence of two of the α points of a subset; if d be the number of coincidences of a pair of points in all the sets of the involution $I_1{}^\alpha$, the coincidences of this kind will be d in number; their aggregate may be denoted by D; but, (2), a coincidence, in fact α coincidences, in the linear series on C, may arise, in which two of the subsets entirely coincide, in consequence of a coincidence of two of the k points of a set of the original series on C'; the aggregate of all such coincidences is exactly that denoted by tJ'. The canonical series on C is expressible by the coincidences in the linear series thereon; namely, if G denote a set of this linear series, we have

$$K \equiv (D + tJ') - 2G, \text{ or } K \equiv D + tJ' - 2tG'.$$

Hence we have $K \equiv tK' + D$; *or, the canonical series on the curve C is defined by sets equivalent to, (1), the aggregate of the sets of α points arising by direct transformation of the points of a set of the canonical series on the representative curve C', with, (2), the addition of the set of coincidences in the involution $I_1{}^\alpha$ which is represented by the curve C'.* This leads to the numerical equation $2p - 2 = \alpha\,(2p' - 2) + d$, which gives the value of d originally stated. When $p' = 0$, the general equivalence is $K \equiv D - 2I$, where I denotes a set of the involution $I_1{}^\alpha$. For in this case all these sets belong to the same linear series. (This is clear because, on the rational curve C', the parameter θ is rational in (x', y'), and hence rational in the (x, y) of a set of the curve C which corresponds to (x', y'); so that the involution on C is given by an equation $\theta U\,(x, y) + V\,(x, y) = 0$.)

The numerical equation $d = 2\alpha + 2p - 2 - 2\alpha p'$ was remarked by Zeuthen, to whom also is due the result to which we now proceed*.

Suppose that there is an (α', α) algebraic correspondence between the curves C', C; we shew that it is possible to define a curve Γ, such that (C', Γ) are in $(1, \alpha)$ correspondence, while (Γ, C) are in $(\alpha', 1)$ correspondence. To any point of C' correspond then α points of Γ, each of which corresponds to a single point of C; and to any point of C correspond α' points of Γ, each of which corresponds to a single point of C'. Thus, a coincidence of two points of a set of α' points on C' which correspond to the same point of C, requires a coincidence of two points of a set of α' points on Γ which correspond to the same point of C. Let d' be the number of such coincidences which exist; they result from d' coincidences in the $(\alpha', 1)$ correspondence of Γ and C. Similarly, let d be the number of existing coincidences of two points of a set of α points on C which corresponds to a point of C'; these result from d coincidences in the $(1, \alpha)$ correspondence of C' and Γ. If D' be the set of coincidences on Γ in the first case referred to, and D the set of coincidences on Γ in the second case, we have $tK \equiv \Lambda - D'$, $t'K' \equiv \Lambda - D$, where K, K' denote canonical sets respectively on C and C', and t refers to the $(1, \alpha')$ correspondence (C, Γ), and t' to the $(1, \alpha)$ correspondence (C', Γ), while Λ is a canonical set on Γ. From these equivalences we infer that, on Γ, $tK - D \equiv t'K' - D'$; this leads to the numerical equation

$$\alpha' (2p - 2) - d = \alpha (2p' - 2) - d',$$

including the former result for a $(1, \alpha)$ correspondence.

A simple way of defining a curve Γ, desired for the argument, is to suppose the curves C', C to lie in different spaces, which both lie in another space but have no point common. (Two spaces $[r]$, $[r']$, of respective dimensions r, r', will not generally intersect when they lie in a space $[r + r' + 1]$.) Then form a ruled surface by joining every point of the curve C' to all the α points of the curve C, to which it corresponds, and, also, every point of C to all the α' points of C', to which it corresponds; to each point of C' will then correspond α generators of the ruled surface, and to each point of C will correspond α' generators. A section of this ruled surface by a prime space of the space containing both C' and C will then give such a curve Γ as desired. If the curves C', C be plane curves in the same plane, we may also obtain such a curve Γ by joining one arbitrary point O, of the plane, to every point of C, and the corresponding point of C' to another arbitrary point O', and taking the locus of the intersection of the corresponding joining lines.

---

* See the exhaustive investigation in Zeuthen, *Lehrbuch d. abzähl. Geometrie*, 1914, pp. 104–7.

Ex. 7 a. We may consider the $(\nu, 1)$ correspondence of two (plane) curves, $f=0, f'=0$, in a more direct analytical way, and so obtain additional results. Let $x'=g(x, y, z)$, $y'=h(x, y, z)$, $z'=k(x, y, z)$, be polynomials of order r in the coordinates x, y, z, which transform the curve $f'(x', y', z')=0$, of order n', into the curve $f(x, y, z)=0$, of order n; there will then be an identity $f'(x', y', z')=Mf(x, y, z)$, where M is a polynomial in x, y, z. If α, β, γ be arbitrary numbers, this leads, on $f=0$, to the equations

$$M\left(\alpha\frac{\partial f}{\partial x}+\beta\frac{\partial f}{\partial y}+\gamma\frac{\partial f}{\partial z}\right)=\alpha\left(\frac{\partial f'}{\partial x'}\cdot\frac{\partial x'}{\partial x}+\frac{\partial f'}{\partial y'}\cdot\frac{\partial y'}{\partial x}+\frac{\partial f'}{\partial z'}\cdot\frac{\partial z'}{\partial x}\right)+\ldots+\ldots$$

$$=\alpha'\frac{\partial f'}{\partial x'}+\beta'\frac{\partial f'}{\partial y'}+\gamma'\frac{\partial f'}{\partial z'},$$

where α', β', γ' are given by

$$\alpha'=\alpha\frac{\partial x'}{\partial x}+\beta\frac{\partial x'}{\partial y}+\gamma\frac{\partial x'}{\partial z}, \text{ etc.,}$$

or, say, to
$$M\left(\alpha\frac{\partial f}{\partial x}\right)=\left(\alpha'\frac{\partial f'}{\partial x'}\right).$$

Hence, also, if $x_1'=\partial x'/\partial x$, $x_2'=\partial x'/\partial y$, $x_3'=\partial x'/\partial z$, etc., we find

$$\begin{vmatrix} \alpha', & x', & dx' \\ \beta', & y', & dy' \\ \gamma', & z', & dz' \end{vmatrix}=\frac{1}{r}\begin{vmatrix} x_1', & x_2', & x_3' \\ y_1', & y_2', & y_3' \\ z_1', & z_2', & z_3' \end{vmatrix}\begin{vmatrix} \alpha, & x, & dx \\ \beta, & y, & dy \\ \gamma, & z, & dz \end{vmatrix},$$

which, if J denote the Jacobian determinant $\partial(x', y', z')/\partial(x, y, z)$, we may write

$$(\alpha', x', dx')=\frac{1}{r}J\cdot(\alpha, x, dx).$$

Wherefore, if $\phi'(x', y', z')$ be any polynomial in x', y', z', we have

$$\frac{\phi'(x', y', z')\cdot(\alpha', x', dx')}{\left(\alpha'\frac{\partial f'}{\partial x'}\right)}=\frac{\phi'(g, h, k)J}{rM}\cdot\frac{(\alpha, x, dx)}{\left(\alpha\frac{\partial f}{\partial x}\right)},$$

of which the former member is expressed by x', y', z', and the latter member by x, y, z.

Now it is a familiar fact, easy to establish, that the fraction

$$(\alpha, x, dx)\Big/\left(\alpha\frac{\partial f}{\partial x}\right)$$

has a value, on $f=0$, which is independent of α, β, γ; and similarly

$$(\alpha', x', dx')\Big/\left(\alpha'\frac{\partial f'}{\partial x'}\right)$$

on $f'=0$; also, it is known that, if $\phi'(x', y', z')$ be an adjoint polynomial of order $n'-3$ for the curve $f'=0$, the left side is the differential of an everywhere finite integral for the curve $f'=0$, and that, conversely, every such integral is capable of this form. And it is obvious that an everywhere finite integral for $f'=0$ is transformed into an everywhere finite integral. Thus, for the curve $f=0$, there is an adjoint polynomial $\phi(x, y, z)$, of order $n-3$, such that, on $f=0$, we have

$$\frac{\phi'(g, h, k).J}{M} = r\phi(x, y, z).$$

This identity involves certain consequences. For it shews that any zero of M on $f=0$, that is, any intersection of the curve $M=0$ with $f=0$, is a zero either of $\phi'(g, h, k)$ or of J; and we can shew that the latter is the right alternative. Indeed, the identity holds for all the p' adjoint polynomials $\phi'(x', y', z')$ of $f'=0$, and these are known not to have a common zero on $f'=0$; whence the p' polynomials in x, y, z, expressed by $\phi'(g, h, k)$, have no common zero on $f=0$; a zero of M, not being a common zero of these $\phi'(g, h, k)$, must then be a zero of J (of at least equal order of multiplicity). Other zeros of J are obvious; for any point (x, y, z), of $f=0$, at which there is a coincidence of two of the points of the set of ν points on $f=0$ which correspond to a single point (x', y', z') of $f'=0$ (in virtue of the equations $x'=g$, $y'=h$, $z'=k$), is a zero, on $f=0$, of the polynomial J. Indeed, from $\rho x'=g(x, y, z)$, etc., $\sigma x'=g(x+dx, y+dy, z+dz)$, etc., we deduce $x_1'\,dx+x_2'\,dy+x_3'\,dz=0$, and so on. Conversely, from $f'=Mf$, if $J=0$, we deduce an equation of the form

$$M(af_1+bf_2+cf_3)=0,$$

and can hence infer that a zero of J, when not a zero of M, is such a point of coincidence, of two points of a set on $f=0$, in general. Thus we have the consequence that, on $f=0$, the zeros of J consist of the zeros of M together with the set of coincidences of the correspondence.

We infer, therefore, that these points of coincidence on $f=0$ are zeros of all the particular adjoint polynomials, of order $n-3$, for $f=0$, which arise, as here, by direct transformation of the p' existing polynomials ϕ' of $f'=0$ (and $p \geqslant p'$).

Now let the p' everywhere finite integrals of $f'=0$ be denoted by $w_1, \ldots, w_{p'}$; and let the everywhere finite integrals of $f=0$ obtainable as here by direct transformation of these be denoted by $u_1, \ldots, u_{p'}$; further, let u_i $(i>p')$ be one of the $p-p'$ other everywhere finite integrals of $f=0$. Also, let $(x_1), \ldots, (x_\nu)$ be the places of $f=0$ corresponding to a place (x') of $f'=0$; and, in particular, when (a') is a fixed place of $f'=0$, let $(a_1), \ldots, (a_\nu)$ be the corresponding places of $f=0$. Then, from the algebraic character of the correspondence, it is clear that the sum $u_i{}^{x_1, a_1} + \ldots + u_i{}^{x_\nu, a_\nu}$ is equal to an everywhere finite integral of $f' = 0$ (its differential being symmetric in $(x_1), \ldots, (x_\nu)$). This integral of $f' = 0$ is thus of the form (where $c_{i,1}, \ldots, c_{i,p'}$ are constants)

$$c_{i,1} w_1{}^{x', a'} + \ldots + c_{i,p'} w_{p'}{}^{x', a'};$$

and, as we have seen, this transforms directly into the sum, on $f = 0$,

$$c_{i,1} u_1{}^{x, a} + \ldots + c_{i,p'} u_{p'}{}^{x, a},$$

where, for (x) and (a), respectively, we may take any one of the points corresponding to (x') and (a'). If then we take an integral on $f = 0$ defined by

$$v_i{}^{x, a} = u_i{}^{x, a} - \frac{1}{\nu}(c_{i,1} u_1{}^{x, a} + \ldots + c_{i,p'} u_{p'}{}^{x, a}), \qquad i = p' + 1, \ldots, p,$$

we have $v_i{}^{x_1, a_1} + \ldots + v_i{}^{x_\nu, a_\nu} = 0$; thus, the $p - p'$ integrals v_i have the property that the sum of the values of any one of these, at the ν places corresponding to a point of $f' = 0$, is independent of this point. The p integrals of $f = 0$ consist of these $p - p'$ integrals, together with the p' integrals which arise by direct transformation of the integrals of $f' = 0$. The fact that, when $p' = 0$, the set $(x_1), \ldots, (x_\nu)$ belongs to a linear series on $f = 0$, which we have already proved, arises again here, by the converse of Abel's Theorem (Vol. v, pp. 146, 176).

We suppose that neither of the curves $f = 0, f' = 0$ is hyperelliptic; then these can be represented by the canonical curves, say γ and γ', of respective orders $2p - 2$, $2p' - 2$, lying in spaces Π, Π', of dimensions $p - 1$, $p' - 1$, of which we suppose the space Π' to be contained in Π. For coordinates in Π' we may use $z_1, \ldots, z_{p'}$; and, for coordinates in Π, we may use $x_1, \ldots, x_{p'}, x_{p'+1}, \ldots, x_p$, of which $x_1, \ldots, x_{p'}$ are proportional, on the canonical curve, to those ϕ-polynomials of $f = 0$ which arise by direct transformation of the ϕ-polynomials of $f' = 0$. Thus we have, on the canonical curves,

$$\frac{x_1}{z_1} = \ldots = \frac{x_{p'}}{z_{p'}}.$$

This shews that the canonical curve γ', in the space Π', is obtainable by projection from the canonical curve γ, in the space Π, the centre of projection being the space Σ, of dimension $p - p' - 1$, which is given by $x_1 = 0, \ldots, x_{p'} = 0$. By the argument given, this space Σ contains the d points of the curve γ which are the coincidences of the correspondence between the two curves. Through Σ, and an arbitrary point P of the curve γ, passes a space $[p - p']$, which meets the space Π', of dimension $p' - 1$, in one point (because two spaces, of dimensions $p - p'$ and $p' - 1$, both lying in a space Π

of dimension $p-1$, meet in a point). This point is then on the curve γ', and is the point P' corresponding to P. Through Σ, and an arbitrary point P' of γ', again, passes a space $[p-p']$, which, therefore, intersects the curve γ in the ν points (P) which correspond to P', beside containing the d points of coincidence which lie in Σ. The space, say Ω, through Σ and $p'-1$ points of γ', depending in all on $p-p'+p'-1$, or $p-1$ points, is a prime, or space $[p-2]$, of the space Π, and contains the prime of the space Π' determined by the $p'-1$ points of γ'; this prime space Ω meets γ in a canonical set on γ. This set consists then of the d points of coincidence which lie on Σ, together with $2p'-2$ sets of ν points, each arising from one of the $2p'-2$ points, of γ', lying in the prime space of Π' through which Ω passes. We thus have geometrical interpretations of the equations $K \equiv tK' + D$, $2p-2 = \nu(2p'-2) + d$; and the curve γ has the geometrical property that every space $[p-p']$ through the space Σ, and one point of the curve γ', is a multiple secant of γ, meeting it in ν points. These $[p-p']$, though ∞^1 in aggregate, are not all the spaces of a rational pencil through Σ, but form an aggregate of genus p'; and p' is necessarily $\geqslant 3$, since $f' = 0$ is supposed not to be hyperelliptic. Further, the total aggregate of p primes, of the space Π, consists of the particular p' primes we have considered, passing through the space Σ, together with $p-p'$ others, independent of these. The former correspond to everywhere finite integrals of the curve γ which are reducible to such integrals for the curve γ'; these integrals have thus only $2p'$ independent columns of periods. It will later be shewn to follow that the remaining $p-p'$ integrals of γ have also a defective system of periods, reducing to only $2(p-p')$ columns at most (p. 62, below).

Repeating, in slightly different form, what we have proved: The sets of ν points, of an involution I_1^ν, of genus p', on a canonical curve γ, of genus p, are obtainable by projection of a canonical curve γ', of genus p', from a space $[p-p'-1]$, which contains the points of coincidence, on γ, of pairs of points of a set of the involution. There are canonical sets on γ, each of which contains these points of coincidence, together with the projection of a canonical set of γ'.

Ex. 7 *b.* A simple example may be given (not fully representative, however, as $M=1$). Let the curves $f'=0$ and $f=0$ be given respectively by $x'^4 + y'^4 = z'^4$, $x^8 + y^8 = z^8$, with $x' = x^2$, $y' = y^2$, $z' = z^2$, so that $p' = 3$, $\nu = 4$, $p = \frac{1}{2}(8-1)(8-2) = 21$, and $d = 24$, the points of coincidence being such as $x=0$, $y = z(1)^{\frac{1}{8}}$. Then $J = xyz$. The coordinates for the canonical curve of $f=0$ may be expressed by

$$\frac{x_1}{x^3 yz} = \frac{x_2}{xy^3 z} = \frac{x_3}{xyz^3} = \frac{x_i'}{u_i}, \qquad i = 1, \ldots, 18,$$

u_i being one of the 18 terms $x^l y^m z^n$ for which $l+m+n=5$, other than x^3yz, xy^3z, xyz^3. The canonical curve for $f'=0$ (identical therewith) is given by $z_1/x^2=z_2/y^2=z_3/z^2$. The space Σ, of dimension 17, is given by $x_1=0=x_2=x_3$, and contains the points of coincidence.

Ex. 7 *c.* If U, V, W be homogeneous, of order 2, in x, y, z, the curve of genus 5 in space [4], where the coordinates are x, y, z, t, u, which is given by $\frac{1}{2}t^2+U=0$, $tu+W=0$, $\frac{1}{2}u^2+V=0$, is projected from the line $x=0=y=z$ into a plane quartic curve, to each point of which correspond two points (x, y, z, t, u) and $(x, y, z, -t, -u)$, of the octavic curve in [4]. This remark was made by Mr F. P. White.

Ex. 7 *d.* The curve of contact of a Weddle quartic surface with the enveloping cone drawn from one of the six nodes of the surface, is a septimic curve of genus 5. This curve is projected from any one of the other 5 nodes into an elliptic cubic curve. The septimic curve is thus in (2, 1) correspondence with an elliptic curve in 5 ways, and each of its everywhere finite integrals is reducible to an elliptic integral (see the author's *Multiply-periodic functions*, Cambridge, 1907, pp. 322–6; and Humbert, "Sur un complexe remarquable...," *Journ. École polyt.* LXIV, 1894).

Ex. 8. We have obtained the number of points of coincidence of $r+1$ points in a set of a linear series g_r^n; and the number of coincidences of pairs of points of a set of an irrational involution I_1^ν, determined by a (1, ν) correspondence between a curve, whose genus is that of the involution, and the given curve on which the involution exists. A correspondence (ν', ν), between a curve of genus p', and a curve of genus p, establishes an involution of sets of ν points, of freedom 1, upon the curve of genus p, in which, however, there are ν' sets of the involution which contain a given point of the curve; this is expressed by saying that the involution is of *index* ν'. From the consideration of the correspondence, we have proved Zeuthen's formula, involving the number d of coincidences of a pair of points of a set of the involution I_1^ν, namely

$$\nu' (2p-2) - d = \nu (2p'-2) - d',$$

which gives $d = 2\nu + (2p-2) \nu' - 2\nu p' + d'$. This involves, however, the number, d', of points of the curve (of genus p), for which two of the ν' sets, of ν points, containing a point, coincide with one another.

We have also obtained the number of pairs of points common to two linear series, both of freedom one, g_1^m and g_1^n. We obtain now the number of pairs of points common to a linear series g_1^n and an involution I_1^ν, supposing here that this latter is of index unity.

If, on a ruled surface, of order N, in space of any number of dimensions, which has no multiple points, whose prime section is of genus p', there be a curve of order n and genus p, which meets every generator of the surface in ν points, then the generators determine on this curve an involution $I_1{}^\nu$, of genus p', and index 1. The number, d, of points of simple contact of the curve with a generator, is therefore, by what is proved above, given by

$$d = 2\nu + 2p - 2 - 2\nu p'.$$

This number d can however be evaluated in terms of N (cf. also Ex. 12, p. 216 of Chap. VIII, Vol. V). For consider a pencil of primes, that is, planes passing through an arbitrary general line when the ruled surface is in ordinary space, and primes through an arbitrary general space $[r-2]$, when the ruled surface is in space $[r]$; let two primes of the pencil be regarded as corresponding when they meet a proper generator in two points which are among the ν intersections of this generator with the curve in question. Any prime of the pencil meets the curve in n points, and each of the $(\nu - 1)$ other points of the curve, on any one of the generators through these n points, gives rise to a corresponding prime. The correspondence between the primes of the pencil, which is obviously symmetrical, is thus of indices $n(\nu-1)$, $n(\nu-1)$. Two corresponding primes coincide, (1), when they both pass through a point where the curve touches a generator, that is, in d cases; or (2), when the generator lies in both primes; in this case there are $\frac{1}{2}\nu(\nu-1)$ pairs of points of the curve lying in this generator, and there are N generators which meet the space which is the base of the pencil of primes; so that $N\nu(\nu-1)$ coincidences are obtained. Hence we have $d + N\nu(\nu-1) = n(\nu-1) + n(\nu-1)$, or* $d = 2n(\nu-1) - N\nu(\nu-1)$. Equating this to the former value of d we have

$$\nu + p - 1 - \nu p' = n(\nu-1) - \tfrac{1}{2}N\nu(\nu-1).$$

This may be regarded as determining the genus, p, of a curve of order n, lying on a ruled surface of order N, whose prime section is of genus p', when the curve meets each generator in ν points; and agrees with the result found on p. 216 of Vol. V. The formula may be written however $(n-1)(\nu-1) - p + \nu p' = \frac{1}{2}N\nu(\nu-1)$. Here the right side evidently is the number of pairs of points of the curve which belong, both to one set of the involution $I_1{}^\nu$, and also to one set of the linear series $g_1{}^n$, which is determined on the curve by the primes of the pencil considered. Thus, for this involution, on the curve, determined by the generators, and this linear series, the number of pairs belonging to a set of both, is $(n-1)(\nu-1) - p + \nu p'$, where p' is the genus of the involution. When $p' = 0$ this reduces to the number (p. 11) of pairs common to two linear series $g_1{}^\nu$, $g_1{}^n$.

---

* This value of d must however be diminished by 2δ, if the curve have δ double points at points of the ruled surface where there is a unique generator (cf. Segre, *Math. Ann.* XXXIV, 1889, p. 3).

That this formula remains valid when the sets of $I_1{}^\nu$ do not lie on lines, generating a ruled surface upon which the given curve (n, p) lies, may perhaps be regarded as evident; also, another proof is given below, in Ex. 12. But a proof, that no limitation of the involution $I_1{}^\nu$ is involved in the assumption, is obtained by recurring to the figure considered above (Ex. 7 a), of the case of two canonical curves. In that figure, the sets of $I_1{}^\nu$ were obtained by spaces $[p-p']$, passing through a fixed base space Σ, of dimension $p-p'-1$, and a point of the curve γ', in space $[p'-1]$. If, through this last space, there be drawn a space $[p']$, lying in the space $[p-1]$ which contains the curves γ and γ', this space $[p']$ will be met in a line by every one of the spaces $[p-p']$ determining a set of the involution on γ, and this line will contain the point of the curve γ' which fixes the set of the involution.

Ex. 9. We can use the principle of correspondence to determine, for a ruled surface in ordinary space: (a), the number of pairs of consecutive generators which intersect one another (that is, the number of torsal generators, or of pinch points of the double curve of the surface); or, (b), the genus of the double curve of the surface; or, (c), the number of triple points of this double curve. It is assumed that the surface has no cuspidal curve, and that the double curve is irreducible.

For (a), we may proceed as follows: let the plane sections of the ruled surface be of order n and genus p. Take an arbitrary line, and, through it, two arbitrary plane sections, α and β. To any point, P, of the line correspond $m = 2n + 2p - 2$ points of contact, T, of tangents from P to the section in one plane, α; and the generator through T meets the plane section in the other plane, β, in a point U; let the tangent line of the section β meet the line in a point Q. We thus have an obviously symmetrical correspondence of points P, Q of this line. When P is at one of the n points in which the line meets the surface, the point T coincides with this P, and the tangent plane at U, which is to give Q, contains PU; so that Q coincides with this P. Similarly, when Q is at this point, so is P. If P be on a torsal generator, the tangent plane is the same at all points of this, and P, Q coincide; and conversely. The number of torsal generators is thus $2m - 2n$, or $2 (n + 2p - 2)$. Schubert gives two determinations of this number; for one of these see his paper, *Math. Ann.* XVII. The other is noticed in the following Chapter, dealing with Schubert's methods (p. 102).

A clear deduction of the number of torsal lines, and of the number of generators of the ruled surface which touch the double curve, is obtained also by the representation, referred to in Vol. v, p. 195, by which the generators correspond to points of a curve γ, of order n and genus p, lying on a quadric Ω in space of five dimensions. To the $n-2$ generators which meet a generator h correspond then $(n-2)$ points (Q), of the curve γ, lying on the tangent prime of Ω at a point P of the curve. The lines PQ lie on Ω, and there is a correspondence of indices $(n-2, n-2)$, and valency 2, between P and Q. The torsal generators correspond then to the coincidences of this correspondence, and are in number $2(n-2)+4p$, as before*. For the double curve to touch a generator h, two of the $n-2$ generators which meet h must coincide. The number of contacts is, therefore, the number of coincidences of the direct lateral correspondence derived from the correspondence (P, Q) on the curve γ. For this correspondence $s=r=n-2$ and $\gamma=2$; and from the formula (above, p. 11) $2s(r-1)+2p(s-\gamma^2)$ we thence derive, for the number, ζ, of contacts of the double curve with a generator, $\zeta=2(n-2)(n-3)+2p(n-6)$.

(b) Now let P be the genus of the double curve of the ruled surface; and consider the $(n-2, 2)$ correspondence between the double curve, and a plane section, wherein, to any point, L, of the double curve, correspond the two points, M, of the plane section in which this is met by the generators through L, and, to any point M of the plane section, correspond the $n-2$ intersections, L, of the double curve with the generator through M. The coincidences, in this correspondence, which lie on the double curve, are the ζ contacts of this curve with generators; the coincidences on the plane section are the η intersections of this with the torsal generators (where $\eta=2n+4p-4$). By Zeuthen's formula, obtained above, we therefore have $2(2P-2)-\zeta=(n-2)(2p-2)-\eta$, leading to

$$2P-2=(n-5)(n+2p-2).$$

We may also notice that the rank $(2b+2P-2)$ of the double curve, of order b, the number of its tangents which meet an arbitrary line, is equal to ζ, the number of its tangents which are generators of the surface. It may perhaps also be remarked (see below, p. 32) that, for a curve of order n and genus p, for which the numbers of cusps, inflexions, double points and double tangents are respectively κ, i, δ, τ, the number of the tangents of the curve which meet the curve again is $2(n-2)(n-3)+2p(n-6)-(n\kappa+2i+4\delta+4\tau)$.

(c) To determine the number of triple points of the double curve, consider the correspondence of pairs of planes of a pencil, drawn

---

* If η be the number of torsal generators, and b, $=\frac{1}{2}(n-1)(n-2)-p$, be the order of the double curve, we have $8b+2\eta=4n(n-2)$, in accordance with Salmon's remark (*Solid Geometry*, 1882, p. 591) that the Hessian of a ruled surface (of order $4n(n-2)$) meets the surface in the double curve (which is 4-ple on the Hessian), and in the torsal generators (which are double on the Hessian). Zeuthen (*Lehrbuch d. abzähl. Geometrie*, 1914, p. 219) obtains the number of torsal generators by allowing p of the generators of the surface to become double, so that the plane section becomes rational.

through an arbitrary axis, l, determined by the condition that two
of these planes correspond when they meet a generator in two points
of the double curve. Any plane through l meets the double curve in
$b \, (=\tfrac{1}{2} \, (n-1)(n-2)-p)$ points, through each of which pass two
generators, each meeting the double curve in $n-3$ other points; the
indices of the correspondence are therefore $2b \, (n-3)$, $2b \, (n-3)$. The
coincidences of corresponding planes arise, either when the two
points of the same generator which determine a pair of corresponding planes, coincide with one another, or when a generator meets the
line l. The former arises in two ways, either because the double
curve touches a generator (ζ cases), or because there are three
generators passing through a point of the double curve; then the
curve, and the surface, have a triple point; and, as each branch of
the double curve is then met by both the other branches, there are
6 coincidences of pairs of corresponding planes, so that, if t be the
number of triple points, the number of coincidences so arising is $6t$.
For the latter, when a generator meets the line l, each of the
$\tfrac{1}{2} \, (n-2)(n-3)$ pairs chosen from the $n-2$ intersections of the
generator with the double curve, gives rise to a double coincidence
of pairs of corresponding planes. On the whole then

$$4b \, (n-3)=\zeta+6t+n \, (n-2)(n-3),$$

which leads to

$$t=(n-4)[\tfrac{1}{6} \, (n-2)(n-3)-p], \;=(n-4)[h-\tfrac{1}{3}n \, (n-2)],$$

if h be the number of apparent double points of the double curve.

It has been assumed that the double curve is irreducible, and
that the surface has no cuspidal curve; in general, consult W. L.
Edge, *Ruled Surfaces* (Cambridge).

Ex. 10. We prove now that, for two curves, in ordinary space,
of orders n_1, n_2, with respectively h_1, h_2 apparent double points,
and i mutual intersections, the number of common chords is

$$h_1h_2+\tfrac{1}{4}n_1n_2(n_1-1)(n_2-1)-i(n_1-1)(n_2-1)+\tfrac{1}{2}i(i-1).$$

This result follows from a known result (Vol. IV, p. 51), that the
number of common lines of two algebraic congruences of lines, of
orders p_1, p_2 and classes q_1, q_2, is $p_1p_2+q_1q_2$ (the order and class of
a congruence being the numbers of its lines respectively through
an arbitrary point and in an arbitrary plane). This result was
stated by Halphen (*Compt. rend.* 1872, p. 41), and is interesting
historically as being one of the first of such linear formulae (cf.
Schubert, *Abzähl. Geom.* pp. 62, 337).

The chords of the first curve being a congruence of order $p_1=h_1$,
and class $q_1=\tfrac{1}{2}n_1 \, (n_1-1)$, the general formula for the number of
common chords is $h_1h_2+\tfrac{1}{4}n_1n_2(n_1-1)(n_2-1)$. This however includes the joins of every two intersections of the curves, in number

$\frac{1}{2}i(i-1)$; it also includes, for each intersection, the common generators of the two cones projecting the curves from this intersection, of which the number, beside those to another intersection, is $(n_1-1)(n_2-1)-(i-1)$. The total number of proper common chords of the two curves, not passing through any common point, is thus

$$h_1h_2+\tfrac{1}{4}n_1n_2\,(n_1-1)(n_2-1)-\tfrac{1}{2}i(i-1)-i[(n_1-1)(n_2-1)-(i-1)],$$

agreeing with the statement made.

But an examination of the problem by the method of the theory of correspondence is interesting. From a point P, of the curve (n_1), draw the h_2 chords of the curve (n_2), and let P' be one of the $2h_2$ extremities of these chords; from P' draw the h_1 chords of the original curve (n_1), and let Q be one of the $2h_1$ extremities of these. Thus from P are obtained $4h_1h_2$ points Q, of the curve (n_1), and we regard these as being in correspondence. To pass back from Q to P, we take one of the $n_2(n_1-1)$ common generators of the cones which project the two curves from Q, denoting by P' the extremity, lying on (n_2), of one such chord; and then take one of the $n_1(n_2-1)$ common generators of the cones which project the two curves from P', denoting by P the extremity, lying on (n_1), of one such chord. There are thus $n_2(n_1-1).n_1(n_2-1)$ positions of P corresponding reversely to Q. The relation between P and Q is evidently an algebraic correspondence. We shall prove that this is of zero valency, the $4h_1h_2$ positions of Q, for an arbitrary P, being a set of a linear series, taken by themselves. The number of coincidences, of P and Q, is thus $4h_1h_2+n_1n_2(n_1-1)(n_2-1)$; each such coincidence involves that PP' is a common chord of the two curves; and, conversely, a common chord gives rise to four such coincidences. Thus, after correction for the common points of the two curves, we reach the same result as before.

To complete this proof, we shew that, if the h chords be drawn from an arbitrary point O, to a non-singular curve of order n and genus p, then the $2h$ extremities of these chords, as O varies, are sets of the same linear series on the curve. They are in fact equivalent with the sets of residual intersections, with the curve, of surfaces of order $n-3$, drawn through any chosen canonical set (of $2p-2$ points) on the curve. For, consider the projection of the curve to a plane, from O. The canonical series on the plane curve is obtained by the curves of order $n-3$ through its double points; and, for general positions of O, the canonical sets on the plane curve are in $(1, 1)$ correspondence with the canonical sets on the space curve. There is thus a cone of order $n-3$, with vertex at O, which contains any particular canonical set on the space curve, and also contains the h chords to the curve from the point O. This proves the statement made. In the application to the problem in hand, the set of $4h_1h_2$ points Q is made up of $2h_2$ sets each by itself belonging to a linear series when P varies. The aggregate set of points Q is thus one of a linear series.

Ex. 10 *a*. Consider the quintic curve which is the residual intersection of a quadric surface with a cubic surface through one generator; this curve is of genus 2, and meets the generator in 3 points. From any arbitrary point O, there can be drawn 4 chords of this curve; also the plane joining O to the fixed generator has two points A, B on the curve (beside 3 on the generator); the four chords from O, and the two lines OA, OB, lie on a quadric cone, with vertex at O (as is proved in Vol. v, Chapter vIII). Now let P be a point of the quintic curve, and l, m be two fixed arbitrary lines; let the transversal from P meet these lines in P' and H', respectively; from P' draw the four chords to the quintic curve, meeting this in $Q_1, Q_1'; Q_2, Q_2'; Q_3, Q_3'; Q_4, Q_4'$; also, from H', draw the four chords $K_1, K_1'; \ldots; K_4, K_4'$. By what has been proved in Ex. 10, the sixteen points (Q), (K) are a set of a linear series, as P varies on the curve.

Ex. 10 *b*. For the number of chords of a curve (n_1, h_1) which meet two arbitrary lines, the formula of Ex. 10 (putting $h_2 = 1$, $n_2 = 2$), gives $h_1 + \frac{1}{2}n_1(n_1 - 1)$. When the lines intersect this is obvious. This is then the order of the ruled surface generated by chords of the curve which meet an arbitrary line; in terms of the genus of the (non-singular) curve, it is $(n_1 - 1)^2 - p_1$.

Ex. 10 *c*. The chords of a curve (n) having simple intersections with two other curves (n_1), (n_2), are in number

$$n_1 n_2 \left[h + \tfrac{1}{2} n(n-1) \right] - (n-1) \left[n_1 i_2 + n_2 i_1 \right] - h i_{12} + i_1 i_2,$$

where i_1, i_2 are the respective numbers of intersections of the curves (n_1), (n_2) with the first curve (n), and i_{12} the number of their mutual intersections.

Ex. 10 *d*. The number of lines having simple distinct intersections with each of four curves (n_1), (n_2), (n_3), (n_4), is

$$2 n_1 n_2 n_3 n_4 - \Sigma n_p n_q i_{rs} + \Sigma i_{pq} i_{rs},$$

where the curves (n_r), (n_s) have i_{rs} intersections, and p, q, r, s are the numbers 1, 2, 3, 4 in some order.

Ex. 10 *e*. For a curve, of order n, class n', and genus p, with k-ple points, of separated tangents, the number of chords each of which lies in the osculating planes of both extremities of the chord, is

$$(n-3, 2) + (n'-3, 2) - \Sigma(k, 2) - 12p,$$

where (λ, μ) means

$$\lambda(\lambda - 1) \ldots (\lambda - \mu + 1)/\mu!.$$

For example, for a rational quartic curve, there are 3 such chords.

References to the literature for these results are given by Berzolari, *Pascal's Repertorium*, II, 2, 1922, p. 906.

Ex. **11.** We consider now the chords having three, or four, intersections with a curve, the trisecants and quadrisecants. Of the former there is in general an infinite number, determining a ruled surface; of the latter a definite finite number.

The trisecant lines which can be drawn, to a curve of order n, and genus p, without multiple points, from any point P of itself, each meeting the curve in two other points, are the same in number as the double points of the curve, of order $n-1$, obtained by projection of the given curve, from P, upon a plane. Assuming that the plane curve is of genus p, this number is $\frac{1}{2}(n-2)(n-3)-p$, which is the same as $h-(n-2)$, if h be the number of apparent double points of the original curve. The points Q, in number $(n-2)(n-3)-2p$, or say ν, where these trisecants from P again meet the curve, are in a correspondence (P, Q) with P, of indices ν, ν. We prove that this correspondence is of valency $n-4$. Take a particular canonical set (of $2p-2$ points) on the plane curve obtained by projection of the given curve from P; let (A) be the set of $2p-2$ points on the space curve projecting from P into this particular canonical set; there is then a cone, of order $n-4$, of vertex P, containing the set (A) and the trisecant chords from P to the space curve (as follows by the definition of canonical sets on the plane curve); but, conversely, if (A_0) be any canonical set of the space curve, these project from a point P of the curve into a canonical set of the resulting plane curve. There is thus a system of cones, of order $n-4$, containing the set (A_0), with vertices at points P of the space curve, each containing the trisecants of the curve drawn from the point P. The points (Q), which are the other intersections of these trisecants with the curve, thus lie with P, taken $(n-4)$ times over, in a set of a linear series upon the curve.

But we found (p. 16, above) that the order of the ruled surface generated by the lines joining points P, Q of the curve, in (s, r) correspondence, of valency γ, is $(n-1)(s+r)-2\gamma p$. Thus, as, in the present case, each trisecant arises from the correspondence in six ways, the trisecants generate a surface of order

$$\tfrac{1}{6}\{(n-1)[2(n-1)(n-3)-4p]-2(n-4)p\}$$

which is $(n-2)[\frac{1}{3}(n-1)(n-3)-p]$. If the curve have a double point, the ruled surface, as so found, will include the cone, of order $n-2$, projecting the curve from this point; and it is the same for a cusp. The residual ruled surface of proper trisecants is then of order $(n-2)[\frac{1}{3}(n-1)(n-3)-p-\delta-\kappa]$, if the curve have δ double points, and κ cusps. This is the same as the number, t_3, given by $t_3=(n-2)[h-\frac{1}{6}n(n-1)]$, where h is the number of apparent double points of the curve. And this form includes the simpler case.

The tangents of the curve which meet the curve again arise by the coincidences of the correspondence (P, Q), and are thus, in general, of number $2v + 2(n-4)p$, namely

$$2(n-2)(n-3) + 2(n-6)p;$$

but, if the curve have i inflexions, κ cusps, δ actual double points, and τ actual double tangents, we must in fact subtract from this the number $2i + n\kappa + 4\delta + 4\tau$. If r denote the rank of the curve, the resulting number, ξ, may also be written

$$\xi = (n-4)\,r + 4h - 2n\,(n-3) - 2i - 4\tau.$$

Save for the term -4τ, this corrected result is given by Zeuthen, *Ann. d. Mat.* III, Milan, 1869–70; and, for historical reasons, it may be interesting to give an indication of Zeuthen's reasoning. Of a pencil of planes through an arbitrary axis, let two be regarded as corresponding when they meet the curve in two points of a trisecant; we thus have a correspondence, between pairs of planes of this pencil, whose indices are $2nt$, $2nt$, where t is the number, found above, of trisecants which can be drawn to the curve from a point of itself. There are coincidences in this correspondence, (1), for a tangent meeting the curve again (say ξ cases); (2), for each inflexion (counting twice); (3), for every proper trisecant of the curve which meets the axis of the pencil of planes (counting six times, giving therefore $6t_3$, in the notation above); (4), for each double point of the curve (counting $2n-8$ times); (5), for each cusp (counting $3n-12$ times); (6), for each double tangent (counting four times). Thus we have $4nt = \xi + 2i + 6t_3 + (n-4)(2\delta + 3\kappa) + 4\tau$; substituting

$$t = h - (n-2), \quad t_3 = (n-2)[h - \tfrac{1}{6}n\,(n-1)], \quad 2\delta + 3\kappa = n(n-1) - 2h - r,$$

we find for ξ the value put down.

We have above deduced the number $h - (n-2)$ of trisecants to the curve, from an arbitrary point P of itself, with the assumption that the plane curve obtained by projection of the given curve is of genus p. Zeuthen deduces this result without this assumption: taking two arbitrary points A, B of the space, and two points Q, Q' of the curve, and excluding the cases when the planes APQ, BPQ' are coincident, the points Q, P, Q' are in line when the plane APQ contains Q', and the plane BPQ' contains Q. By an ingenious argument, equivalent to using the product of two correspondences, it is shewn that the number of cases in which this happens is $(n-2)^2 - p$.

Ex. 11 *a.* We pass now to the problem of finding the number of quadrisecants of a curve. Let A_1, A_2 be two arbitrary fixed points, not lying on the curve. Let P_1 denote one of the $n-3$ intersections,

with the curve, of the plane joining A_1 to a trisecant of the curve
(not on the trisecant), and P_2 one of the $n-3$ intersections, with the
curve (other than on the trisecant), of the plane joining A_2 to the
same trisecant. When P_1 is any point of the curve, the line A_1P_1 is
met by a definite number of trisecants of the curve, and each of
these, joined by a plane to A_2, gives $(n-3)$ points P_2; the number
of trisecants which meet A_1P_1, excluding those through P_1, is t_3-t
(where, as above, $t_3=(n-2)h-\frac{1}{6}n(n-1)(n-2)$ and $t=h-(n-2)$).
Thus the points P_1, P_2 of the curve are in correspondence, with indices both equal to $(n-3)(t_3-t)$. We shew that this correspondence
is of valency k_{12}, where $k_{12}=n-4-t$. Let P_3, P_4 be two of the intersections, with the curve, of a trisecant (not through P_1) which meets
A_1P_1. There is a correspondence from P_1 to P_3, say (P_1, P_3); the
sum of this, and the correspondence (P_1, P_2) under examination, is
a correspondence leading from P_1 to all the points of the curve
which lie on the planes joining A_2 to the trisecants (not through P_1)
which meet A_1P_1; this aggregate of planes does not contain P_1, and
meets the curve in a set of a linear series. Assuming this, we infer
$k_{13}+k_{12}=0$, where k_{13} is the valency of the correspondence (P_1, P_3).
Again, when P_3 is given, the planes joining A_1 to the trisecants of
the curve through P_3, evidently give, as the aggregate of their intersections with the curve, the points arising from P_3 by the sum
correspondence $(P_3, P_1)+(P_3, P_4)$; but they give a linear series on
the curve in which P_3 arises t times in each set $(t=h-n+2)$, that is,
a correspondence of valency t; thus, with a similar notation,
$k_{31}+k_{34}=h-(n-2)$. We have already shewn (p. 31) that $k_{34}=n-4$;
hence
$$k_{12}=-k_{13}=-k_{31}=n-4-(h-n+2), \quad =2(n-3)-h,$$
as stated.

Now consider the coincidences in the correspondence (P_1, P_2);
these arise in two ways: (1), If a trisecant of the curve meet the line
A_1A_2, the planes from A_1 and A_2 to this trisecant coincide in one
plane, and the $n-3$ intersections of this plane with the curve, not
on the trisecant, are coincidences; thus we have $(n-3)t_3$ coincidences; (2), If a trisecant have a fourth intersection with the curve,
the planes joining A_1 and A_2 to this have intersections with the
curve, other than at the three points which define the trisecant,
which coincide at this fourth point. The quadrisecant thus gives 4
coincidences of P_1 and P_2. Assuming that there is no other possibility, if t_4 denote the total number of quadrisecants of the curve,
the whole number of coincidences is thus $(n-3)t_3+4t_4$. The correspondence has been seen to be of equal indices $(n-3)(t_3-t)$, and
of valency $n-4-t$. Thus we have
$$4t_4=2(n-3)(t_3-t)+2p(n-4-t)-(n-3)t_3,$$

and $t_4 = (h-2n+6, 2)-(n-3, 4)$, where (λ, μ) is the binomial $\lambda (\lambda-1) \dots (\lambda-\mu+1)/\mu !$. This is the same as

$$\tfrac{1}{2}h\,(h-4n+11)-\tfrac{1}{24}n\,(n-2)\,(n-3)\,(n-13)$$

or, the curve having no multiple points, as

$$(p, 2)-p\,(n-3, 2)+\tfrac{1}{3}\,(n-2, 2)\,(n-3, 2).$$

The above argument is very slightly modified from Zeuthen (*Lehrbuch*, 1914, p. 257), who also considers the case of multiple points in *Ann. d. Mat.* III, 1869–70. The number t_4 was obtained by Salmon (*Solid Geometry*, 1882, p. 435), Cayley (*Papers*, v, p. 171), Brill (*Math. Ann.* VI, 1873, p. 52), and by others (Pascal, *Repertorium*, II, 2, 1922, p. 905). In Enriques-Chisini, *Teoria Geometrica*, II, p. 303, the result is found by a functional method due to Cayley; and, *ibid.* III, p. 473, by a method indicated in the following Ex. 11 *b*.

Ex. 11 *b*. Let Q be a point, other than P, of a trisecant drawn to a curve in space from a point P of the curve; let R be a point, other than Q, of a trisecant, other than PQ, drawn to the curve from Q. Denote the correspondences, from P to Q, and from P to R, by (P, Q), (P, R). If I denote the identical correspondence, it may be shewn that $(P, Q)^2 = (P, Q)+(P, R)+2tI$ (where $t = h-n+2$). We have proved that the valency of (P, Q) is $n-4$; the correspondence (P, R) is symmetrical, of indices both equal to $4t\,(t-1)$; by the identity remarked its valency is $2t-(n-4)-(n-4)^2$, which is $2\,(n-3)-2p$. The coincidences of this correspondence (P, R) are in number $24t_4+2\zeta$, where ζ is the number of planes which touch the curve at two points whose join is a trisecant. When ζ is found, we can thence find t_4. It may be shewn that

$$\zeta = -4t^2+2t\,(n^2-14)+18\,(n-2, 3).$$

For this proof compare Zeuthen, *Lehrbuch*, 1914, p. 259.

Ex. 11 *c*. Consider three curves in space, whose orders, genera, numbers of apparent double points, and of cusps are n_i, p_i, h_i, κ_i, and the ruled surface generated by a line meeting simply each of these curves. Prove that the genus, P, of the plane section of this ruled surface is given by $2P-2 = 4n_1 n_2 n_3 + \Sigma n_j n_k\,(2p_i-2)$, where i, j, k are 1, 2, 3 in some order.

Also that the genus, P_0, of the plane section of the ruled surface generated by the trisecants of a curve (n, p, h, κ), is given by $2P_0-2 = \tfrac{1}{3}\,(6h-n^2-n)\,(n-4)+(h-2n+6)\,(2p-2)$. The formulae for the order, rank, etc., of a ruled surface generated by the transversals of three curves, by the chords of a curve which meet another curve, and by the trisecants of a single curve, with many references to the literature, are given in Pascal's *Repertorium*, II, 2, 1922, pp. 907–8.

Ex. 12. Suppose that on a curve we have an involution of sets of ν points, of freedom unity and index 1 (so that only one set of the involution contains a given point of the curve); we denote this by I_1^ν, a single set being denoted by I. Suppose also there exists on the curve a linear series, of sets of n points, of freedom r, say g_r^n which is *simple* (so that sets of the series which have one point in common, of general position on the curve, do not thereby have other points in common). Suppose further that $\nu > r$, and that any r points of a set I of the involution form independent conditions for a set of g_r^n, and so determine this set. We proceed to find how many times it happens that the set of the linear series, so determined by r points of a set I, has another point contained in this set I; so that $r + 1$ points are common to a set I and a set of g_r^n. We have already considered the case when $r = 1$, in Ex. 8.

If the linear series g_r^n be given by a system of curves

$$\lambda_0 \phi_0 + \ldots + \lambda_r \phi_r = 0,$$

it is well known (see Vol. v, p. 26) that, by equations of the forms $X_0/\phi_0 = \ldots = X_r/\phi_r$, the curve can be placed in (1, 1) birational correspondence with a curve of order n in space of r dimensions; the sets of the linear series are then represented by the prime sections of this curve in space $[r]$. And there will be an involution I_1^ν upon this curve similar to that on the original curve. Our problem is then to find the number of sets I of this involution which have $r + 1$ coprimal points (where $\nu > r$). Denote this number by $Z_{n,r}$; also denote by d the number of coincidences of a pair of points belonging to the same set of the involution. For brevity, let ρ denote the binomial coefficient $(\nu - 2, r - 1)$.

Take a point P, on the curve in the space $[r]$. This determines a set I, of ν points of the curve, belonging to I_1^ν; a prime can be put through every r points of the $\nu - 1$ points of this set, other than P; and there will be $(\nu - 1, r)$ such primes. Each of these primes meets the curve in $n - r$ points, Q, beside those through which it is described. The aggregate of these $(\nu - 1, r)$ sets of $n - r$ points Q is determined by P; we consider the correspondence between P and a point of this aggregate. This correspondence (P, Q) we denote by T. With this we consider the correspondence between P and a point P', other than P, of the set of ν points I determined by P; this correspondence (P, P') we denote by U. The primes which determine the aggregate of points Q will use every point P' a certain number of times; namely as many times as there are sets of $r - 1$ points in the set I, beside P and P' (which with P' will make up a set of r points determining such a prime); thus each point P' will be used $(\nu - 2, r - 1)$, or ρ times. Whence we see that the sum correspondence denoted by $T + \rho U$ is of zero valency, the points corresponding to P in this being the complete intersection of the curve with $(\nu-1, r)$ primes. The correspondence U between P and one of the $\nu-1$ points P' is obviously symmetrical. Consider the correspondence reverse to T: the primes through a point Q of the curve determine a linear series g_{r-1}^{n-1} thereon; to get back to a point P it is necessary to find the sets of I_1^{ν} which contain r points of a set of this g_{r-1}^{n-1}. If we assume that the number $Z_{n,r}$, when the involution I_1^{ν} is given, is a function only of n and r, the number of sets of I_1^{ν} which contain r points of the linear series g_{r-1}^{n-1}, is $Z_{n-1,r-1}$; and when one such set, I, is found, P may be any one of the $\nu-r$ points of this set I, other than the r points belonging to a set of g_{r-1}^{n-1}. The reverse index of the correspondence T is thus $(\nu-r)Z_{n-1,r-1}$. The direct and reverse indices of the correspondence $T+\rho U$ are thus

$$(n-r)(\nu-1, r)+\rho(\nu-1), \text{ and } (\nu-r)Z_{n-1,r-1}+\rho(\nu-1);$$

we have remarked (p. 6) that the reverse of a sum of correspondences is the sum of their reverses. Now consider the coincidences in the correspondence $T+\rho U$. These arise, (1), when a point Q coincides with P; then there are $(r+1)$ points of a set I which are coprimal, and, therefore, $(r+1)Z_{n,r}$ coincidences of the correspondence; (2), when a point P' coincides with P; and this gives ρd coincidences of the correspondence $T+\rho U$. Thus we have

$$(r+1)Z_{n,r}+\rho d = (n-r)(\nu-1, r)+(\nu-r)Z_{n-1,r-1}+2\rho(\nu-1);$$

since $(\nu-1, r)=(\nu-1)\rho/r$, this may be put into the form

$$(r+1)Z_{n,r}-(\nu-r)Z_{n-1,r-1}=\rho\left[\frac{(n+r)(\nu-1)}{r}-d\right],$$

and we easily compute that, if we put

$$c_{n,r}=\rho\left[\frac{n(\nu-1)}{r}-\tfrac{1}{2}d\right], =(\nu-2, r-1)\left[\frac{n(\nu-1)}{r}-\tfrac{1}{2}d\right],$$

then this is the same as

$$(r+1)(Z_{n,r}-c_{n,r})=(\nu-r)(Z_{n-1,r-1}-c_{n-1,r-1}).$$

To solve this equation, we prove that $Z_{m,1}=m(\nu-1)-\tfrac{1}{2}d, =c_{m,1}$. Recurring to the original plane curve, this is established by the argument we have used in general, though the result has already been obtained above, in Ex. 8. The number $Z_{m,1}$ is that of the sets of I_1^{ν} which contain two points of a set of a linear series g_1^m. Consider a set I, of I_1^{ν}, determined by a point P; let P_0' be another point of this set; let Q_0 be one of the points, other than P_0', of the set of g_1^m determined by P_0'; consider the sum correspondence $(P, Q_0)+(P, P_0')$. For the correspondence reverse to (P, Q_0): there is a set of I_1^{ν} determined by every one of the $m-1$ points, other than Q_0, belonging to the set of g_1^m which contains Q_0; and there are thus $(m-1)(\nu-1)$ points P corresponding reversely to Q_0. Thus, as in the case $r > 1$, we have the equation

$$2Z_{m,1} + d = (m-1)(\nu-1) + (\nu-1)(m-1) + 2(\nu-1), = 2m(\nu-1),$$

as desired.

Wherefore, supposing $m = n - r$, the general difference equation leads to $Z_{n,r} = c_{n,r}$, namely

$$Z_{n,r} = (\nu-2, r-1)\left[\frac{n}{r}(\nu-1) - \tfrac{1}{2}d\right].$$

This result may be generalised. See Segre, *Ann. d. Mat.* xxii, 1894, p. 99, and the references given in Severi, *Trattato di geom. algeb.* i, 1, 1926, p. 253, to whom the proof here given is due.

We have supposed the involution I_1^{ν} to be of index 1. If this be of index ν', the result is at once obtainable from that found by replacing n by $n\nu'$. Or a direct proof, identical with that given, may be followed.

A consequence of the formula

$$Z_{n,r} = (\nu-2, r-1)\left[\frac{n\nu'}{r}(\nu-1) - \tfrac{1}{2}d\right],$$

for the case of an irreducible involution I_1^{ν}, of freedom 1, of sets of ν points, but with index ν', is remarked by Castelnuovo (*Rend. Lincei*, xv, 1906, p. 337), namely:

The number of coincidences of pairs of points of a set of such an involution, cannot exceed $\nu'(2\nu + 2p - 2)$; and, if d is as great as this, all the sets of ν points, of the involution, are sets of the same linear series. We give a sketch of the proof of this.

With $\nu - 1$ points of a set I of such an involution, and p arbitrary points of the curve, define a linear series, of sets of $\nu - 1 + p$ points of freedom $\nu - 1$, $g_{\nu-1}^{\nu-1+p}$. The number of sets of ν points, common to the linear series and the involution, is then $Z_{\nu+p-1,\,\nu-1}$, say Z, given by $Z = \nu'(\nu + p - 1) - \tfrac{1}{2}d$, and, as this cannot be negative, we have the necessary upper bound for d. Suppose, however, if possible, that $d = 2\nu'(\nu + p - 1)$. Then $Z = 0$, and there are, in a general linear series $g_{\nu-1}^{\nu-1+p}$, no sets which, having $\nu - 1$ points taken in a set of the involution, contain the remaining point. The argument by which we found $Z_{n,\,r}$, however, as in all similar cases, assumes that this number is finite. Thus if we construct a linear series $g_{\nu-1}^{\nu-1+p}$ with the property that, a set of this determined by $\nu - 1$ points of a particular set of the involution contains all points of this set, then the conclusion will be that the same is true for every set of the involution; namely, that all the sets of the involution belong to the same linear series. To construct such a linear series, first form the linear series $g_\nu^{\nu+p}$ of which one set consists of a particular set I of the involution taken with p arbitrary points of the curve, say $A_1, ..., A_p$; then take the linear series $g_{\nu-1}^{\nu-1+p}$ formed by all the sets of $g_\nu^{\nu+p}$ which contain the point A_i (other than this). As, by what has been said, every set of the involution consists of points belonging to one set of $g_{\nu-1}^{\nu-1+p}$, we infer that the sets of $g_\nu^{\nu+p}$ which contain sets of the involution all contain A_i; hence, by parity of reasoning, all these sets of $g_\nu^{\nu+p}$ contain $A_1, A_2, ..., A_p$. Thus the involution consists of sets of the linear series obtained by the sets of $g_\nu^{\nu+p}$ which contain $A_1, ..., A_p$.

We may remark that, for general values of n and r, when $d = \nu'(2\nu + 2p - 2)$, the value of $Z_{n,r}$ is

$$\nu'(\nu-1, r)\left[n - r - \frac{rp}{\nu-1}\right],$$

which, when $\nu = r + 2$, is $\nu'[(n-r)(r+1) - rp]$.

Castelnuovo has proved further, as a Corollary, due to Severi (*Ann. d. Mat.* xii, 1906, p. 55), that, if the sets of $\nu'\nu$ points, consisting of all the sets of the involution I_1^ν which contain a point P of the curve (including this point itself reckoned ν' times), belong to a linear series, so do the separate sets of ν points of the involution.

Ex. 13. For the developable surface formed by the tangents of a curve, beside the cuspidal curve which is constituted by the original curve, there is a double curve, the *nodal curve*, of order ν, such that $p = \frac{1}{2}(r-1)(r-2) - n - \nu - \tau - i$ (see p. 191 of Vol. v). This double curve has triple points whose number is

$$\tfrac{1}{3}\left[-r^3 + 13r^2 - 42r + 8n' + \nu(3r-26) - 2\tau\right].$$

Cf. Zeuthen, *Ann. d. Mat.* iii, 1869–70; Pascal, *Repertorium*, ii, 2, 1922, p. 899. The triple tangent planes of the curve are of number obtainable from this by replacing n' by n, and ν by the class ν' of the developable surface formed by the planes which have two contacts with the curve. For example, for the sextic intersection of a quadric and a cubic surface, we have $n = 6$, $r = 18$, $\nu' = 96$, $\tau = 0$; and the number of triple tangent planes is 120. (See also, below, Ex. 14 *b*.)

Find the genus of the double curve of the developable surface formed by the tangent lines of the curve.

Many particular results will be found in Zeuthen's paper referred to, and in the Pascal. See also Cayley's paper, *Papers*, v, p. 172.

Ex. 14. The curves which determine the sets of a linear series g_r^n, of freedom r, upon a given plane curve, contain r variable parameters. We expect then that there is a determinate number of such curves having $r+1$ coincident intersections with the given curve, if the point of contact is not assigned—and we have found this number (p. 10, above). The conditions to be satisfied may be formulated algebraically, by expressing that the general curve giving a set of the linear series, and the r further curves obtained by differentiation of this, have a common point on the given curve. Similarly, the conditions that, among the intersections of a variable line with a given plane curve, there should be two points of contact, are two in number; we expect then a finite number of bitangents of the curve; though this problem, as formulated algebraically, will have as solutions the lines joining two cusps of the curve, and the tangents to the curve from the cusps. Somewhat similarly, if we consider an algebraic curve in space of four dimensions, since the conditions for a line to meet such a curve are two in number, and there are ∞^2 chords of the curve, we expect to find a finite number of chords meeting the curve in a third point, that is, of trisecants of the curve. Many such problems have been solved by the principle of correspondence. We give here first, with an indication of the proof, a very general formula due to de Jonquières. This determines, in a given linear series g_r^n upon a given plane curve, the number of sets in which there are several coincidences, or contacts, at unassigned points of the curve, of such character that the total number of conditions required for these coincidences is r. And we add, without proof, some formulae for problems of incidence such as that referred to, of the number of trisecants of a curve in space [4].

We consider then, first, upon a given plane curve, a given linear series g_r^n, not necessarily complete, of which the sets have no points in common. We consider the possibility of a set, of n points, of this linear series, consisting of α_0 simple points, of α_1 points each of which is a coincidence of r_1+1 points, of α_2 points at each of which there is a coincidence of r_2+1 points, and so on, finally of α_p points at each of which there is a coincidence of r_p+1 points; so that $\alpha_0+\alpha_1(r_1+1)+...+\alpha_p(r_p+1)=n$. For a coincidence of r_i+1 points at an unassigned point of the curve, r_i conditions are necessary. We suppose the total number of conditions imposed for the set to be r; that is $\alpha_1 r_1+...+\alpha_p r_p=r$. We put $k=n-r$, and so have $\alpha_0+\alpha_1+...+\alpha_p=k$, this being the number of distinct points in the set. We expect that the linear series contains a finite number of sets of the character described; our object is to find this number. If p be the genus of the curve, we put p_i, k_i for the binomial coefficients, $p_i=p!/i!(p-i)!$, $k_i=k!/i!(k-i)!$, and denote by s_m the coefficient of t^m in the ascending expansion of $(1+r_1t)^{\alpha_1}(1+r_2t)^{\alpha_2}...(1+r_pt)^{\alpha_p}$, so that s_m is the sum of the products, m together, without repetitions, of

$$r_1^{(1)}, r_1^{(2)}, \ldots, r_1^{(\alpha_1)}; r_2^{(1)}, r_2^{(2)}, \ldots, r_2^{(\alpha_2)}; \ldots\ldots r_\rho^{(1)}, r_\rho^{(2)}, \ldots, r_\rho^{(\alpha_\rho)},$$

wherein

$$r_1^{(1)} = r_1^{(2)} = \ldots = r_1^{(\alpha_1)} = r_1, \ldots, r_\rho^{(1)} = r_\rho^{(2)} = \ldots = r_\rho^{(\alpha_\rho)} = r_\rho.$$

Thus s_m is zero when $m > \alpha_1 + \ldots + \alpha_\rho$, or $m > k - \alpha_0$. We also use two symbols, M and f, the former as a multiplier, the latter a series of $1 + \alpha_1 + \ldots + \alpha_\rho$ terms, defined by

$$M(0^{\alpha_0}, r_1^{\alpha_1}, \ldots, r_\rho^{\alpha_\rho}) = (k!/\alpha_0!\alpha_1!\ldots\alpha_\rho!)(1 + r_1)^{\alpha_1}\ldots(1 + r_\rho)^{\alpha_\rho},$$

$$f(0^{\alpha_0}, r_1^{\alpha_1}, \ldots, r_\rho^{\alpha_\rho}) = 1 + (p_1/k_1)s_1 + (p_2/k_2)s_2 + \ldots + (p_{k-\alpha_0}/k_{k-\alpha_0})s_{k-\alpha_0},$$

and denote the product of these by ψ, so that

$$\psi(0^{\alpha_0}, r_1^{\alpha_1}, \ldots, r_\rho^{\alpha_\rho}) = M(0^{\alpha_0}, r_1^{\alpha_1}, \ldots, r_\rho^{\alpha_\rho}).f(0^{\alpha_0}, r_1^{\alpha_1}, \ldots, r_\rho^{\alpha_\rho}).$$

The result is that the number of sets of g_r^n satisfying the prescribed conditions is this number ψ.

To explain the notation, first take simple examples: (1), For the number of sets of g_r^n which have a single coincidence of $r+1$ points, we have $\alpha_0 = n - r - 1$, $\rho = 1$, $\alpha_1 = 1$, $r_1 = r$, $k = n - r$, and

$$M = [(n-r)!/(n-r-1)!](1+r), \ = (n-r)(1+r),$$

$$f = 1 + [p/(n-r)]r, \ = 1 + rp/(n-r),$$

so that $\psi = Mf = (r+1)(n-r+rp)$, as was formerly found (Ex. 2, p. 10); (2), For the number of lines meeting a plane curve of order n in two points of coincidence each of two points, and in $n-4$ simple points, we have $\alpha_0 = n - 4$, $\alpha_1 = 2$, $r_1 = 1$, $k = n - 2$, $\rho = 1$, and

$$M = 2^2(n-2)!/(n-4)!2!, \ = 2(n-2)(n-3),$$

so that
$$f = 1 + 2p/(n-2) + p_2/(n-2)_2,$$

$$\psi = 2(n-2)(n-3) + 4(n-3)p + 2p(p-1).$$

This can be verified to be the same as

$$\tau + \tfrac{1}{2}\kappa(\kappa-1) + \kappa[n(n-1) - 2\delta - 3\kappa - 3],$$

where δ, κ, τ denote the number of double points, cusps, and double tangents of the curve; so that ψ is the number of proper double tangents, together with the number of joins of a pair of cusps, and the number of tangents to the curve drawn from the cusps. As a third example, (3), consider how many sets are contained in the canonical series g_{p-1}^{2p-2} which consist of $p-1$ coincidences of two points. For this case, $\alpha_0 = 0$, $\alpha_1 = p-1$, $r_1 = 1$, $\rho = 1$, $k = p-1$ and $n = 2p-2$; thus

$$M = [(p-1)!/(p-1)!]\,2^{p-1},$$

$$f = 1 + [p_1/(p-1)_1]\,(p-1)_1 + [p_2/(p-1)_2]\,(p-1)_2 + \ldots$$
$$+ [p_{p-1}/(p-1)_{p-1}]\,(p-1)_{p-1},$$

or $\qquad f = 1 + p_1 + p_2 + \ldots + p_{p-1} = (1+1)^p - 1, \; = 2^p - 1;$

thus the number is $2^{p-1}(2^p - 1)$, as is well known.

To prove the general formula we proceed by induction; we assume that it is proved for linear series $g_s{}^n$, of freedom $s < r$, and thence deduce it for freedom r. This will be sufficient because the result is known for series of freedom 1. For, then, the only possibility is that for which $\rho = 1$, with $\alpha_1 = 1$, $r_1 = 0$; and the number of Jacobian points of a linear series $g_1{}^n$ is known.

Take a definite point, P, of the fundamental curve; and consider the sets of the series $g_r{}^n$ which have r_1 coincident points at P. The variable intersections then belong to a series, of sets of $n - r_1$ points, with freedom $r - r_1$, $g_{r-r_1}^{n-r_1}$. In this series consider the sets which have, save at P, precisely the same coincidences as those sought, namely have $(\alpha_1 - 1)$ points of $(r_1 + 1)$-fold coincidence, α_2 points of $(r_2 + 1)$-fold coincidence, and so on, finally α_ρ points of $(r_\rho + 1)$-fold coincidence, and therefore $(\alpha_0 + 1)$ points of simple intersection; such a simple intersection we denote by Q. By the assumption made, the number of such sets is $\psi\,(0^{\alpha_0+1}, r_1{}^{\alpha_1-1}, r_2{}^{\alpha_2}, \ldots, r_\rho{}^{\alpha_\rho})$. The simple points Q, of these sets, are determined when P is assigned, and the coincidences in question are fulfilled. We can therefore regard the points Q as arising from P by a correspondence on the curve; and a coincidence in this correspondence, when one of the points Q coincides with P, will give rise to a set of the original series $g_r{}^n$ with the coincidences required; conversely, every such set may be supposed to arise in this way. We desire then to obtain the index of the reverse correspondence, from Q to P, and the valency of this correspondence. Now, to pass from Q to P, we should consider all sets of the original series $g_r{}^n$ having Q as a given point, namely a series g_{r-1}^{n-1}; and therein the sets having one r_1-fold coincidence, at P, with $(\alpha_1 - 1)$ points of $(r_1 + 1)$-fold coincidence, α_2 points of $(r_2 + 1)$-fold coincidence, and so on, finally α_ρ points of $(r_\rho + 1)$-fold coincidence, and therefore α_0 simple points. As the freedom of this series is $r - 1$, the number of such sets, by the assumption made, is $\psi\,(0^{\alpha_0}, r_1 - 1, r_1{}^{\alpha_1-1}, r_2{}^{\alpha_2}, \ldots, r_\rho{}^{\alpha_\rho})$; this is then the reverse index of the correspondence (P, Q). As there are $(\alpha_0 + 1)$ points Q in each of the contemplated sets of the series $g_{r-r_1}^{n-r_1}$, the direct index is $(\alpha_0 + 1)\,\psi\,(0^{\alpha_0+1}, r_1{}^{\alpha_1-1}, r_2{}^{\alpha_2}, \ldots, r_\rho{}^{\alpha_\rho})$. But the determination of the valency of the correspondence (P, Q) is less easy. For this vital step we refer the reader to Zeuthen, *Lehrbuch*, 1914, p. 246, and, especially, Torelli, *Rend. Palermo*, XXI, 1906, p. 58 (see also, de Jonquières, *Crelle*, LXVI, 1866, p. 289; Brill, *Math. Ann.* VI, 1873, p. 47, and Cayley, *Papers*, VII, p. 41). This valency is, in fact, γ, given by $\gamma = r_1 \psi_{p-1} (0^{\alpha_0}, r_1^{\alpha_1-1}, r_2^{\alpha_2}, \ldots, r_\rho^{\alpha_\rho})$, where ψ_{p-1} is formed in accordance with the definition above given for ψ, except in one respect, namely that $p-1$ is substituted for p in the series f; another form for γ is given by

$$p\gamma = \frac{\alpha_1}{r_1+1} \psi(0^{\alpha_0}, r_1^{\alpha_1}, \ldots, r_\rho^{\alpha_\rho}) - (\alpha_0+1)\,\psi(0^{\alpha_0+1}, r_1^{\alpha_1-1}, r_2^{\alpha_2}, \ldots, r_\rho^{\alpha_\rho}).$$

If Z be the number required, of sets of the series g_r^n, which have the specified coincidences (α_1 of (r_1+1)-fold coincidence, ..., α_ρ of $(r_\rho+1)$-fold coincidence, and α_0 simple points), then the number of coincidences of the correspondence (P, Q) will be $\alpha_1 Z$, since P may be at any one of the α_1 points of (r_1+1)-fold coincidence. Thus, to deduce the formula stated for Z, it is necessary to prove that the sum of the two indices above, with $2p\gamma$, is equal to

$$\alpha_1 \psi(0^{\alpha_0}, r_1^{\alpha_1}, \ldots, r_\rho^{\alpha_\rho}).$$

If we denote the portions of these symbols which relate to (α_1-1) points of (r_1+1)-fold coincidence, α_2 points of (r_2+1)-fold coincidence, ..., α_ρ points of $(r_\rho+1)$-fold coincidence, simply by S, the necessary equation is

$$\alpha_1 \psi(0^{\alpha_0}, r_1, S)$$
$$= (\alpha_0+1)\,\psi(0^{\alpha_0+1}, S) + \psi(0^{\alpha_0}, r_1-1, S) + 2pr_1 \psi_{p-1}(0^{\alpha_0}, S),$$

of which the algebraic verification is not difficult.

The result stated in regard to the valency is so remarkable that it is perhaps worth a separate statement in more general terms. The sets of a series of given grade (n) and freedom (r), are to satisfy certain conditions of coincidence, of prescribed multiplicity at each of a prescribed number of points; the number of conditions imposed is to be equal to the freedom of the series, a coincidence of multiplicity r_i+1 being counted as requiring r_i conditions for the sets sought. There will be a certain number, α_0, of simple points in each such set, depending on the grade of the series; these impose no condition. Consider apart, for distinctness, one of the coincidences, of multiplicity r_1+1, say at P_1, speaking of the remaining coincidences, with their prescribed character, as imposing conditions S. Now, take, in the sets of the original linear series, those which satisfy the conditions S, but have, at an assigned point P, a coincidence of multiplicity r_1; this imposes as many (r_1) conditions as did the ascription of a coincidence of multiplicity r_1+1 at an un-

Correspondence; elementary methods 43

assigned point P_1; there will then be a finite number of such sets; but in each of these there will be $\alpha_0 + 1$ simple points, say Q. We consider the correspondence between P and all the sets of $\alpha_0 + 1$ points Q which exist satisfying, beside the coincidence at the assigned point P, the conditions S at the other unassigned points. A coincidence in this correspondence, in which a point Q coincides with P, gives a set satisfying the original conditions, of $(r_1 + 1)$-fold coincidence at P, beside the conditions S.

Then all these points Q arising from P, taken with P counted $r_1 \psi_{p-1}$ times, form a set of a linear series on the original curve; where ψ_{p-1} is the number, on a curve of genus $p-1$, of sets (in a linear series $g^{n-r_1-1}_{r-r_1}$), whose coincidences are those of the conditions S, and have beside α_0 simple points; that is, of sets characterised precisely by the coincidences of the original sets to be enumerated, other than the $(r_1 + 1)$-fold coincidence at P_1.

Ex. 14*a.* If on the canonical curve of order $2p-2$ in space of $p-1$ dimensions, we consider primes touching the curve in $(p-2)$ points, the two remaining intersections of such a prime with the curve are in symmetrical correspondence, with indices both equal to $2^{2p-4}(2p-2) - 2^{p-2}(2^{p-1}-1)$, and the valency of this correspondence is equal to $2^{p-2}(2^{p-1}-1)$. It can be proved in fact that, if $q_i = q!/i!(q-i)!$, then

$$1 + \frac{p_1}{(p-1)_1}(p-2)_1 + \frac{p_2}{(p-1)_2}(p-2)_2 + \dots + \frac{p_{p-2}}{(p-1)_{p-2}}(p-2)_{p-2}$$
$$= \frac{2^{p-1}p - (2^p - 1)}{p-1}.$$

A direct proof, by the theory of correspondence, for the number, $2^{p-1}(2^p-1)$, of primes having $(p-1)$ simple contacts with the canonical curve, is given by Welchman, *Proc. Camb. Phil. Soc.* XXVI, 1930, p. 453.

Ex. 14*b.* The general formula for the number of planes having 3 contacts with a curve of order n and genus p, and $n-6$ simple intersections, is found to be

$$8\left[(n-3,3) + (n-4,2)p + (n-5)(p,2) + (p,3)\right],$$

where (λ, μ) denotes the binomial coefficient. Identify this with the result given in Ex. 13 (p. 38) above, explaining the discrepancy (as in the case of the double tangents of a plane curve on p. 40). Either formula, as in all these cases, supposes that the curve has only a finite number of such triply touching planes; for example, a curve, lying on a developable surface, if it meets each generator in three points, has every tangent plane of the developable as a triple tangent plane.

44 *Chapter I*

Ex. 15. Closely connected with the problem considered in Ex. 14 is that of determining the number of linear spaces which satisfy conditions of incidence (or contact) with one, or more, curves; it is supposed, of course, that the number of conditions is equal to the number of parameters on which such a linear space depends. We do not enter into a complete account; but we quote the result for one case.

Consider, in space $[r]$, a curve of order n, and genus p, with $n \geqslant p + r$. The number of conditions for a space of dimension k, say $[k]$, in this space, to meet a line, is $r - k - 1$; we assume that this is also the number of conditions necessary for such a space to meet an algebraic curve. The number of conditions necessary for the space to have i intersections with the curve is then $i(r - k - 1)$; and a space $[k]$, in space $[r]$, depends on $(k+1)(r-k)$ parameters. These two numbers will be equal if $i(r-k-1) = (k+1)(r-k)$. We suppose this equality to hold. It requires $i > k+1$; we denote $i-k-1$ by j; we also put $m = n - p - r$, and $l = m - j$; and also $s = r - k - 1$. Thus the numbers j, m, l, s are perfectly definite.

We take all possible, zero or positive, integers $\lambda, \rho_0, \rho_1, \ldots, \rho_s$, whose sum is p, supposing $\lambda < p$; we denote the product of the squares of the differences of the numbers $\rho_0, \rho_1 + 1, \ldots, \rho_s + s$, by Δ; also we denote the number

$$\frac{(m + \rho_t + t)!}{(\rho_t + t)!(j + t)!(l + \rho_t + t)!},$$

for $t = 0, 1, \ldots, s$, by n_t; and the product $n_0 n_1 \ldots n_s$ by $n(\rho_0 \ldots \rho_s)$. With these notations, the number of spaces $[k]$ which have i intersections with the curve, is given by

$$p! \sum_{\lambda=0} \frac{(-s)^\lambda}{\lambda!} \Sigma \Delta . n(\rho_0 \ldots \rho_s), \quad 0 \leqslant \rho_0 \leqslant \rho_1 \leqslant \ldots \leqslant \rho_s.$$

For this formula we refer to Giambelli, *Mem....Torino*, LIX, 1908, p. 433, quoted by Segre, *Enzykl. Math. Wiss.* III, C 7, p. 888.

To test the meaning of the formula, apply it to the case of the trisecant lines of a curve in space of four dimensions. We thus obtain (writing μ, ν, σ for ρ_0, ρ_1, ρ_2)

$$p! \Sigma \frac{(-2)^\lambda}{\lambda!} (\nu + 1 - \mu)^2 (\sigma + 2 - \mu)^2 (\sigma + 1 - \nu)^2$$

$$\times \frac{(n-p-4+\mu)(n-p-4+\nu+1)(n-p-4+\sigma+2)}{\mu!(\nu+1)!(\sigma+2)!\,1!\,2!\,3!}$$

for all values of $\lambda, \mu, \nu, \sigma$ subject to

$$\lambda + \mu + \nu + \sigma = p, \quad 0 \leqslant \lambda < p, \quad 0 \leqslant \mu \leqslant \nu \leqslant \sigma,$$

for which $n - p - 4 + \mu > 0$. For example, if $n = 10$ and $p = 6$, there are seven terms in the sum, corresponding to the values of λ, μ, ν, σ given by

$$(3, 1, 1, 1), (2, 1, 1, 2), (1, 1, 1, 3), (1, 1, 2, 2), (0, 1, 1, 4),$$
$$(0, 1, 2, 3), (0, 2, 2, 2),$$

and computation leads to 20, as the number of trisecant chords of the curve. Simpler formulae were given for particular cases by Castelnuovo (*Rend. Palermo*, III, 1889, p. 27; *Rend. Lincei*, V, 1889, p. 130):

(*a*), In space $[r]$, the number of primes having r contacts with a curve of order n and genus p is the coefficient of t^r in the ascending expansion of $2^r (1+t)^{n-r-p} (1+2t)^p$. Or, again, in space of an even number of dimensions, with spaces $[k]$ for which $k = \frac{1}{2}r - 1$, to meet the curve in the number i, $= (k+2)$, of points which follows from the formula $(k+1)(r-k) = i(r-k-1)$, the number of such spaces is the coefficient of t^{k+2} in the ascending expansion of
$$(1+t)^{n-k-1}[1 - t^2/(1+t)^2]^p.$$

Thus the number of trisecants of a curve (n, p) in space $[4]$ is $(n-4)[\frac{1}{6}(n-2)(n-3)-p]$, which is $\frac{1}{3}(n-4)d$, where d is the number of accidental double points of a ruled surface of order n with prime section of genus p, in space $[4]$, as will be seen in a subsequent chapter. For $n=10$, $p=6$, this gives the number 20 found from Giambelli's formula. Or again, the number of quadrisecant planes of a curve (n, p) in space $[6]$ is

$$(n-3, 4) - p(n-5, 2) + (p, 2),$$

where (λ, μ) is the binomial coefficient. A similar, but slightly more complicated formula gives, without requiring r to be even, the number of spaces $[r-2]$ which have $2(r-1)$ intersections with a curve (n, p), in the form

$$\sum_{\lambda=0} \frac{(-1)^\lambda}{r-\lambda} (n-r-\lambda, r-\lambda-1)(n-r-\lambda+1, r-\lambda-1)(p, \lambda),$$

which, for example, leads to the result found, in Ex. 11*a* (p. 32 above), for the number of quadrisecants of a curve in space $[3]$.

(*b*), When the curve is in space $[r]$, with $r = n - p$, the spaces $[k]$ with i intersections, where $i(r-k-1) = (k+1)(r-k)$, are of number

$$\frac{p!}{(p-i)!} \frac{\{r-k-1\}\{i-k-2\}}{\{r+i-2k-2\}},$$

where $\{m\}$ denotes $1! \, 2! \, \ldots m!$, with $\{0\} = 1$. This again gives 20 for $r=4$, $n=10$, $p=6$, $k=1$, $i=3$.

For more general results cf. Severi, *Mem. Torino*, L, 1901, p. 81, LI, 1902, p. 103, where formulae are given for the number of spaces

[k] which meet a curve of order n and genus p in space [r] with ν_1-point intersection at one point, ν_2-point intersection at another point, and so on, finally with ν_t-point intersection at a t-th point, subject to $(r-k)\Sigma\nu_i - t = (r-k)(k+1)$. See also Schubert, *Mitt. d. math. Ges. Hamburg*, III, 1891, p. 12.

Part II. Transcendental methods. The application of Riemann's methods to the theory of algebraic correspondence leads to new results, as was first shewn by Hurwitz, *Math. Ann.* XXVIII, 1887. In particular it is possible to shew that, on a single curve of general moduli, for its genus, every such correspondence is of the kind we have considered hitherto, having a valency; but that, on a curve of particular moduli, there may exist a correspondence in which, instead of the single number, the valency, there arises a matrix of numbers. We explain the matter so as to include both possibilities.

Consider a curve C, of genus p, whose normal integrals of the first kind are denoted by $u_1, \ldots, u_p$. Denote the periods of u_i by $(1)_{ij}, \rho_{ij}$, where $(1)_{ij} = 0$ unless when $j = i$, and $(1)_{ii} = 1$. Thus we speak of these integrals as having the period scheme $(1, \rho)$, where 1 denotes the unit matrix of type (p, p), and ρ denotes the matrix (ρ_{ij}), also of type (p, p). These periods are those corresponding to *crossings* of period ovals on the associated Riemann surface, these ovals being denoted by (γ, γ'); thus in particular ρ_{ij} is the excess of the value of u_i on the left of γ_j' over its value on the right of this oval (see Chap. VI of Vol. V, p. 141). Consider also a curve D, of genus q, with normal integrals of the first kind denoted by $v_1, \ldots, v_q$, whose period scheme is $(1, \sigma)$, over ovals, denoted by (δ, δ'), on the associated Riemann surface. Let an elementary normal integral, of the curve C, of the *third* kind be denoted by $\Pi_{z,c}^{x,a}$, as in the chapter referred to; and a similar integral of the curve D by $\Phi_{z,c}^{x,a}$. For definiteness, we suppose $p \leqslant q$.

Now suppose there exists an algebraic (s, r) correspondence between these curves, whereby to the place (x), of C, correspond the places $(z_1), \ldots, (z_r)$ of the curve D, it being the case that any rational symmetric function of $(z_1), \ldots, (z_r)$ is expressible rationally by (x); while to the place (z), of D, correspond the places $(x_1), \ldots, (x_s)$ of C, with a like property. We denote particular positions of $(x), (x_1), \ldots, (x_s)$ by $(a), (a_1), \ldots, (a_s)$; and particular positions of $(z), (z_1), \ldots, (z_r)$ by $(c), (c_1), \ldots, (c_r)$. From these assumptions it follows that the sum of the values of v_j at the places $(z_1), \ldots, (z_r)$, of D, depends uniquely on the place (x) of C, and is an everywhere finite integral thereon. Hence there exist q equivalences of the forms

$$v_j^{z_1, c_1} + \ldots + v_j^{z_r, c_r} \equiv M_{j,1} u_1^{x,a} + \ldots + M_{j,p} u_p^{x,a}, \quad j = 1, 2, \ldots, q,$$

where the coefficients $M_{j,k}$ are constants; and it will follow that there are similar equivalences

$$u_i^{x_1, a_1} + \ldots + u_i^{x_s, a_s} \equiv N_{i,1} v_1^{z,c} + \ldots + N_{i,q} v_q^{z,c}, \quad i = 1, 2, \ldots, p;$$

here the integrations are on the respective Riemann surfaces, dissected along the ovals (γ, γ') and (δ, δ'), and the signs $\equiv$ denote the omission of *definite* constant aggregates of periods, all the integrals being single valued on the respective dissected Riemann surfaces.

The constants $M_{j,k}, N_{i,l}$, which enter here form two matrices, respectively of types (q, p) and (p, q). We proceed to express these matrices in terms of the relations which arise, in virtue of the correspondence, between the period ovals on the two Riemann surfaces representing the curves C and D. Denote any algebraic integral on the curve C by I; and, on the curve D, by J. When (x) describes a closed curve on the Riemann surface C, which we may suppose not to cross any of the $2p$ period ovals (γ, γ'), the set $(z_1), \ldots, (z_r)$ will vary on the surface D, and come back to their original positions, with (possibly) interchange among themselves. We can suppose these paths of $(z_1), \ldots, (z_r)$ also taken so as not to cross the period ovals (δ, δ'), a passage of, for example, δ_i, being avoided by a detour along δ_i'. In virtue of such an identity as

$$J^{z_2, k_1} + J^{z_1, k_2} = J^{z_2, z_1} + J^{z_1, k_1} + J^{z_1, z_2} + J^{z_2, k_2},$$

we can suppose that the aggregate of the paths described by $(z_1), \ldots, (z_r)$ is equivalent with an aggregate of closed paths. Now any closed path, on the surface D, dissected along the $2q$ period loops, (the $q-1$ connecting traverses being ineffective, see Vol. v, p. 141) is equivalent, so far as its effect in adding a period increment, to an algebraic integral, is concerned, to a certain number of circuits of the positive (left) rims of the $2q$ period ovals. And a circuit, in the negative direction, of the left rim of the period oval δ_i', is equivalent to a crossing of the associated δ_i; while a circuit, in the positive direction, of the positive rim of δ_i is equivalent to a crossing of the associated δ_i'. Hence we may express the paths on D, arising by the correspondence, 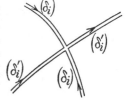 corresponding to circuits on C, by saying that the crossing of an oval γ_i by (x) gives rise, for an integral sum $J^{z_1, c_1} + \ldots + J^{z_r, c_r}$ on D, to a certain number of crossings of the $2q$ oval cuts made along (δ, δ'); and, precisely, that it is equivalent to

$$\alpha_{1,i}, \alpha_{2,i}, \ldots, \alpha_{q,i}; \alpha'_{1,i}, \alpha'_{2,i}, \ldots, \alpha'_{q,i}, \quad i = 1, \ldots, p,$$

crossings of the respective oval cuts

$$\delta_1, \delta_2, \ldots, \delta_q; \delta_1', \delta_2', \ldots, \delta_q';$$

and, similarly, that the crossing of γ_i' by (x) is equivalent to, respectively,

$$\beta_{1,i}, \beta_{2,i}, \ldots, \beta_{q,i}; \beta'_{1,i}, \beta'_{2,i}, \ldots, \beta'_{q,i}$$

crossings of the same oval cuts. The crossing is intended to be from the right (or negative) rim to the positive rim, in each case, corresponding to the ascription of the periods; a crossing in the other direction will be called a recrossing. Later we shall modify these symbols, expressing crossings by circuits, of associated ovals.

We thus have four matrices, say $\alpha, \alpha', \beta, \beta'$, each of q rows and p columns, of which all elements, such as $\alpha_{j,i}$, etc., are integers, positive or negative or zero; and, in accordance with what is said above, the crossing, by (x), on the surface C, of the loop γ_i, corresponds to the crossings, for the integral sum on D, of the $2q$ period ovals δ, δ', which are given by taking the column (i) of the matrices α, α'; while, when (x) crosses γ_i', the corresponding crossings on D are similarly defined by the i-th column of the matrices β, β'. There will be a similar set of matrices when we employ the reverse transformation to pass from D to C; the values of these will be expressed in what follows.

The coefficients $M_{j,k}$ in the equivalences

$$v_j^{z_1, c_1} + \ldots + v_j^{z_r, c_r} \equiv M_{j,1} u_1^{x,a} + \ldots + M_{j,p} u_p^{x,a}, \qquad j = 1, \ldots, q,$$

can be evaluated at once in terms of the matrices $\alpha, \alpha', \beta, \beta'$. For, first, when (x) crosses γ_i, the right side in this suffers the increment $M_{j,i}$, while the left side has the increment

$$(1)_{j,1}\alpha_{1,i} + \ldots + (1)_{j,q}\alpha_{q,i} + \sigma_{j,1}\alpha'_{1,i} + \ldots + \sigma_{j,q}\alpha'_{q,i},$$

where $(1)_{j,k}$ is the period of v_j for crossing of δ_k, ($= 1$ or zero), and $\sigma_{j,k}$ the period for crossing δ_k'. Thus we have

$$M_{j,i} = \alpha_{j,i} + \sigma_{j,1}\alpha'_{1,i} + \ldots + \sigma_{j,q}\alpha'_{q,i}, \qquad \begin{cases} i = 1, \ldots, p, \\ j = 1, \ldots, q, \end{cases}$$

which we write, in matrix notation,

$$M = \alpha + \sigma\alpha', \qquad\qquad (A),$$

M being, like α, α', of type (q, p), while σ is of type (q, q). Similarly, when (x) crosses γ_i', the right side of the equivalence under consideration obtains the increment

$$M_{j,1}\rho_{1,i} + \ldots + M_{j,p}\rho_{p,i},$$

and the left side has the increment

$$(1)_{j,1}\beta_{1,i} + \ldots + (1)_{j,q}\beta_{q,i} + \sigma_{j,1}\beta'_{1,i} + \ldots + \sigma_{j,q}\beta'_{q,i};$$

thus, with the notation of matrices,

$$M\rho = \beta + \sigma\beta', \qquad\qquad (B).$$

Eliminating M from the equations (A), (B), we have

$$(a + \sigma\alpha')\,\rho - \beta - \sigma\beta' = 0,$$

which is equivalent to

$$(1,\,\sigma)\begin{pmatrix} -\,\beta,\,\alpha \\ -\,\beta',\,\alpha' \end{pmatrix}\begin{pmatrix} 1 \\ \rho \end{pmatrix} = 0,$$

where the three matrices involved are respectively of types $(q,\,2q)$, $(2q,\,2p)$, $(2p,\,p)$, the symbol 1 denoting, in the first and last of these, the unit matrix respectively of types $(q,\,q)$ and $(p,\,p)$. Similarly the two relations (A), (B) are both expressible by the single equation

$$M\,(1,\,\rho) = (1,\,\sigma)\begin{pmatrix} \alpha,\,\beta \\ \alpha',\,\beta' \end{pmatrix}, \qquad (C),$$

wherein the types of the four matrices are in turn $(q,\,p)$, $(p,\,2p)$, $(q,\,2q)$, $(2q,\,2p)$.

It may happen that the sets (z_1), ..., (z_r), for different points (x), all belong to a linear series on the surface D; in this case, by Abel's Theorem, the matrix M is zero; and also, $v_j^{z_1,\,c_1}$, ..., $v_j^{z_r,\,c_r}$ being single valued on the dissected Riemann surface, every matrix α, α', β, β' consists of zeros. It will follow from the succeeding theory that, in this case, also (x_1), ..., (x_s) are a set of a linear series on the surface C; and this is clear in an elementary way, the correspondence being then representable by a single polynomial equation $\phi\,(x,\,z) = 0$, as can be easily proved.

We now consider the sum $\Phi_{z,\,c}^{z_1,\,c_1} + \ldots + \Phi_{z,\,c}^{z_r,\,c_r}$, on the surface D, wherein (z), (c) are arbitrary positions, and (z_1), ..., (z_r) correspond as before to a place (x) of the surface C, while (c_1), ..., (c_r) correspond to a place (a) of this surface; the function $\Phi_{z,\,c}^{\zeta,\,\gamma}$, as before said, is the elementary normal integral of the third kind. The sum itself we may, for brevity, denote by $(\Phi_{z,\,c}^{z_\nu,\,c_\nu})$.

This sum, being symmetrical with respect to (z_1), ..., (z_r), and also in regard to (c_1), ..., (c_r), may be regarded as a function of the places (x), (a) of the surface C, as well as of the independent places (z), (c) of D.

When one of the places (z_1), ..., (z_r), of D, which correspond to (x) on C, is in the neighbourhood of (z), the function is logarithmically infinite, with multiplier $+1$. This will happen if, and only if, (x) be in the neighbourhood of one of the places (x_1), ..., (x_s), of C, which correspond to the place (z) of D; and, from the analytical character of the correspondence between C and D, the function, regarded as a function of (x) on C, will have the like logarithmic behaviour in this case. There is similarly logarithmic infinity, with multiplier -1, when one of (z_1), ..., (z_r), on D, is in the neighbourhood of (c); and, therefore, when (x) is in the neighbourhood of one of the places $(a_1), \ldots, (a_s)$ of C which correspond to (c). The function depends on (x) only through $(z_1), \ldots, (z_r)$, and hence has no other infinities than those enumerated. Moreover, the function is single valued (and analytical) in (x), upon the surface C, dissected along the period ovals γ, γ'. Now, on this surface C, make the $(p-1)$ traverses, connecting the p pairs of ovals (γ_i, γ_i'), so as to obtain a single boundary of the surface C, with a direction of description of this consistent with that employed for γ_i and γ_i' (see Vol. v, pp. 131, 141). This being done, and the surface thereby rendered simply connected, consider the integral $\int u_m{}^{x,a} d\,(\Phi_{z,c}^{z_\nu,c_\nu})$, taken, as a function of (x), along the single boundary explained. The result will be equal to the sum of the values obtained by taking the same integral in turn round the infinities of the function; this sum is

$$2\pi i\,(u_m{}^{x_1,a_1} + \ldots + u_m{}^{x_s,a_s}),$$

where the paths of integration are on the dissected surface (and supposed not to intersect one another). Consider the value obtained by taking the integral round the boundary. This is the sum of the values obtained for the p pairs (γ_k, γ_k'), the $(p-1)$ traverses furnishing no contribution. For the pair γ_k, γ_k', the whole path may be considered by taking together opposite rims of the cut γ_k, and also taking together opposite rims of the cut γ_k'. For clearness of description, if (x) denote a point on the negative (right) side of γ_k or γ_k', the opposite point, on the positive (left) side, may be denoted by $(\bar{x})$. The two sides of γ_k thus give a contribution

$$\int u_m{}^{\bar{x},a} d_{\gamma_k}(\Phi_{z,c}^{\bar{z}_\nu,c_\nu}) - \int u_m{}^{x,a} d_{\gamma_k}(\Phi_{z,c}^{z_\nu,c_\nu}),$$

where, in *both* integrals, d_{γ_k} corresponds to an increment of (x) along the positive direction of γ_k, and $(\bar{z}_1), \ldots, (\bar{z}_r)$ are the points of the surface D corresponding to $(\bar{x})$. But we may put $u_m{}^{\bar{x},a} = u_m{}^{x,a} + H_{m,k}$, where $H_{m,k}$, the period of u_m, is independent of the position of (x) on the rim of γ_k. Thus, from the two integrals, the expression under the integral sign is $u_m{}^{x,a} d_{\gamma_k}(\Phi_{z,c}^{\bar{z}_\nu,z_\nu}) + H_{m,k} d_{\gamma_k}(\Phi_{z,c}^{\bar{z}_\nu,c_\nu})$. Herein, however, in the sum $(\Phi_{z,c}^{\bar{z}_\nu,z_\nu})$, the integration, on D, from the set $(z_1), \ldots, (z_r)$ to the set $(\bar{z}_1), \ldots, (\bar{z}_r)$, is obtainable (as in the preceding work), with the integers from the matrices α, α', in terms of crossings of the period ovals (δ, δ') on D, using the k-th columns of these matrices, and the periods of the function (of ζ) $\Phi_{z,c}^{\zeta,\gamma}$, on D. These periods all vanish at the ovals (δ); for the ovals (δ') they are $2\pi i v_1^{z,c}, \ldots, 2\pi i v_q^{z,c}$. Thus the sum is $2\pi i\,(\alpha'_{1,k}v_1^{z,c}+\ldots+\alpha'_{q,k}v_q^{z,c})$. This is independent of (x), so that $\int u_m^{x,a}d_{\gamma_k}(\Phi_{z,c}^{\bar{z}\nu,z\nu})=0$. We have then to consider $H_{m,k}\int d_{\gamma_k}(\Phi_{z,c}^{\bar{z}\nu,c\nu})$.

Now the complete circuit, by (x), of the positive rim of the oval γ_k, which is equivalent to a crossing of γ_k', will correspond to an aggregate of circuits on the surface D, whose expression, in terms of crossings of the period ovals δ, δ', is given by the k-th columns of the matrices β, β'. Wherefore the integral in question is given by $2\pi i H_{m,k}(\beta'_{1,k}v_1^{z,c}+\ldots+\beta'_{q,k}v_q^{z,c})$. Here, in fact, $H_{m,k}=0$ unless $k=m$; and then $=1$.

We may similarly examine the contribution to the integral $\int u_m d(\Phi_{z,c}^{z\nu,c\nu})$ which arises from the two rims of the period oval γ_k'. There is the difference that the complete circuit by (x), of the positive rim of the oval γ_k', is equivalent to a recrossing, instead of a crossing, of the oval γ_k. The result is therefore

$$-2\pi i K_{m,k}(\alpha'_{1,k}v_1^{z,c}+\ldots+\alpha'_{q,k}v_q^{z,c}),$$

where $K_{m,k}$, the period of $u_m^{x,a}$ for a crossing of γ_k', is $\rho_{m,k}$.

Adding together the results for the p pairs (γ_k, γ_k'), we thus have

$$2\pi i \sum_{k=1}^{p}\sum_{j=1}^{q}(H_{m,k}\beta'_{j,k}-\rho_{m,k}\alpha'_{j,k})\,v_j^{z,c},$$

which is expressible as

$$2\pi i \sum_{j=1}^{q}(H\beta'-\rho\bar{\alpha}')_{m,j}\,v_j^{z,c},$$

in which H denotes the matrix unity of p rows and columns, and $\bar{\mu}$ denotes the matrix obtained by interchange of rows and columns of a matrix μ, the so-called transposition of μ.

Equating this to the value found above, by integration round the infinities of the function, we have

$$u_m^{x_1,a_1}+\ldots+u_m^{x_s,a_s}\equiv N_{m,1}v_1^{z,c}+\ldots+N_{m,q}v_q^{z,c}, \quad m=1,\ldots,p,$$

where N, a matrix of type (p, q), is given by

$$N=\bar{\beta}'-\rho\bar{\alpha}', \qquad (A').$$

By the transposition of an identity obtained above, however, we have $\bar{\rho}\,(\bar{\alpha}+\bar{\alpha}'\bar{\sigma})-\bar{\beta}-\bar{\beta}'\sigma=0$, or, since ρ and σ are symmetrical $(\bar{\rho}=\rho, \bar{\sigma}=\sigma)$, $\rho\bar{\alpha}-\bar{\beta}=(\bar{\beta}'-\rho\bar{\alpha}')\sigma$. Hence we also have

$$N\sigma=-\bar{\beta}+\rho\bar{\alpha}, \qquad (B').$$

Further, comparing

$$M=\alpha+\sigma\alpha', \quad M\rho=\beta+\sigma\beta',$$

with

$$N=\bar{\beta}'-\rho\bar{\alpha}', \quad N\sigma=-\bar{\beta}+\rho\bar{\alpha},$$

we see that the matrix associated with the correspondence from C to D, say

$$\Delta = \begin{pmatrix} \alpha, & \beta \\ \alpha', & \beta' \end{pmatrix},$$

is replaced, in the reverse transformation from D to C, by

$$\nabla = \begin{pmatrix} \bar{\beta}', & -\bar{\beta} \\ -\bar{\alpha}', & \bar{\alpha} \end{pmatrix},$$

which is of type $(2p, 2q)$, just as Δ is of type $(2q, 2p)$. We also have

$$N(1, \sigma) = (1, \rho)\,\nabla.$$

Further, using $(0)_p$, $(1)_p$ for zero and unit matrices of type (p, p), etc., and putting

$$\epsilon_{2p} = \begin{pmatrix} (0)_p, & -(1)_p \\ (1)_p, & (0)_p \end{pmatrix}, \quad \epsilon_{2q} = \begin{pmatrix} (0)_q, & -(1)_q \\ (1)_q, & (0)_q \end{pmatrix}$$

we have $\quad \epsilon_{2p}\nabla\epsilon_{2q} = \begin{pmatrix} \bar{\alpha}', & -\bar{\alpha} \\ \bar{\beta}', & -\bar{\beta} \end{pmatrix}\epsilon_{2q} = \begin{pmatrix} -\bar{\alpha}, & -\bar{\alpha}' \\ -\bar{\beta}, & -\bar{\beta}' \end{pmatrix} = -\bar{\Delta},$

and, similarly $\epsilon_{2q}\Delta\epsilon_{2p} = -\bar{\nabla}$; and we may also write

$$\epsilon_{2q}\Delta = -\overline{\epsilon_{2p}\nabla}, \quad \Delta\epsilon_{2p} = -\overline{\nabla\epsilon_{2q}}.$$

We may now express the relations between *circuits* on the surfaces C, D, arising by the direct and reverse correspondences; hitherto we have used only crossings. In the direct correspondence we had (p. 47)

a crossing of γ_k leads to

k-th column of α crossings $\delta + k$-th column of α' crossings δ',

a crossing of γ_k' leads to

k-th column of β crossings $\delta + k$-th column of β' crossings δ';

now, a crossing of γ_k' is obtainable by a positive circuit of (the positive rim of) γ_k, and a crossing of γ_k by a negative circuit of (the positive rim of) γ_k'; and similarly for δ and δ'. Hence, taking the facts written down in inverse order, and denoting positive circuits of γ_k, γ_k' by (γ_k), (γ_k'), etc., we may say that

(γ_k) leads to $\beta'_{1,k}(\delta_1) + \ldots + \beta'_{q,k}(\delta_q) - \beta_{1,k}(\delta_1') - \ldots - \beta_{q,k}(\delta_q')$,

(γ_k') leads to $-\alpha'_{1,k}(\delta_1) - \ldots - \alpha'_{q,k}(\delta_q) + \alpha_{1,k}(\delta_1') + \ldots + \alpha_{q,k}(\delta_q')$;

these may be represented by saying that the circuits

$$[(\gamma_1), \ldots, (\gamma_p), (\gamma_1'), \ldots, (\gamma_p')],$$

lead to $\quad \begin{pmatrix} \bar{\beta}', & -\bar{\beta} \\ -\bar{\alpha}', & \bar{\alpha} \end{pmatrix}[(\delta_1), \ldots, (\delta_q), (\delta_1'), \ldots, (\delta_q')]$, or, still more briefly, by saying that $[(\gamma), (\gamma')]$ lead to $\nabla [(\delta), (\delta')]$. We have remarked that $\epsilon_{2p}\nabla = \bar{\Delta}\epsilon_{2q}$; thus we also have

$$\epsilon_{2p} [(\gamma), (\gamma')] \equiv \bar{\Delta}\epsilon_{2q} [(\delta), (\delta')],$$

which merely recovers the original definition of Δ in terms of crossings. By the reverse transformation we similarly have

$$[(\delta), (\delta')] \equiv \Delta [(\gamma), (\gamma')].$$

Suppose now that the second curve D coincides with the first. Then we have an (s, r) correspondence on the curve C, with relations

$$u_i^{z_1, c_1} + \ldots + u_i^{z_r, c_r} \equiv M_{i,1} u_1^{x,a} + \ldots + M_{i,p} u_p^{x,a}, \quad i = 1, \ldots, p,$$

in which (replacing ρ by the more usual notation τ for the period matrix) $M = \alpha + \tau\alpha'$, $M\tau = \beta + \tau\beta'$; and, also, relations

$$u_i^{x_1, a_1} + \ldots + u_i^{x_s, a_s} \equiv N_{i,1} u_1^{z,c} + \ldots + N_{i,p} u_p^{z,c}$$

with $N = \beta' - \tau\bar{\alpha}'$, $N\tau = -\beta + \tau\bar{\alpha}$.

The existence of the correspondence thus requires the equation (equivalent to p^2 equations) $\tau\alpha'\tau + \alpha\tau - \tau\beta' - \beta = 0$. This equation may express actually existing relations, with numerical coefficients, quadratic in the elements of the period matrix τ; or, it may be satisfied, if the numerical matrices α, α', β, β' be properly chosen, without any limitation for the matrix τ. We first enquire as to the possibility of the second alternative. The (m, n)th of the p^2 relations is

$$\sum_{i,j} \tau_{m,i}\alpha'_{i,j}\tau_{j,n} + \sum_i (\alpha_{m,i}\tau_{i,n} - \tau_{m,i}\beta'_{i,n}) - \beta_{m,n} = 0, \quad i, j = 1, \ldots, p;$$

for these p^2 relations to hold without restriction of the periods $\tau_{i,j}$, it is necessary and sufficient that

$$\beta_{m,n} = 0, \quad \alpha'_{i,j} = 0; \quad \alpha_{m,i} = 0, \quad \beta'_{i,n} = 0, \quad i \neq m, \ i \neq n; \quad \alpha_{m,m} = \beta'_{n,n},$$

for all values of m, n, i, j $(1, \ldots, p)$. Thus the matrices β, α' consist of zeros, and the matrices α, β' are both equal to the same diagonal matrix, of equal elements; namely using γ for a single integer, we have

$$\beta = 0, \quad \alpha' = 0, \quad \alpha = \beta' = -\gamma, \quad \Delta = -\begin{pmatrix} \gamma, & 0 \\ 0, & \gamma \end{pmatrix} = -\gamma, \quad \nabla = \Delta = -\gamma,$$

in accordance with $\epsilon^2_{2p} = -1$. And the two sets of p equivalences, of which we have deduced the second from the first, become, in this case,

$$u_i^{z_1, c_1} + \ldots + u_i^{z_r, c_r} + \gamma u_i^{x,a} \equiv 0, \quad u_i^{x_1, a_1} + \ldots + u_i^{x_s, a_s} + \gamma u_i^{z,c} \equiv 0,$$
$$i = 1, \ldots, p;$$

here γ may be any positive or negative integer, or zero. By the meaning of these equivalences (in virtue of the converse of Abel's Theorem), γ is what we have called the valency; and is the same for the reverse and direct correspondences.

Thus we have the result, first proved by Hurwitz (*Math. Ann.*
XXVIII, 1887, p. 561), that, unless the periods of the matrix τ be
connected by quadratic relations, of appropriate form, with
numerical coefficients, every existing (s, r) correspondence, of a
curve with itself, is a correspondence with valency. When such
quadratic relations exist for the periods τ, the " valency coefficients "
M, or N, are not arbitrary, but related to the conditions connecting
the periods.

We can prove, however, that the number of coincidences for *any*
(s, r) correspondence upon a single curve is given by

$$s+r-(\alpha_{11}+\ldots+\alpha_{pp}+\beta'_{11}+\ldots+\beta'_{pp}),$$

(of which, for a valency correspondence, the particular form is
$s+r+2\gamma p$, as found before).

For this, we consider the function of (x) expressed by

$$\Phi \equiv \Pi_{x,c}^{z_1,c_1}+\ldots+\Pi_{x,c}^{z_r,c_r},$$

where Π denotes the elementary normal integral of the third kind,
and $(z_1), \ldots, (z_r)$ are the places corresponding to the place (x). Here
$(c_1), \ldots, (c_r)$ are the places corresponding to a place (a), but (c) is
an arbitrary place.

With the notation employed above (p. 50), to distinguish two
places of the Riemann surface which are opposite to one another
on two rims of a period cut, we have

$$\Pi_{\bar{x},c}^{z_i,c_i}=\Pi_{\bar{x},c}^{z_i,z_i}+\Pi_{\bar{x},c}^{z_i,c_i}=\Pi_{x,c}^{z_i,z_i}+\Pi_{\bar{x},x}^{z_i,c_i}+\Pi_{x,c}^{z_i,c_i},$$

so that $$\Pi_{\bar{x},c}^{z_i,c_i}-\Pi_{x,c}^{z_i,c_i}=\Pi_{\bar{x},c}^{z_i,z_i}+\Pi_{z_i,c_i}^{\bar{x},x};$$

here, the second term on the right is the period, of the function
$\Pi_{z_i,c_i}^{\zeta,c}$ (with ζ current), arising by the crossing of the period cut on
which (x), $(\bar{x})$ lie; its value is therefore zero, or $2\pi i v_k^{z_i,c_i}$, for $k=1$,
or $2, \ldots,$ or p. Thus we see that, for a crossing by (x) of a period cut
(from the right rim to the left), the increment of the function Φ is
$\sum\limits_{i=1}^{r} \Pi_{\bar{x},c}^{z_i,z_i}+\sum\limits_{i=1}^{r} \Pi_{z_i,c_i}^{\bar{x},x}$; and further, by what was explained before,
if, with the usual notation, this period cut be (a_k), this increment is
$2\pi i (\alpha'_{1,k}v_1^{\bar{x},c}+\ldots+\alpha'_{p,k}v_p^{\bar{x},c})$, while, if this period cut be (b_k), the
increment is

$$2\pi i \left(\beta'_{1,k}v_1^{\bar{x},c}+\ldots+\beta'_{p,k}v_p^{\bar{x},c}+\sum\limits_{i=1}^{r} v_k^{z_i,c_i}\right),$$

namely, is

$$2\pi i (\beta'_{1,k}v_1^{\bar{x},c}+\ldots+\beta'_{p,k}v_p^{\bar{x},c}+M_{k,1}v_1^{x,c}+\ldots+M_{k,p}v_p^{x,c}).$$

We express now that the total increment of the function Φ, considered on the dissected Riemann surface, when taken round the

aggregate of the contours of the p period cut pairs (a_k), (b_k), is equal to its increment taken round its infinities.

The two rims of the (a_k) cut, as contribution to $\int d(\Phi)$, will give the value of $\int d(\bar{\Phi} - \Phi)$ taken once along the positive rim of (a_k); this circuit is equivalent to a crossing of (b_k), and gives, therefore, by what we have seen, the corresponding period of the function

$$2\pi i \,(\alpha'_{1,k}v_1^{\bar{x},c} + \ldots + \alpha'_{p,k}v_p^{\bar{x},c}),$$

namely that at (b_k), which is

$$2\pi i \,(\alpha'_{1,k}\tau_{1,k} + \ldots + \alpha'_{p,k}\tau_{p,k}),$$

and this, as $\tau_{m,k} = \tau_{k,m}$, is the product of $2\pi i$ by the (k, k)-th element of the matrix $\tau\alpha'$, or, say, $2\pi i\,(\tau\alpha')_{k,k}$. Taking next the two rims of the (b_k) cut, these give $\int d(\bar{\Phi} - \Phi)$, taken once along the positive rim of (b_k); this circuit is equivalent to a recross of (a_k); thus, by what we have seen for $\bar{\Phi} - \Phi$, we obtain

$$-2\pi i\,(\beta'_{k,k} + M_{k,k}), \quad \text{or} \quad -2\pi i\,(\beta' + \alpha + \tau\alpha')_{k,k}.$$

Adding this to the former contribution, and taking all pairs (a_k), (b_k), the whole result is

$$-2\pi i \sum_{k=1}^{p} (\alpha + \beta')_{k,k}.$$

Now consider the infinities of the function

$$\Phi = \Pi^{z_1, c_1}_{x, c} + \ldots + \Pi^{z_r, c_r}_{x, c},$$

regarded as depending on (x). There is a logarithmic infinity, with multiplier unity, when (x) is at any one of the places $(z_1), \ldots, (z_r)$ which correspond to (x); say, at the C coincidences of the correspondence. There is a logarithmic infinity, with negative unity as multiplier, when (x) is at any of the places $(c_1), \ldots, (c_r)$, which correspond to (a); and there is such an infinity when (c) is at any of the places $(z_1), \ldots, (z_r)$ which correspond to (x), namely, when (x) is at any of the places $(a_1), \ldots, (a_s)$ which correspond reversely to the place (c). The total increment of Φ, round the infinities, is thus $2\pi i\,(C - r - s)$. Equating this to the former result, we have the anticipated formula

$$C = r + s - \sum_{k=1}^{p} (\alpha_{k,k} + \beta'_{k,k}),$$

where the last part is the sum of the diagonal elements of the matrix Δ, or, equally, of the matrix ∇.

It is easy to prove that if we have two correspondences T, T_1 upon the curve C, and form the product $T_1 T$, whereby to (x) first correspond $(z_1), \ldots, (z_r)$ by T, and to (z_m) then correspond, say, $(z_{m.1}), \ldots, (z_{m.r_1})$, by T_1, then the matrices M, and Δ, for this product correspondence, are, respectively, M_1M and $\Delta_1\Delta$; and thus the number of coincidences, for this product correspondence, is given by $rr_1 + ss_1 - \Sigma(\Delta_1\Delta)_{k,k}$. Replacing Δ, by ∇_1, considering the product $T_1^{-1}T$, we have, for the number of pairs of places corresponding both for T and for T_1, $rs_1 + r_1s - \Sigma(\nabla_1\Delta)_{k,k}$, which generalises the formula, due to Brill, for valency correspondences, given above (Ex. 3, p. 10).

We may also make an extension to the case of the correspondences previously considered between two curves C, D. Suppose we have two such correspondences, T and T', whereby, respectively to (x), of C, correspond (z_1), ..., (z_r), of D, and then (z_1'), ..., $(z'_{r'})$, of D. We may enquire, when does it happen that one of (z_1), ..., (z_r) coincides with one of (z_1'), ..., $(z'_{r'})$?

For this, take first the place (x_k), of C, which is one of the set (x_1), ..., (x_s) corresponding to the place (z_1) of D by the reverse correspondence T^{-1} from D to C; then the places $(z'_{k,1})$, ..., $(z'_{k,r'})$ of D, corresponding to the place (x_k), of C, by the correspondence T'; and then consider, on D, the correspondence by which, from (z_1), arises the aggregate of the sets $(z'_{1,1})$, ..., $(z'_{1,r'})$; ...; $(z'_{s,1})$, ..., $(z'_{s,r'})$. Namely, we consider, on D, the product correspondence $T'T^{-1}$, and consider the coincidences therein. The indices of this correspondence will be $(s'r, sr')$, and the associated matrix will be $\Delta'\nabla$. The number of times it happens that both T and T' lead from the point of C to the same point of D will thus be

$$rs' + r's - \Sigma(\Delta'\nabla)_{k,k}, \qquad\qquad k = 1, ..., 2q.$$

It is easy to see that the latter term may be replaced by any of

$$\Sigma(\nabla'\Delta)_{h,h}, \ \ \Sigma(\nabla\Delta')_{h,h}, \ \ \Sigma(\Delta\nabla')_{k,k} \qquad h = 1, ..., 2p.$$

Remark. It is clear from the fundamental equations, and the theory of the so-called inversion problem (Vol. v, p. 145, Ex. 4) that there exists on any curve a correspondence for which the valency coefficients M have arbitrary values, and one of the indices (r) of the correspondences is the genus, p, of the curve. By the use of the theta functions, Hurwitz (*Math. Ann.* xxviii) obtains the value of the other index, and the number of coincidences, which, in the notation used here, is $\frac{1}{2}\Sigma[(1-\Delta)(1-\nabla)]_{k,k}$.

The theory of the correspondence between two curves has been further developed in a very interesting way, first by Severi, and then by Lefschetz, by means of the conception of a surface, of which every point represents the aggregate of a point of one curve and a point of the other curve. Such a surface may be realised geometrically by supposing the two curves to be in different independent spaces $[C]$, $[D]$, both lying in a space $[N]$, then joining, by a line, every point of one curve to every point of the other, and then taking the section of the resulting ∞^3 manifold by a prime of the space $[N]$. The resulting surface may be called the *product* of the two curves C, D. Upon this surface, say Σ, there will correspond, to every point (x) of the curve C, a curve K_x, the prime section of the cone of lines joining (x) to the curve D; and the aggregate of such curves on Σ will form a system of which one curve passes through an arbitrary point of Σ (we assume that the construction is so general that, from every point of Σ, there passes only one line to meet both C and D, determining thereby a single point on each curve); an algebraic system of curves, on a surface, of this character (that one curve of the system passes through any point of the surface) is called a *pencil* of curves. Each curve of the pencil will have the genus, q, of the curve D; and, as these curves correspond $(1, 1)$ to the points of the curve C, the pencil itself is said to have the genus, p, of the curve C. The surface Σ possesses, also, another pencil of curves, K_z, each of genus p, this other pencil having genus q. Every curve of either pencil meets every curve of the other pencil in one point; and there is a single curve of each pencil through an arbitrary point of the surface Σ.

A correspondence between the curves C, D, in which, to each point (x) of C correspond r points of D, and to each point (z) of D correspond s points of C, say a (s, r) correspondence, gives rise to a ruled surface, formed by joining every (x) to the r corresponding points of D, and every (z) to the s corresponding points of C; and the section of this ruled surface by the prime, in which Σ lies, gives rise to a curve on this surface. This curve *represents* the correspondence, every point of the curve leading to a point of C and a point of D, associated with one another in the correspondence; conversely every such pair of associated points leads to a point of this curve. With every such point (x, z), of this curve, ϑ, we may take the $(r-1)$ other points $(x, z_2), \ldots, (x, z_r)$, where $(z), (z_2), \ldots, (z_r)$ are the points of D which correspond to the point (x) of C; and we may consider the aggregate of r curves, of the pencil K_z, which pass through the points $(x, z), (x, z_2), \ldots, (x, z_r)$, denoting this aggregate by ϑ_z. The aggregate ϑ_z, regarded as one composite curve, defines a linear system of curves on the surface Σ (see below, Chap. v). Similarly, with any point (x, z) of the curve ϑ, we may take $(s-1)$ other points $(x_2, z), \ldots, (x_s, z)$, wherein $(x), (x_2), \ldots, (x_s)$ are the points of C which correspond to the point (z) of D, and we may consider the aggregate, ϑ_x, of curves of the pencil K_x, which pass through the s points $(x, z), (x_2, z), \ldots, (x_s, z)$. This aggregate ϑ_x, regarded as a single composite curve, defines another linear system of curves on the surface Σ. The curve ϑ itself likewise defines a linear system on the surface.

A correspondence, T, between the curves C, D, may have the

particular character that the sets of points $(z_1), \ldots, (z_r)$ of D, which
arise from the points (x), all belong to the same linear series on D;
in that case the valency coefficients M, of the foregoing theory, and
the integers of the matrices α, β, α', β', are all zero; then the points
$(x_1), \ldots, (x_s)$ of C, which correspond to a point (z) of D, are also a
set of a linear series on C. Such a correspondence may be called a
correspondence of zero valency between C and D. For such a corre-
spondence it may be shewn that the linear system $|\vartheta|$, on Σ,
defined by the curve which represents the correspondence, is the
sum of the linear systems $|\vartheta_z|$, $|\vartheta_x|$ defined respectively by the
curves ϑ_z and ϑ_x. Or we may say that there is the equivalence
$\vartheta \equiv \vartheta_z + \vartheta_x$, or, also, that the linear system $|\vartheta - \vartheta_z - \vartheta_x|$ is *null*.
This result we shall assume here; such a correspondence is repre-
sented by a single equation $\phi(x, z) = 0$, which is a polynomial in
rational polynomials in the coordinates (x) of the curve C, and also
in rational polynomials in the coordinates (z) of the curve D. Conversely, any curve ϑ, of the surface Σ, represents a correspondence
between the curves C and D, in which, to (x), of C, correspond
points (z) of D, represented by the curves K_z, forming ϑ_z, which
pass through the points where the curve K_x meets the curve ϑ. If
$\vartheta \equiv \vartheta_z + \vartheta_x$, as K_x does not meet any of the K_x curves of ϑ_x, the
points $(x, z_1), \ldots, (x, z_r)$, where ϑ is met by K_x, are on the aggregate
ϑ_z, and form a set of a linear series when (x) varies. A curve ϑ, for
which $\vartheta \equiv \vartheta_z + \vartheta_x$, thus represents a correspondence of zero valency.
This geometrical relation is thus equivalent to the analytical fact
that, for the correspondence represented by the curve ϑ, in this case,
the matrices M, N, Δ, ∇ are all zero; and conversely.

More generally, if $\lambda_1, \ldots, \lambda_t$ be integers, a correspondence represented by a curve on the surface Σ which is of the same linear
system as the aggregate of curves given by $-\vartheta + \lambda_1 \vartheta^{(1)} + \ldots + \lambda_t \vartheta^{(t)}$,
wherein $\vartheta^{(1)}, \ldots, \vartheta^{(t)}$ are, like ϑ, curves on Σ, is of zero valency if

$$\vartheta \equiv \vartheta_z + \vartheta_x + \lambda_1 [\vartheta^{(1)} - \vartheta_z^{(1)} - \vartheta_x^{(1)}] + \ldots + \lambda_t [\vartheta^{(t)} - \vartheta_z^{(t)} - \vartheta_x^{(t)}];$$

and this is consequently the same as the statement that, for the
corresponding matrices Δ, we have $\Delta = \lambda_1 \Delta_1 + \ldots + \lambda_t \Delta_t$, there
being similar equations, therefore, for the matrices M, N, ∇.

Now, the arithmetic theorem can be proved that, any matrix of
integers Δ is expressible in a form $\lambda_1 \Delta_1 + \ldots + \lambda_t \Delta_t$, where $\Delta_1, \ldots, \Delta_t$
are appropriate particular matrices, and t is a limited number
$(\leqslant 2pq)$. Thus we can infer that any curve on the surface Σ can be
expressed as equivalent to at most $2pq$ curves $\vartheta_1, \vartheta_2, \ldots$, together
with an aggregate of curves of the pencils K_x, K_z. This is a wide
generalisation of a theorem already illustrated in the preceding
volume (p. 216), in the case of a ruled surface (and a quadric, and
cubic surface); and it is itself capable of much wider development

(associating itself indeed with the existing so-called Picard integrals of a surface which have curves of logarithmic infinity). In the present case, the immediate application is to the expression of any possible correspondence between the curves C, D in terms of a finite number of such correspondences. Also, surfaces such as Σ, representing the pairs of points of two curves, have very interesting properties, considered further below, in Chapter VII (p. 282).

The following particular remark, dealing with the number of intersections of two curves ϑ, ϑ', on the surface Σ, may also be made. Suppose that the correspondences represented by the curves $-\vartheta+\lambda_1\vartheta^{(1)}+\ldots+\lambda_t\vartheta^{(t)}$, $-\vartheta'+\lambda_1'\vartheta^{(1)}+\ldots+\lambda_t'\vartheta^{(t)}$ are both of zero valency. Then we have seen that $\Delta'=\lambda_1'\Delta_1+\ldots+\lambda_t'\Delta_t$, and, also, from the same equation for ϑ, $\nabla=\lambda_1\nabla_1+\ldots+\lambda_t\nabla_t$; hence

$$\Delta'\nabla=\Sigma\lambda_i'\lambda_j(\Delta_i\nabla_j),$$

of which, if $d(\mu)$ be used for the sum of the diagonal elements of a square matrix μ, a particular consequence is $d(\Delta'\nabla)=\Sigma\lambda_i'\lambda_j d(\Delta_i\nabla_j)$; on the other hand, denoting by i the number of intersections (ϑ, ϑ'), of the curves ϑ, ϑ', the intersections of $\vartheta-\vartheta_z-\vartheta_x$ and $\vartheta'-\vartheta_z'-\vartheta_x'$, are given by

$$(\vartheta-\vartheta_z-\vartheta_x, \vartheta'-\vartheta_z'-\vartheta_x')=(\vartheta, \vartheta')-(\vartheta, \vartheta_z')-(\vartheta, \vartheta_x')- \text{ etc.},$$

and, if (s, r) be the indices for the correspondence ϑ, while (s', r') belong to ϑ', this is at once found to be $i-rs'-r's$. By a result given above (p. 56), however, this is the same as $-d(\Delta'\nabla)$. Thus we have $(\vartheta-\vartheta_z-\vartheta_x, \vartheta'-\vartheta_z'-\vartheta_x')=-d(\Delta'\nabla)$.

The preceding remarks, slight as they are, may serve to lead the reader to the remarkable paper of Severi, *Mem....Torino*, LIV, 1903, p. 32. In the present place only a still slighter indication is possible of a most striking paper by S. Lefschetz, *Annals of Maths.* XXVIII, 1927, wherein the expression, $rs'+r's-d(\Delta'\nabla)$, for the number of coincidences of two correspondences, is obtained by the methods of Analysis Situs. The comparison of this paper with the foregoing theory is full of interest and suggestion.

Part III. Correspondence and defective integrals. We have already seen (p. 22) that the existence of a $(1, \nu)$ correspondence between two curves involves the existence, on one of the curves, of a batch of integrals, of the first kind, in number less than the whole number of such integrals existing for that curve, whose periods depend likewise upon a defective number of independent periods. The equation we have obtained, for the general correspondence between two curves, $M(1, \rho)=(1, \sigma)\Delta$, and the similar equation for the correspondence between points of the same curve, give rise to further interesting results, as to the existence of systems of algebraic integrals, of the first kind, upon a curve, which have defective period systems. As such integrals are of great importance, we give now an introductory account of such results.

Some preliminary remarks should be made, which are of interest in themselves. Thinking first of the case of a single curve, of genus p, for which the normal integrals, $v_1, \ldots, v_p$, have a period system denoted as before by $(1, \tau)$, we consider a space $[2p-1]$, of dimension $2p-1$, which we denote by R_{2p}; and consider, in this space, a space of $p-1$ dimensions $[p-1]$, denoted by $[\tau]$, which is based on the p points of the space R_{2p} whose coordinates are, respectively,

$$(\tau_{11}, \ldots, \tau_{1p}; -1, 0, \ldots, 0), (\tau_{21}, \ldots, \tau_{2p}; 0, -1, 0, \ldots, 0), \ldots,$$
$$(\tau_{p1}, \ldots, \tau_{pp}; 0, \ldots, 0, -1).$$

Using $(x_1, \ldots, x_p; x_{p+1}, \ldots, x_{2p})$ for the homogeneous coordinates in R_{2p}, the space $[\tau]$ is thus the intersection of the p primes

$$x_s + \tau_{s1}x_{p+1} + \ldots + \tau_{sp}x_{2p} = 0, \qquad s = 1, \ldots, p.$$

We also consider, in R_{2p}, the *null system*, say ϵ, in which the polar prime of a point $(y_1, \ldots, y_{2p})$ is that given by the equation

$$x_1 y_{p+1} - x_{p+1}y_1 + \ldots + x_p y_{2p} - x_{2p}y_p = 0;$$

this equation may also be expressed, in matrix notation, by

$$\begin{pmatrix} 0, & -1 \\ 1, & 0 \end{pmatrix} (x_1, \ldots, x_{2p})(y_1, \ldots, y_{2p}) = 0,$$

where, in the matrix of type $(2p, 2p)$, which in future we shall denote by ϵ, 0 and 1 denote respectively the zero and the unit matrices of p rows and columns. It is then at once clear that, the polar prime, in this Null system, of any point lying in the space $[\tau]$, wholly includes this space; and, conversely, that the pole, in the system ϵ, of any prime containing the space $[\tau]$, is a point of this space. This we may express by saying that the space $[\tau]$ is self-polar in regard to the null system ϵ. In fact,* if c denote $(c_1, \ldots, c_p)$, a general point of $[\tau]$ has, in R_{2p}, the coordinates

$$((\tau c)_1, \ldots, (\tau c)_p, -c_1, \ldots, -c_p),$$

and the polar prime of this, in the system ϵ, has the equation

$$c_1 x_1 + \ldots + c_p x_p + (\tau c)_1 x_{p+1} + \ldots + (\tau c)_p x_{2p} = 0,$$

which, if x' denote $(x_{p+1}, \ldots, x_{2p})$, is the same as

$$c_1 [x_1 + (\tau x')_1] + \ldots + c_p [x_p + (\tau x')_p] = 0,$$

so that this prime is one of the so-called star of primes containing $[\tau]$.

The space $[\tau]$, in virtue of Riemann's results, has the curious property of not containing any point whose coordinates $(x_1, \ldots, x_{2p})$

---

* If x, y denote rows $(x_1, \ldots, x_n)$, $(y_1, \ldots, y_m)$, and a be a matrix of type (m, n), then $(ax)_s$ denotes $a_{s1}x_1 + \ldots + a_{sn}x_n$; and axy denotes $\Sigma(ax)_s y_s$.

are all real. For if $\tau_{sm} = \rho_{sm} + i\sigma_{sm}$, where ρ_{sm}, σ_{sm} are real, the equations of the primes defining τ shew that such a point would need to satisfy the p equations $\sigma_{s1}x_{p+1} + \ldots + \sigma_{sp}x_{2p} = 0$, and would thus have $x_{p+1} = \ldots = x_{2p} = 0$, because the determinant $|\sigma_{sm}|$ is known not to vanish (Vol. v, p. 145). From the equations of these primes, we should then also have $x_1 = \ldots = x_p = 0$.

By the polarity just remarked, or by simple direct proof, the space $[\tau]$ has also the property of not lying in any prime whose equation has wholly real coefficients. More generally, a space $[2p - 1 - n]$, or, say, R_{2p-n}, common to any n primes containing $[\tau]$, $(n \leqslant p)$, cannot contain more than $2p - 2n$ *independent* points, for each of which all the coordinates are rational numbers—a point being dependent on others when its $2p$ coordinates are linearly expressible each by the like coordinates of these others, with co-efficients the same for all the coordinates. The polar of this fact, with respect to the system ϵ, may also be stated. Analytically this property has the following enunciation: *let M be any matrix of p columns and (we may suppose) of p rows, whose rank is n ($n \leqslant p$); let ξ be a matrix, whose elements are integers, with $2p$ rows and (we may suppose) of $2p$ columns. Let the identities expressed by the matrix equation $M (1, \tau) \xi = 0$ hold. Then, the rank, r, of ξ, is equal to, or less than, $2p - 2n$. It is thus sufficient to regard ξ as having only r independent columns.*

This result is important; we state the proof here in the simplest way, in a form often applied. In the first place, it is a consequence of the fact that the determinant of the imaginary parts of the periods τ_{sm} is not zero that, there exists no integral of the first kind of which all the $2p$ periods are real. Next, if $x_1, \ldots, x_{2p}$ be integers, and we consider the closed curve, on the Riemann surface representing the curve under consideration, which is equivalent to x_s crossings of the period oval (a_s), and x_{p+s} crossings of the conjugate oval (b_s), for $s = 1, \ldots, p$, then the equation $M (1, \tau) (x_1, \ldots, x_{2p}) = 0$ expresses that, for each of the n independent integrals $V_1, \ldots, V_n$ contained in the formula

$$V_s = M_{s1}v_1 + \ldots + M_{sp}v_p, \qquad s = 1, \ldots, p,$$

the period, obtained by circuit of the closed curve spoken of, vanishes. Now consider an integral V given by

$$V = (\lambda_1 + i\mu_1) V_1 + \ldots + (\lambda_n + i\mu_n) V_n,$$

in which $\lambda_1, \ldots, \mu_n$ are all real; let $P_1 + iQ_1, \ldots, P_n + iQ_n$ denote the periods, for $V_1, \ldots, V_n$ respectively, obtained by circuit of any specified closed curve. The period of V, for this circuit, will then be real if $\lambda_1 Q_1 + \mu_1 P_1 + \ldots + \lambda_n Q_n + \mu_n P_n = 0$, so that $\lambda_1, \ldots, \mu_n$ can be chosen to make the periods of V real for any $2n - 1$, or fewer, specified circuits. On the other hand, if r be the rank of the matrix

ξ, there being r independent columns in this matrix, the periods of all of $V_1, \ldots, V_n$ are zero (and hence real), for the r circuits corresponding to these columns. As there are $2p$ independent circuits on the Riemann surface, there will be $2p - r$ other circuits independent of these r. This number must then be greater than the number $2n - 1$ of circuits mentioned above, since else the integral $\Sigma(\lambda_s + i\mu_s)V_s$ would have all its periods real. Wherefore $2n - 1 < 2p - r$, or $r \leqslant 2p - 2n$; which is the theorem to be proved.

Another statement of the result is that the period scheme, for a system of n independent integrals of the first kind, must have at least 2n independent columns, in which not every period is zero. When this number is exactly 2n, the system of integrals is called regular.

It will now be proved that, when there exists one such regular defective system, of n independent integrals, with $2n$ independent columns of not all zero periods, then there also exists another regular defective system, consisting of $p - n$ integrals, independent of one another and of the integrals of the former system. This latter system, with a period scheme of $2n$ columns, may be called complementary to the former, the two systems together making up a set of p independent integrals.

If a general one of the former system of integrals, expressed in terms of the p normal integrals, be

$$\lambda_{s1}v_1 + \ldots + \lambda_{sp}v_p, \qquad s = 1, \ldots, n,$$

there will exist a matrix ξ, of $2p - 2n$ columns, each of $2p$ integers, such that $\lambda(1, \tau)\xi = 0$, these equations merely expressing that there are $2p - 2n$ independent closed circuits for which all the integrals $(\lambda v)_s$ have zero periods. We prove then that there exists a matrix, μ, of $p - n$ rows and p columns (of rank $p - n$), and a matrix η, of $2n$ columns, each of $2p$ integers, such that $\mu(1, \tau)\eta = 0$, there being no linear combination of the second system of integrals, that is, of the integrals $\mu_{t1}v_1 + \ldots + \mu_{tp}v_p, \qquad t = 1, \ldots, p - n,$ which is the same as any linear combination of the integrals of the first system. In other words, the matrices λ, μ are such that there exists no set, c, of p quantities $c_1, \ldots, c_p$, other than all zero, for which the p equations $(\bar{\lambda}, \bar{\mu})c = 0$ are satisfied.

We can, in fact, regard the $2p - 2n$ columns of the matrix ξ as defining $2p - 2n$ points of the space R_{2p}, with rational coordinates; the $2p - 2n$ polar primes of these points, in the Null system ϵ, then intersect in a space $[2n - 1]$, which can be regarded as based upon $2n$ rational points (that is, points of rational coordinates). In other words, we can determine a matrix, η, of integers, of $2p$ rows and $2n$ columns, such that $\xi\epsilon\eta = 0$, a typical one, of the $2p - 2n$ equations to be satisfied by a column of η, being

$$-\xi_{1s}\eta_{p+1, t} - \ldots - \xi_{ps}\eta_{2p, t} + \xi_{p+1, s}\eta_{1t} + \ldots + \xi_{2p, s}\eta_{pt} = 0,$$

for $s=1, \ldots, 2p-2n$, t varying from column to column of η. The $2p$ elements $\eta_{1t}, \ldots, \eta_{pt}, \eta_{p+1,t}, \ldots, \eta_{2p,t}$, of a column of η, are thereby subject to $2p-2n$ equations, when, for the $\xi_{1s}, \ldots, \xi_{2p,s}$, we put, in turn, all the columns of ξ. There are thus $2n$ independent columns for the matrix η; or the polar space is based on the $2n$ points given by these columns. It will be proved presently that the spaces $[\xi]$, $[\eta]$, thus explained, have no point in common.

Of the space $[\xi]$, based on the $2p-2n$ points given by the columns of ξ, there will be an intersection with the space $[\tau]$; this intersection is a space which, as lying in $[\xi]$, can be described as based upon a certain number of points, of each of which the coordinates are given by a column of a matrix of the form ξa, wherein a is a matrix of $2p-2n$ rows, and a certain number of columns; the number of such (independent) columns will be the number of independent points of the space which is the intersection in question. We prove that this number is $p-n$. For, first, it is an obvious remark that, if A be a matrix of type (p, q), and x a matrix of type (q, r), such that $Ax=0$, then the rank of x is q diminished by the rank of A; because, if α be the rank of A, the equations for the elements of a column of x which are expressed by $Ax=0$ are equivalent to α independent equations, and the q elements of a column of x are then expressible by $q-\alpha$ of them; so that x has $q-\alpha$ independent columns. This being noticed, in the equations $\lambda\,(1, \tau)\,\xi=0$, satisfied by ξ, as λ is of type (n, p), and of rank n, the rank of $(1, \tau)\,\xi$ is $p-n$; and, by the same remark, as $(1, \tau)\,\xi$ is of type $(p, 2p-2n)$, the rank of a matrix a, such that $(1, \tau)\,\xi a=0$, is $p-n$. The equations expressed by $(1, \tau)\,\xi a=0$ give, as the columns of ξa, the basic points of the space, say $[\tau\xi]$, of dimension $p-n-1$, in which the space $[\xi]$ meets the space $[\tau]$.

Now put $\xi=\begin{pmatrix} h \\ k \end{pmatrix}$, where h, k are matrices both of the type $(p, 2p-2n)$. Then, from $(1, \tau)\,\xi a=0$, we have the consequences

$$(h+\tau k)\,a=0, \quad ha+\tau ka=0, \quad \bar{a}\bar{h}+\bar{a}\bar{k}\tau=0;$$

wherefore, from $\bar{\xi}\epsilon\eta=0$, using the last of these, we have

$$\bar{a}\bar{\xi}\epsilon\eta=0, \quad \bar{a}\,(\bar{h}, \bar{k})\,\epsilon\eta=0, \quad \bar{a}\bar{k}\,(-\tau, 1)\,\epsilon\eta=0,$$

and this is the same as $\bar{a}\bar{k}(1, \tau)\,\eta=0$; here $\bar{a}$ is of type $(p-n, 2p-2n)$, and $\bar{k}$ of type $(2p-2n, p)$, so that $\bar{a}\bar{k}$ is of type $(p-n, p)$. We put then $\mu=\bar{a}\bar{k}$, and have consequently the equations $\mu\,(1, \tau)\,\eta=0$; and these express that the $p-n$ integrals $(\mu v)_t$, or

$$\mu_{t1}v_1+\ldots+\mu_{tp}v_p, \qquad t=1, \ldots, p-n$$

form a defective system; all these in fact have zero columns of periods for circuits given by the matrix η. And further, as is expressed by $\bar{a}\,(\bar{h},\bar{k})\,\epsilon\eta=0$, the space $[\eta]$, based on the points given by the columns of the matrix η, lies in the polar space, in regard to the Null system ϵ, of the space $\begin{pmatrix} h \\ k \end{pmatrix} a$, or $[\xi a]$, which is the meet $[\tau\xi]$.

In the same way, for the matrix λ which determines the original system of defective integrals, we may substitute a matrix obtained by considering the meet $[\tau\eta]$. For let this meet be given by the columns of ηb, where b is, as above, of type $(2n, n)$, and is to be determined from $(1, \tau)\,\eta b=0$. Putting $\eta=\begin{pmatrix} l \\ m \end{pmatrix}$, where l, m are matrices both of the type $(p, 2n)$, we have $\bar{b}\bar{l}+\bar{b}\bar{m}\tau=0$, and the equations $\bar{\eta}\epsilon\xi=0$ lead to

$$\bar{b}\bar{\eta}\epsilon\xi=0,\quad \bar{b}\,(\bar{l},\bar{m})\,\epsilon\xi=0,\quad \bar{b}\bar{m}\,(-\tau,1)\,\epsilon\xi=0,\quad \bar{b}\bar{m}\,(1,\tau)\,\xi=0;$$

if, however, we regard ξ as given, and the n equations expressed by $\lambda\,(1,\tau)\,\xi=0$ as equations for the p elements of a row of λ, then there will be n such rows, because $(1,\tau)\,\xi$ is of rank $p-n$. We may therefore take $\bar{b}\bar{m}v$ as the original system of defective integrals; and the space $[\xi]$ lies in the polar, with respect to ϵ, of the meet $[\tau\eta]$.

It is important, however, to shew that the $p-n$ integrals $\bar{a}\bar{k}v$, and the n integrals $\bar{b}\bar{m}v$, are linearly independent. Geometrically, as we first shew, this is the statement that the spaces $[\tau\xi]$, $[\tau\eta]$, both within $[\tau]$, have no common point. As $[\tau]$ is of dimension $p-1$, and these spaces are of respective dimensions $p-n-1$ and $n-1$, this would be exceptional. To prove this, remark that, if there be a set of $p-n$ quantities x, and a set of n quantities y, which, used as multipliers respectively for the $p-n$ integrals $\bar{a}\bar{k}v$, and the n integrals $\bar{b}\bar{m}v$, give a zero sum, then we shall have $kax+mby=0$; and hence, as $ha+\tau ka=0$, $lb+\tau mb=0$, we shall also have $hax+lby=0$; and these lead to $\xi ax+\eta by=0$; when this is so, the point $(ha, ka)x$, or ξax, of the space $[\tau\xi]$, is the same as the point $(lb, mb)\,y$, or ηby, of the space $[\tau\eta]$. To disprove the possibility contemplated, we invoke the theorem, found by Riemann (cf. Vol. v, Chap. vi, p. 145), that, if the matrix τ be written as $\rho+i\sigma$, where ρ, σ are matrices of real elements, and $n_1, \ldots, n_p$ be p real numbers, then the quadratic form σn^2 has a necessarily positive value (and is not zero). We have $(1,\tau)\,\xi a=0$; and hence $(h+\tau k)\,ax=0$, with an arbitrary set x. Thus the $2p$ quantities ξax, or (hax, kax), can be expressed in the form

$$(-\tau kax,\ kax),\quad \text{or}\quad \xi ax=\begin{pmatrix} -\tau \\ 1 \end{pmatrix}z,$$

when z denotes the p quantities kax. Also we have

$$0=\bar{\eta}\epsilon\xi,\quad 0=\bar{\eta}\epsilon\xi\,(ax)\,(b_0 y_0)=\epsilon\,(\xi ax)\,(\eta by)_0,$$

where b_0, y_0 are the conjugate complexes of b, y (and η is real). Thus, using the matrix rule $\mu x \cdot vy = \bar{v}\mu\, xy$, we have from $\xi ax + \eta by = 0$

$$0 = \epsilon\, (\xi ax)(\xi ax)_0$$

$$= \epsilon\left[\binom{-\tau}{1} z\right]\left[\binom{-\tau_0}{1} z_0\right]$$

$$= (-\tau_0, 1)\, \epsilon\, \binom{-\tau}{1} zz_0, \ = (1, \tau_0)\binom{-\tau}{1} zz_0, \ = (\tau_0 - \tau)\, zz_0,$$

which, if $\tau = \rho + i\sigma$, is $-2i\sigma zz_0$. By Riemann's theorem quoted this cannot vanish.

Thus, the integrals $\bar{a}kv$, $\bar{b}\bar{m}v$ are linearly independent, and form, in all, the p independent integrals.

Having then proved that the spaces $[\tau\xi]$, $[\tau\eta]$ have no common point, their polar spaces, with regard to the Null system ϵ, will not lie in a prime. These polar spaces, we have shewn, include, respectively, the spaces $[\eta]$ and $[\xi]$; hence these do not lie in a prime. As they are of respective dimensions $2n-1$, $2p-2n-1$, they can, therefore, have no common point. Namely, the spaces $[\xi]$, $[\eta]$, defined by points whose (integer) coordinates give the period ovals for which the two systems of defective integrals have zero periods, are independent; and the $(2p-2n)+(2n)$ basic points of these give $2p$ independent period ovals.

In regard to the space $[\tau]$, we may add the further simple remark: the equation, $M(1, \tau) = (1, \tau)\Delta$, which arises for a transformation of a single curve into itself, shews that the linear transformation, in the space R_{2p}, with integer coefficients, which is expressed by $x' = \Delta x$, transforms any point of the space $[\tau]$ into another point of the same space. For it leads, from $(1, \tau)\, x = 0$, to $(1, \tau)\, x' = 0$. Conversely, for the existence of a transformation, $x' = ax$, in which the matrix a is of type $(2p, 2p)$, which changes the space $[\tau]$ into itself, the equations $(1, \tau)\, x' = 0$ must be a consequence of $(1, \tau)\, x = 0$; or $(1, \tau)\, ax = 0$ must be a consequence of $(1, \tau)\, x = 0$; thus we infer the existence of a matrix M such that $M(1, \tau) = (1, \tau)a$.

We pass now to consider the equations, which arose in considering the correspondence between two curves, $M(1, \rho) = (1, \sigma)\Delta$, $N(1, \sigma) = (1, \rho)\nabla$. It is clear, from consideration of the processes, of transposition, and interchange of rows and columns, by which ∇ may be obtained from Δ, or, more directly, from the identity $\epsilon_{2q}\Delta\epsilon_{2p} = -\bar{\nabla}$, that the rank of ∇ is the same as that of Δ. Denote this rank by δ, and the ranks of M and N respectively by μ and ν.

Then, if ξ be such a matrix of $2p$ rows that $\Delta\xi = 0$, this matrix will (by a remark made above) be of rank $2p - \delta$; and, by the former of the equations put down, we shall have $M(1, \rho)\, \xi = 0$. Hence, by a result proved, $2\mu + 2p - \delta \leqslant 2p$. Again, M being of type (q, p), if

L be such a matrix that $LM = 0$, so that L is of rank $q - \mu$, the equation $M (1, \rho) = (1, \sigma) \Delta$ leads to $L (1, \sigma) \Delta = 0$; and this, by the same result, shews that $2 (q - \mu) + \delta \leqslant 2q$. By the comparison of these two consequences we infer that $\delta = 2\mu$. Wherefore, in an equation $M (1, \rho) \xi = 0$, the rank of ξ is $2p - 2\mu$. By similar consideration of the equation $N (1, \sigma) = (1, \rho) \nabla$ we infer that $\delta = 2\nu$, so that $\mu = \nu$; also, in an equation $N (1, \sigma) \eta = 0$, the matrix η is of rank $2q - 2\nu$. We shall denote μ, or ν, by n. The q integrals Mu (where $u_1, \ldots, u_p$ are as above the normal integrals for the curve C) are then linear functions of n integrals, and these form, by what has been proved, a regular defective system on the curve C, with $2p - 2n$ columns of zero periods; while the p integrals Nv are also linear functions of n integrals, likewise forming a regular defective system on the curve D, with $2q - 2n$ columns of zero periods. Further, since a matrix L for which $LM = 0$ is of a rank $q - n$, it follows, from the equations

$$v_j^{z_1, c_1} + \ldots + v_j^{z_r, c_r} \equiv M_{j1} u_1^{x, a} + \ldots + M_{jp} u_p^{x, a}, \quad j = 1, \ldots, q,$$

that there are $q - n$ independent integrals for the curve D, namely the integrals Lv, whose values, at the points $(z_1), \ldots, (z_r)$ which correspond to the point (x) of C, have a sum which is independent of (x) (namely the same as when (x) is at (a)). By parity of reasoning, there are $p - n$ integrals on C whose sum values, at a set corresponding to (z) of D, are independent of (z). We have reached this conclusion in a particular case (p. 22 above).

The question, which we do not conclusively answer, naturally arises, whether the n defective integrals Nv, on D, and the $q - n$ integrals Lv, on D, where L is a matrix for which $LM = 0$, are always independent. The former have been shewn to form a regular system; the latter also form a regular system, since, from $M(1, \rho) = (1, \sigma) \Delta$ and $LM = 0$, we have $L(1, \sigma) \Delta = 0$; and Δ is of rank $2n$. An equation $(1, \rho) \zeta = 0$, in which ζ consists of integers, is impossible, since this would involve the existence of circuits for which all q integrals v have zero periods; hence an equation $N (1, \sigma) \eta = 0$, with η integral, being the same as $(1, \rho) \nabla \eta = 0$, involves $\nabla \eta = 0$. Thus the columns of zero periods for the system Nv are given by a matrix of integers, η, such that $\nabla \eta = 0$; and the columns of zero periods for the system Lv are given by Δ; here η has $2q - 2n$ columns, and Δ has $2n$ columns. To find whether these systems form q independent integrals, we may take account of the equation $\bar{\Delta} \epsilon_{2q} \eta = 0$, which follows from $\bar{\Delta} \epsilon_{2q} = \epsilon_{2p} \nabla$. When the answer to the question raised is affirmative, we shall have two complementary regular systems of defective integrals on both curves C and D.

Another remark, also only suggestive, dealing with the correspondence on a single curve, may be made. The equation $M (1, \tau) = (1, \tau) \Delta$

leads to $M^2 (1, \tau) = M (1, \tau) \Delta = (1, \tau) \Delta^2$; and, in general, if $\phi (M)$ denote any integral polynomial in the matrix M, with coefficients which are single quantities, we have $\phi (M) (1, \tau) = (1, \tau) \phi (\Delta)$.
A very particular case, θ being a single number, is

$$(M - \theta) (1, \tau) = (1, \tau) (\Delta - \theta),$$

with the corresponding equation for the reverse transformation

$$(N - \theta) (1, \tau) = (1, \tau) (\nabla - \theta).$$

It can be easily shewn that the roots of the algebraic equation in θ expressed by $|N - \theta| = 0$ are the conjugate complexes of the roots of $|M - \theta| = 0$, and that $|\Delta - \theta| = |M - \theta||N - \theta|$. For we have, if τ_0 be the conjugate complex of τ, etc.,

$$\begin{pmatrix} 1, & \tau \\ 1, & \tau_0 \end{pmatrix} \Delta = \begin{pmatrix} M, & M\tau \\ M_0, & M_0\tau_0 \end{pmatrix} = \begin{pmatrix} M, & 0 \\ 0, & M_0 \end{pmatrix} \begin{pmatrix} 1, & \tau \\ 1, & \tau_0 \end{pmatrix},$$

and this shews that $|\Delta - \theta| = |M - \theta||M_0 - \theta|$; while, recalling

$$\Delta = \begin{pmatrix} \alpha, & \beta \\ \alpha', & \beta' \end{pmatrix}, \quad M = \alpha + \tau \alpha', \quad M\tau = \beta + \tau \beta',$$

$$\nabla = \begin{pmatrix} \beta', & -\beta \\ -\bar{\alpha}', & \bar{\alpha} \end{pmatrix}, \quad N = \beta' - \tau \bar{\alpha}',$$

we have

$$\begin{pmatrix} 1, & \tau \\ 0, & 1 \end{pmatrix} \Delta \begin{pmatrix} 1, & -\tau \\ 0, & 1 \end{pmatrix} = \begin{pmatrix} M, & M\tau \\ \alpha', & \beta' \end{pmatrix} \begin{pmatrix} 1, & -\tau \\ 0, & 1 \end{pmatrix} = \begin{pmatrix} M, & 0 \\ \alpha', & \beta' - \alpha'\tau \end{pmatrix} = \begin{pmatrix} M, & 0 \\ \alpha', & \bar{N} \end{pmatrix};$$

and, for a single quantity θ, we have

$$\begin{pmatrix} 1, & \tau \\ 0, & 1 \end{pmatrix} \theta \begin{pmatrix} 1, & -\tau \\ 0, & 1 \end{pmatrix} = \theta \begin{pmatrix} 1, & \tau \\ 0, & 1 \end{pmatrix} \begin{pmatrix} 1, & -\tau \\ 0, & 1 \end{pmatrix} = \theta \begin{pmatrix} 1, & 0 \\ 0, & 1 \end{pmatrix} = \theta,$$

thus $|\Delta - \theta| = |M - \theta||\bar{N} - \theta|$. The two equations

$$|\Delta - \theta| = |M - \theta||M_0 - \theta| = |M - \theta||\bar{N} - \theta|$$

establish the facts stated (cf. the author's *Abel's Theorem*, Camb. 1897, pp. 632, 638, and the references there given).

We may also consider derived correspondences associated with polynomials $\phi (\Delta)$.; and the consequences of the fact that the matrix Δ satisfies a polynomial equation.

We have already (pp. 58, 59, Part II above) touched the question of the number of independent correspondences existing upon a curve, a correspondence being said to be dependent upon others when the matrix Δ, belonging thereto, is linearly expressible by the matrices Δ' belonging to the others. It can be shewn, from the equation $\beta + \tau \beta' = (\alpha + \tau \alpha') \tau$, that certainly there are not more than $2p^2$ independent correspondences; for, if x, y both denote sets of p

integers, it is impossible to solve the equations $x + \tau y = 0$; thus, in the equations quoted, the assumption of values for the $2p^2$ integers contained in α and α' leaves no ambiguity in the values of β and β'. But, in fact, it can be shewn that all existing so-called symmetrical correspondences on the curve, for which the matrices ∇, Δ are identical, are dependent on at most p^2 symmetrical correspondences; and all skew correspondences (for which $\nabla = -\Delta$) are likewise dependent on p^2 skew correspondences at most. We may prove these results in connexion with the linear transformation of the space $[\tau]$, considered above, remarking, from $\epsilon\Delta = \bar{\nabla}\epsilon, = -\epsilon\bar{\nabla}$, that, for a symmetrical correspondence, $\epsilon\Delta$ is a skew symmetric matrix (and a symmetric matrix for a skew correspondence).

For the theory of defective integrals many references may be given. The consideration of such integrals arises imperatively in the study of multiply periodic functions (cf. the writer's volume, Cambridge, 1907, Chap. VIII); it arises also in the study of the Picard integrals associated with an algebraic surface, these integrals giving rise to a defective system of integrals upon a prime section of the surface (cf. Poincaré, *Ann. d. l'École norm.* XXVII, 1910, and *Sitzber. Berliner Math. Gesell.* X, 1911; also Severi, *Rend. Lincei*, XXX, 1921, p. 163, etc.). And the consideration arises also, we have seen, in the multiple correspondence of one, or of two, curves. The foregoing summary of this connexion has been suggested by the papers of Rosati referred to below. The theorem, that the existence of one regular defective system of integrals on a curve involves the existence of a complementary system, was first given in general by Poincaré, *Amer. J. of Math.* VIII, 1886, p. 289; for particular cases of this, references are given in the writer's *Abel's Theorem*, p. 659; a proof of the theorem was given in the writer's *Multiply Periodic Functions*, p. 240. The proof here given, in terms of Rosati's geometrical point of view, seems very simple.

The papers of Rosati referred to are: *Ann. di Mat.* XXV, 1916, p. 1; *Atti...Torino*, L, 1915, p. 685 (see Scorza, *Rend. Lincei*, XXIII–XXV, 1914–16); *Atti...Torino*, LI, 1916, p. 991; *ibid.* LIII, 1918, p. 5; also *Ann. di Mat.* XXVIII, 1919, p. 35. Many references are also given in a paper by Scorza, *Rend. Palermo*, XLI, 1916, pp. 263–379.

CHAPTER II

SCHUBERT'S CALCULUS. MULTIPLE CORRESPONDENCE

Part I. Schubert's methods. Preliminary, as to notations. Characters of a manifold. The theory of correspondence so far developed deals with ∞^1 couples of corresponding points, and the number of coincidences of the points of a couple. Some extension is possible to cases when we have ∞^r couples of corresponding points, or, more generally, an infinite aggregate of corresponding linear *spaces*, of which we determine the number of couples of coinciding elements. This, and many other problems of enumeration, can be treated with the help of a symbolic calculus elaborated by Schubert; and some account of this calculus seems necessary, because, in intricate cases, its power has been shewn by its application to problems for which ordinary methods have failed to give a solution. The calculus is founded on the notion of representing a condition, to which a geometrical entity is to be subject, by an algebraic symbol. Then the alternative imposition of one *or* other of two independent conditions, is appropriately represented by the sum of the two chosen symbols, in either order; and the simultaneous imposition of both conditions is represented by the product of the two symbols, in either order. This multiplication then obeys the distributive law; while the associative law holds both for the sum and product of such symbols. The symbols, in the two operations, thus obey the ordinary laws of algebra. Subtraction of symbols may arise in equations between condition symbols, as a brief expression of additions; but division of the symbols does not arise. Schubert, in his remarkable book (*Kalkül der abzählenden Geometrie*, Teubner, 1879) ascribes the first traces of a calculus of conditions to Halphen, in his work for the characteristics of conics (*Compt. rend.* LXXVI, 1873, p. 1074; and *Bull. d. l. Soc. math.* I, 1873, p. 130), and to his own early work (*Gött. Nachr.* 1874, 1875). To Schubert is due also, beside the notation $[r]$ for a linear space of dimension r, a notation for a condition to which a linear space may be subject, which includes many familiar cases. For convenience we explain this notation at starting. It will be understood that the account of Schubert's methods given here is intended only to be introductory and illustrative; it is necessarily very incomplete.

It is implied generally that every geometrical entity for which conditions are postulated in the theory, is determined by the values of a certain number of parameters (and that the conditions are algebraic). A class of entities, of the same description, of which each depends on the values of m parameters, is often said to be of freedom m, or the aggregate of members of the class is said to be ∞^m; when i (independent) conditions are assigned for the parameters, the entities of the class which satisfy these conditions form a sub-class of freedom $m-i$. In particular, when $i=m$ this subclass consists, in general, of a finite number of entities.

These conceptions are well illustrated by classes of linear spaces, say of dimension k, lying in space $[r]$, of dimension r; and the results, which are of frequent application, can be verified by detailed algebra.

(*a*) The spaces $[k]$, of dimension k, in space $[r]$, are of freedom $(r-k)(k+1)$, or of aggregate $\infty^{(r-k)(k+1)}$. If, in $[r]$, we denote by k' the dimension of a linear space $[r-k-1]$, *dual* to the space $[k]$, the freedom in question is $(k+1)(k'+1)$. For such a space necessarily meets (in one point) an arbitrary space $[r-k]$, and is determined by $(k+1)$ points of the space $[r]$. Hence, it is determined by its points of intersection with $(k+1)$ spaces $[r-k]$; and each of these spaces contains ∞^{r-k} points.

(*b*) In order that the space common to p primes of the space $[r]$ should be, not merely a space $[r-p]$, but a space $[m]$ of greater dimension, $m, \geqslant r-p$, it is necessary that these primes should satisfy $(m+1)(m-r+p)$ conditions. For $r-m$ general primes intersect in an $[m]$; the condition required here is that, choosing $r-m$ primes from among the p, the remaining $p-r+m$ primes should all contain the $m+1$ points which are necessary to define the space $[m]$.

(*c*) If two spaces $[h]$, $[k]$, in space $[r]$, do not intersect, the $h+1$ and $k+1$ points, on which they are respectively based, are independent; and the least-dimensional space which contains both $[h]$ and $[k]$ is an $[h+k+1]$. If however $[h]$ and $[k]$ have a common space $[m]$, which we may call their *meet*, then the least-dimensional space containing both $[h]$ and $[k]$, or, say, their *join*, is a $[j]$, wherein $m+j=h+k$. This is proved as in the case $m=0$, by considering the numbers of independent points on which the spaces are based. In order that the meet of an $[h]$ and $[k]$ should be an $[m]$, with $m \geqslant h+k-r$, the spaces must satisfy $(m+1)(r-j)$, or $(m+1)(m-h-k+r)$, conditions. For, in order that the $r-h$ and $r-k$ primes, by which the spaces $[h]$ and $[k]$ may be defined, should have, for their intersection, a space $[m]$, requires, by (*b*) preceding, a number of conditions equal to $(m+1)(m-r+r-h+r-k)$, which is $(m+1)(m-j)$.

This is also the number of conditions in order that a space $[j]$, which can be defined by $r-j$ independent primes, should contain a given space $[m]$. A particular case is when $j=r-1$; then the number of conditions is $m+1$.

(*d*) The spaces $[k]$ which contain a given space $[m]$, when $k \geqslant m$, are of freedom $(r-k)(k-m)$. For, any such space $[k]$ meets an arbitrary $[r-k]$ in a point, and the $[r-k]$ contains ∞^{r-k} points; while, a $[k]$ depends on $(k+1)$ points, of which $m+1$ may be taken in the $[m]$ through which the $[k]$ passes. Such a $[k]$ is then determined by its intersections with $k-m$ spaces $[r-k]$.

(e) If the space $[m]$ be not given, but lie anywhere in a given $[h]$, the result is different: The spaces $[k]$ which meet a given $[h]$ in spaces $[m]$ are of freedom $(r-k)(k-m)+(m+1)(h-m)$; for the freedom of spaces $[m]$ contained in a given $[h]$ is $(m+1)(h-m)$.

The same result is obtained by remarking that the spaces $[k]$ which meet a given $[h]$ in an $[m]$, may be found by taking the spaces $[j]$ which contain the $[h]$, of which the freedom is $(j-h)(r-j)$, by (d)'preceding, with $j=h+k-m$; and then taking the spaces $[k]$ which lie in such a $[j]$, of which the freedom is $(j-k)(k+1)$. We easily verify that

$$(r-k)(k-m)+(m+1)(h-m)=(j-h)(r-j)+(j-k)(k+1),$$

the right side being $(k-m)(r-j)+(h-m)(k+1)$.

We now consider a certain composite condition, which may be required of a $[k]$ lying in the given space $[r]$. For this, we suppose a series of spaces, $[a_0]$, $[a_1]$, ..., $[a_k]$, to be given, of which each lies in the following space of this series. And the composite condition for $[k]$ is that it should meet $[a_0]$ in a point, $[a_1]$ in a line, ... $[a_h]$ in an $[h]$, ..., finally should lie in $[a_k]$. It is supposed that $a_i < a_{i+1}$, for $i < k$.

A $[k]$ meets an $[r-k]$ in a point, but meets an $[a_0]$ for which $a_0 > r-k$ in a line at least. For the $[k]$ to meet the $[a_0]$ in a point, we must therefore have $a_0 \leqslant r-k$; thus we put $a_0 = r-k-\beta_0$, supposing $\beta_0 \geqslant 0$. Similarly, a $[k]$ will meet an $[a_1]$ in more than a line if $a_1 > r-k+1$; thus we put $a_1 = r-k+1-\beta_1$, supposing $\beta_1 \leqslant 0$. In general, we put $a_i = r-k+i-\beta_i$, with $\beta_i \leqslant 0$. Then $a_i < a_{i+1}$ involves $i-\beta_i < i+1-\beta_{i+1}$, or $\beta_i > \beta_{i+1}-1$; so that

$$\beta_0 \geqslant \beta_1 \geqslant \beta_2 \geqslant ... \geqslant \beta_k \geqslant 0,$$

with $\beta_0 \leqslant r-k$. The composite condition described may be denoted by $(a_0, a_1, ..., a_k)$, or also by $\{\beta_0, \beta_1, ..., \beta_k\}$; and when necessary we may describe the former as the *dimensions symbol*, and the latter as the *condition symbol*. We proceed to prove that the composite condition is equivalent in all to $\beta_0 + \beta_1 + ... + \beta_k$ conditions for the space $[k]$.

To state the proof of this clearly, we may write

$$[k] = [r-a_0][a_0+k] = [r-a_1+1][a_1+k-1]$$
$$= [r-a_2+2][a_2+k-2] = ...,$$

meaning thereby that the $[k]$ may be regarded as the intersection of a space $[r-a_0]$ with a space $[a_0+k]$, or, in general, of a space $[r-a_i+i]$ with a space $[a_i+k-i]$. Then the space $[r-a_0]$ meets $[a_0]$ in a point, and, for $[k]$ to meet the $[a_0]$ in a point, we require the space $[a_0+k]$ to pass through this point. This requires $r-a_0-k$, or β_0, conditions. Suppose these to be satisfied, and $[k]$ to meet $[a_0]$ in a point. Then, similarly, the space $[r-a_1+1]$ meets $[a_1]$ in a line; and the $[r-a_1+1]$ was taken to contain the $[k]$, which now contains a point of $[a_0]$, and $[a_0]$ lies in $[a_1]$. Thus, the line in which the space $[r-a_1+1]$ meets $[a_1]$ contains already a point of $[a_1]$. We require that the $[a_1+k-1]$ should contain the line; and the $[a_1+k-1]$, like the $[r-a_1+1]$, contains already a point of $[a_1]$ lying on this line. Thus, for the $[a_1+k-1]$ to contain this line, it is sufficient that it contains some other point of this line. This requires $r-(a_1+k-1)$, or β_1, conditions, additional to the β_0 already found. Suppose these additional conditions also satisfied, and $[k]$ to meet $[a_1]$ in å line. Then, for the next step, the $[r-a_2+2]$ and the $[a_2+k-2]$ both contain $[k]$, and so contain this line; and this line, as lying in $[a_1]$, lies in $[a_2]$. The $[r-a_2+2]$ meets $[a_2]$ in a plane; we require that the $[a_2+k-2]$ should contain this plane. But this $[a_2+k-2]$, we have said, contains a line of $[a_2]$; it is thus only necessary that this $[a_2+k-2]$ should contain a further point of $[a_2]$. This requires $r-(a_2+k-2)$, or β_2, conditions. Thus the satisfaction of the partial condition which we may denote by $(a_0,\ a_1,\ a_2)$ requires, in all, $\beta_0+\beta_1+\beta_2$ conditions. The argument may be continued, with the result we have enunciated for the whole composite condition.

In particular, if $\beta_{s+1}=0$, $\beta_{s+2}=\ldots=\beta_k=0$; namely, if

$$a_{s+1}=r-k+(s+1),\ a_{s+2}=r-k+(s+2),\ \ldots,\ a_k=r-k+k,\ =r,$$

then the number of conditions required is $\beta_0+\beta_1+\ldots+\beta_s$. In this case we may denote the composite condition simply by $(a_1, a_2, \ldots, a_s)$; though it is convenient to have an entry of $k+1$ indices, in order to indicate that we are expressing a condition for a $[k]$. We may also, in this case, denote the condition by $\{\beta_1, \ldots, \beta_s, 0^{k-s}\}$.

We may remark that the number of symbols $\{\beta_0, \beta_1, \ldots, \beta_k\}$, including the null symbol $\{0, 0, \ldots, 0\}$, is the binomial coefficient, $(r+1, k+1)$, which is the same as the number of homogeneous coordinates of a space $[k]$ in space $[r]$. For recalling

and putting $$r-k \geqslant \beta_0 \geqslant \beta_1 \geqslant \ldots \geqslant \beta_k \geqslant 0,$$

$$\beta_k=x_k,\ \beta_{k-1}=\beta_k+x_{k-1},\ \beta_{k-2}=\beta_{k-1}+x_{k-2},\ \ldots,\ \beta_0=\beta_1+x_0,$$

so that $x_0, x_1, \ldots, x_k$ are zero or positive integers, all values are possible for which $\beta_0 \leqslant r-k$, or $x_0+x_1+\ldots+x_k \leqslant r-k$; the number of sets of values, equal to the number of terms in a homogeneous polynomial of order $r-k$ in $1+(k+1)$ variables, is thus

$$(r-k+k+1,\ k+1),\ \text{or}\ (r+1,\ k+1).$$

For instance, for a line in $[r]$, the number of possible conditions which may be imposed is $\frac{1}{2}r(r+1)$.

As examples of the notation, the condition that a line, in ordinary space [3], should lie in a given plane, is (1, 2) or {1, 1}, requiring two conditions; or, the condition that a line should pass through a given point of a given plane, and lie in this plane, is (0, 2) or {2, 1}, and is three-fold; or, again, the condition for a line to pass through a given point is (0, 3) or {2, 0}, and is two-fold. Beside the ineffective condition (2, 3), the other two possible conditions that can be required of a line are (1, 3), or {1, 0}, the one-fold condition for the line to meet a given line, and the four-fold condition (0, 1), or {2, 2}, for the line to pass through a given point of a given line, and coincide with this given line.

Ex. Among the $(r+1, k+1)$ conditions which a $[k]$, lying in an $[r]$, may be required to satisfy, find how many of these conditions are ρ-fold $(\rho = 0, 1, \ldots, (k+1)(r-k))$; for example, for a line in [3], there are two two-fold conditions, but the other four are all simple. For a plane in [4], the conditions {1, 1, 0}, {2, 0, 0} are two-fold, the conditions {1, 1, 1}, {2, 1, 0} are three-fold, and the conditions {2, 2, 0}, {2, 1, 1} are four-fold; but the other four conditions are all simple.

The notation may be usefully combined with another notation which will be employed below. For instance we may denote the order of the curve of contact of the tangents drawn from a given point to a given surface in ordinary space [3], by the symbol $\epsilon \{2, 0\}\,(2)$, or $\epsilon\,(0, 3)\,(2)$, where the line symbol {2, 0} or (0, 3) refers to the condition that the line passes through a given point, the symbol ϵ denotes that two of the intersections of the line with the surface are coincident, and the point symbol (2) refers to the condition that this point of coincidence lies in an arbitrary plane. In a similar way, the number of tangents of the surface which pass through a given point, and lie in a given plane through this point, may be denoted by $\epsilon \{2, 1\}\,(3)$, or $\epsilon\,(0, 2)\,(3)$, where the line symbol {2, 1} or (0, 2) refers to the fact that the line passes through a given point and lies in a given plane containing this point, the symbol ϵ expresses that two of the intersections of the line with the surface are coincident, and the point symbol (3) merely expresses that the point of contact is subject to no further condition. The use of the symbol ϵ to refer to a coincidence of points is customary (after Schubert); and the symbol of a condition which is satisfied only in a finite number of cases is often used to express this number. More generally, for any algebraic manifold of k dimensions, say M_k, lying in space $[r]$, we may consider two conditions, both referring to the composite entity consisting of the tangent k-fold of M_k and its point of contact with M_k; these are the conditions ρ_i, ω_j respectively expressed by

$$\rho_i = \epsilon \{1^i,\ 0^{k+1-i}\}\,(r-k+i), \quad \omega_j = \epsilon \{j,\ 0^k\}\,(r-k+j),$$
$$(0 \leqslant i \leqslant k;\ 0 \leqslant j \leqslant k,\ 0 \leqslant j \leqslant r-k).$$

Here the symbol $\{\ \ \}$ refers, in both cases, to the $[k]$ which is the tangent space, and the symbol $(\ \ )$ to its point of contact; and ϵ indicates that it is a point of contact which is under consideration. In the first case, the symbol $\{1^i, 0^{k+1-i}\}$ is an i-fold condition for the ∞^k tangent spaces of the manifold M_k; so that the points of contact for which this is satisfied are an aggregate ∞^{k-i}. And, of such an aggregate, a finite number of points satisfy the point condition $(r-k+i)$, of lying in a space $[r-k+i]$. Similarly in the second case, with j for i. Both symbols therefore are satisfied by a finite number of points, and can be used to denote these numbers.

In the symbol ρ_i, the $\{1^i, 0^{k+1-i}\}$ can also be written as

$$(r-k-1, r-k, \ldots, r-k+i-2; r-k+i, r-k+i+1, \ldots, r)$$

where, to indicate the application to a space $[k]$, there are written $k+1$ indices, though those beginning with $r-k+i$ represent intersections which are satisfied without conditions. The index $r-k+i-2$ expresses that the tangent space, $[k]$, meets a given $[r-k+i-2]$ in a space $[i-1]$, namely lies in a prime, $[r-1]$, through this $[r-k+i-2]$. This condition, which is equivalent by itself to i conditions, if satisfied, ensures that the conditions denoted by the preceding indices of the symbol are all satisfied*. Some simple examples of the symbol ρ_i, for a surface $(k=2)$, in ordinary space $(r=3)$, will shew that it is familiar. For $i=0$, the tangent plane of the surface need satisfy no conditions, but its point of contact lies on a line; thus ρ_0 is the order of the surface. For $i=1$, the tangent plane must pass through a given point, and the point of contact lie on a given plane; thus ρ_1 is the order of the curve of contact of the tangent cone drawn to the surface from an arbitrary point, or the order of this cone; commonly called the rank of the surface. For $i=2$, the tangent plane of the surface must pass through a given line, the point of contact being anywhere in the space $[3]$; thus ρ_2 is the number of tangent planes of the surface which can be drawn through an arbitrary line, or the class of the surface.

In the symbol ω_j, the symbol for the tangent space $\{j, 0^k\}$ expresses that this meets a given $[r-k-j]$ in a point, and the factor $(r-k+j)$ that the point of contact lies on a given $(r-k+j)$. For the symbol to have meaning, j must be at most equal to the less of k and $r-k$. As a simple example of the meaning of ω_j, take the case of a surface $(k=2)$, in ordinary space $(r=3)$; then for $j=0$, the factor $\{j, 0^k\}$ implies no condition for the tangent plane, and the point

---

* In general, the number of conditions for a $[k]$ to meet an $[a_s]$ in an $[s]$, is $(s+1)\beta_s$, and implies the whole condition $(a_0, a_1, \ldots, a_s)$ if every space of the series $[a_0], [a_1], \ldots, [a_s]$ is a prime of the following space, or $a_i = a_{i-1}+1$; in this case $\beta_0 = \beta_1 = \ldots = \beta_s$.

factor $(r-k+j)$ only that the point of contact lies on a line. Thus ω_0 is the order of the surface, equal to ρ_0; but, for $j=1$, the tangent plane is to pass through a given point, and its point of contact is to lie on a given plane; so that ω_1 is the rank of the surface, equal to ρ_1.

More generally, as an example of the notation, we compute ρ_i and ω_j for the manifold M_k, in $[r]$, which is the complete intersection of $r-k$ primals, given by equations $f_1=0, \ldots, f_{r-k}=0$. Let $x_0, \ldots, x_r$ denote the homogeneous variables, and denote $\partial f_m/\partial x_n$ by $f_{m,n}$. The condition that the tangent space at the point (x) of the manifold, which, with $y_0, \ldots, y_r$ as current coordinates, is given by the $r-k$ equations

$$y_0 f_{m,0} + \ldots + y_r f_{m,r} = 0, \qquad m=1, \ldots, (r-k),$$

should meet the space $[r-k+i-2]$ which is given by the $k-i+2$ equations

$$y_0 c_{n,0} + \ldots + y_r c_{n,r} = 0, \qquad n=1, \ldots, (k-i+2),$$

in a space $[i-1]$, is that this aggregate of $r-i+2$ equations for $y_0, \ldots, y_r$ should be equivalent only to $r-i+1$ equations (this being the number determining an $[i-1]$). Namely the condition is that, in the matrix of $r-i+2$ rows and $r+1$ columns

$$\left\| \begin{array}{c} f_{m,0}, \ldots\ldots, f_{m,r} \\ \ldots\ldots\ldots\ldots \\ c_{n,0}, \ldots\ldots, c_{n,r} \\ \ldots\ldots\ldots\ldots \end{array} \right\|,$$

all determinants of order $r-i+2$ should vanish. We are required to find the order of the manifold common to all the primals whose equations are given by the vanishing of these determinants, in which $x_0, \ldots, x_r$ are regarded as current variables.

For this we invoke a theorem, of frequent application, included in those noted at the end of this chapter (p. 108). Suppose we have a matrix of $p+1$ rows and $q+1$ columns $(q \geqslant p)$, in which the element in the (i,j)-th place, that is, in the i-th row and the j-th column, is of order r_i+c_j in the $r+1$ homogeneous coordinates $x_0, \ldots, x_r$; we can express the dimension, and the order, of the manifold upon which all the determinants, of $p+1$ rows and columns, in this matrix, vanish. The dimension is $r-(q+1-p)$, so that there is no common intersection when this number is negative, and the intersection is a set of points when this number is zero. The order can be expressed as the coefficient of t^{q+1-p} in the ascending expansion of a certain rational function of t, and, precisely, may be denoted by

$$\left[\frac{(1+c_1 t)(1+c_2 t) \ldots (1+c_{q+1} t)}{(1-r_1 t) \ldots (1-r_{p+1} t)} \right]_{t^{q+1-p}};$$

for brevity, we shall generally refer to this as $[(c)/(-r)]_{q+1-p}$. In the case now under discussion, we have $q=r$, $p=r-i+1$; and we suppose $i \geqslant 1$, so that $q+1-p$, or i, is >0; further the orders of the elements of the matrix are the same in any row, being, for the first $r-k$ rows, respectively one less than the orders of $f_1, \ldots, f_{r-k}$, and zero for all succeeding rows; we may thus suppose all of $c_1, \ldots, c_{q+1}$ to be zero, and, of $r_1, \ldots, r_{p+1}$, all but $r_1, \ldots, r_{r-k}$ to be zero. The dimension of the manifold required is thus $r-i$, and, with the proper values of $r_1, \ldots, r_{r-k}$, the order* is

$$[(1-r_1t)^{-1} \ldots (1-r_{r-k}t)^{-1}]_{t^i}.$$

The points of the original M_k which we seek are the intersection of the M_k with the manifold thus determined. The dimension of this intersection is then $r-i-(r-k)$, or $k-i$; further, if n_m be the order of f_m, we have $r_m = n_m - 1$; so far as the term in t^k we can thus put

$$\rho_0 (1-r_1t)^{-1} \ldots (1-r_{r-k}t)^{-1} = \rho_0 + \rho_1 t + \ldots + \rho_k t^k,$$

where ρ_0, $= n_1 n_2 \ldots n_{r-k}$, is the order of M_k. And this determines $\rho_1, \ldots, \rho_k$.

The determination of ω_j for the complete intersection M_k is similar. In this case we are to express that the tangent space meets a given $[r-k-j]$ in a point; in other words, the $(r-k)+(k+j)$, or $r+j$, equations in $y_0, \ldots, y_r$,

$$y_0 f_{m,0} + \ldots + y_r f_{m,r} = 0, \qquad m=1, \ldots, (r-k),$$
$$y_0 c_{n,0} + \ldots + y_r c_{n,r} = 0, \qquad n=1, \ldots, (k+j),$$

are to be capable of solution. The determinant of the coefficients in every $r+1$ of these equations must therefore vanish. This we express by

$$\left\| \begin{array}{ccc} f_{m,0}, & \ldots, & c_{n,0}, \ldots \\ \ldots\ldots\ldots\ldots\ldots\ldots \\ f_{m,r}, & \ldots, & c_{n,r}, \ldots \end{array} \right\| = 0,$$

in which there are $p+1$, $=r+1$ rows, and $q+1$, $=r+j$ columns; thus $q+1-p=j$, and we suppose $j \geqslant 1$; further, the order, in $x_0, \ldots, x_r$, of all the elements of any column of the matrix is the same; thus, in the notation used above, we may take $r_1, r_2, \ldots, r_{p+1}$ all zero, and all of $c_1, \ldots, c_{q+1}$ zero except $c_1, \ldots, c_{r-k}$; of these latter, however, $c_m = n_m - 1$. The rule then requires us to consider

$$[(1+c_1t) \ldots (1+c_{r-k}t)]_{t^j}.$$

The dimension of the intersection of M_k with the construct, of dimension $r-j$, which is given by the matrix, is $k-j$; thus the

---

* We follow the usual notation of r for the dimension of the space considered; and we use here also r, the initial letter of the word *row*, for ease of memory. Confusion is not to be feared.

numbers ω_1, ω_2, ..., so far as ω_k (supposing $k \leqslant r - k$), are defined by

$$\omega_0 [1 + (n_1 - 1) t] \dots [1 + (n_{r-k} - 1) t] = \omega_0 + \omega_1 t + \omega_2 t^2 + \dots,$$

where ω_0, $= \rho_0$, $= n_1 \dots n_{r-k}$.

In this case of a manifold M_k which is a complete intersection, the numbers ω_j are immediately expressible by the numbers ρ_i, or conversely. For, by what was found above,

$$\rho_0 [1 + (n_1 - 1) t]^{-1} \dots [1 + (n_{r-k} - 1) t]^{-1} = \rho_0 - \rho_1 t + \rho_2 t^2 - \dots,$$

and hence, so far as the terms in t^k in the two factors,

$$\omega_0 \rho_0 = (\rho_0 - \rho_1 t + \rho_2 t^2 - \dots)(\omega_0 + \omega_1 t + \omega_2 t^2 + \dots),$$

leading, as $\omega_0 = \rho_0$, to

$$0 = \omega_1 - \rho_1,$$

$$0 = \omega_2 \rho_0 - \omega_1 \rho_1 + \omega_0 \rho_2,$$

$$0 = \omega_3 \rho_0 - \omega_2 \rho_1 + \omega_1 \rho_2 - \omega_0 \rho_3,$$

$$\dots\dots\dots\dots\dots\dots\dots\dots\dots,$$

where $\omega_j = 0$ for $j > r - k$.

Ex. 1. Consider the complete intersection of two quadrics in space [4], shewing that $\rho_0 = 4$, $\rho_1 = 8$, $\rho_2 = 12$, $\omega_1 = 8$, $\omega_2 = 4$, and obtain a direct geometrical verification.

Ex. 2. For the surface which is the complete intersection of three quadrics in [5], obtain $\rho_0 = 8$, $\rho_1 = \omega_1 = 24$, $\rho_2 = 48$, $\omega_2 = 24$; for the M_3 which is the intersection of two quadrics in [5], obtain $\rho_0 = 4$, $\rho_1 = \omega_1 = 8$, $\rho_2 = 12$, $\rho_3 = 16$, $\omega_2 = 4$, and verify these results geometrically.

The calculus of conditions applied to a line. A linear space [k] in space [r] may be required to satisfy any one of the Schubert conditions $(a_0, a_1, \dots, a_k)$. It may however be required to satisfy, simultaneously, two such conditions, provided the total number of elementary conditions so involved is not greater than the number, $(k+1)(r-k)$, of parameters upon which a [k] depends. When this total number of elementary conditions (say $\beta_0 + \beta_1 + \dots + \beta_k$ for one factor of the product condition, and $\beta_0' + \beta_1' + \dots + \beta_k'$ for the other factor) is equal to $(k+1)(r-k)$, there will be a finite number of spaces [k] satisfying the two conditions. This total number is (it is believed), the sum of the numbers of spaces [k] which satisfy certain alternative conditions which are all of the form $(a_0, a_1, \dots, a_k)$. For instance, in ordinary space, the lines which satisfy the two conditions, of meeting two given intersecting lines, consist of those which pass through the point in which these intersect, together with those which lie in their common plane. In this way arises the problem of expressing the product of two (simultaneous) conditions, of the form $(a_0, a_1, \dots, a_k)$, as a sum of (alternative) conditions, all of this form. We deal now with this problem in the case of lines, in space $[r]$.

Suppose that a line, in space $[r]$, is to satisfy both the conditions (a_0, a_1) and (b_0, b_1), or say, the composite condition $(a_0, a_1)(b_0, b_1)$. In condition symbols this is $\{r-1-a_0, r-a_1\}\{r-1-b_0, r-b_1\}$, and is equivalent to a number of elementary conditions equal to $2(2r-1)-a_0-a_1-b_0-b_1$. But a line in space $[r]$ depends on $2r-2$ parameters; the previous number must therefore not exceed $2r-2$. Hence it is necessary that $a_0+a_1+b_0+b_1 \geqslant 2r$.

We consider now the problem of expressing the product, or composite, condition, as a sum of (alternative) conditions all of the form (c_0, c_1). To obtain this expression we apply a notion which, in simple cases, is familiar: we assume that the number of solutions of an algebraic problem, when finite and definite, remains unaltered by such changes of the values of the constants, entering algebraically in the algebraic relations which express the problem, as leave the number of solutions, of the changed problem, still finite and determinate. In both problems it is to be understood that multiple solutions are to be counted an appropriate number of times. The underlying reason for the assumption is that the change in the values of the constants can be made in a continuous way, thereby effecting only a continuous change in the number of solutions; whereas one integer number cannot change continuously into another integer number.

As an example of this notion, consider the problem, in space of three dimensions, of finding the number of lines of an algebraic congruence which meet two given arbitrary lines. If p, q denote the order and class of a congruence (the numbers of its lines respectively passing through an arbitrary point and lying in an arbitrary plane), we already know (e.g. Vol. IV, p. 51) that the number of lines common to two congruences, (p, q) and (p', q'), is $pp'+qq'$. The lines meeting two arbitrary lines form a particular congruence with $p'=q'=1$. Thus the number of lines of (p, q), which meet two arbitrary lines, is $p+q$. This is the sum of the numbers of lines of the congruence which satisfy one or other of the two conditions, of passing through an arbitrary point, or of lying in an arbitrary plane. In order to obtain this result by an application of the notion under consideration, we seek such a special position of the two lines that either of these alternative conditions could be satisfied in the same figure. This is obviously so if the lines intersect; the respective numbers would then be p and q. The assumption made is that, the number of lines of the congruence (p, q) which intersect two *skew* lines, is also $p+q$.

The argument employed in this case leads to a particular conclusion for the product of two conditions, for lines in ordinary space.

As above, denote the condition for a line to pass through a given point, by $(0, 3)$; also the condition, for the line to lie in a given plane, by $(1, 2)$, where the first index in the bracket is inserted in order that the symbol may be understood to refer to a line; and, similarly, denote the condition for the line to meet a given line by $(1, 3)$. Then the argument and conclusion we have reached may be summarised by the equation $(1, 3)(1, 3) = (0, 3) + (1, 2)$; in this equation it is to be understood, unless the contrary is stated in any case, that the lines indicated by the first index, 1, in the two multiplied symbols on the left, are not assumed to be the same.

Any appearance of novelty in the argument is lessened by considering the representation of lines of the space [3] by the points of a quadric Ω in space [5] (cf. Vol. IV, as referred to). The given congruence is then represented by an algebraic surface lying in the quadric Ω; and we are to find the number of points of this surface which lie in the tangent spaces [4] at both of two arbitrary points of Ω; these lie in the solid, a space [3], in which these two tangent spaces intersect. The number is the order of the surface which represents the congruence, and is the same for any solid of general position in the space [5]. If, in particular, we take a solid which meets Ω in two planes, respectively of the first and second kind upon Ω, which meet in a line of Ω (of which the solid is the polar with respect to Ω), then, the intersections in question, since the surface lies wholly on Ω, must be in one or other of these two planes. The points of one of the planes represent lines through a point in the original space [3], the points of the other represent lines in a plane. The total number of points of the surface which lie in a solid is thus obviously $p + q$.

By an argument which is almost the same we prove now, in space [r], under the limitations $a < r-1$, $b < r-1$, $a+b \geqslant r-1$, the symbolical equation

$$(a, r)(b, r) = (a+b-r+1, r) + (a, \overline{r-1})(b, \overline{r-1}),$$

where the upper bar in $\overline{r-1}$ is to indicate that the two spaces $[r-1]$ are the same space. In words this equation is expressed by saying that, the condition for a line to meet both of two arbitrary spaces $[a]$ and $[b]$, is of a kind which may be satisfied alternatively, either by the line meeting a space $[a+b-r+1]$; or by the line lying in a space $[r-1]$, and meeting both of two spaces $[a]$ and $[b]$ which lie therein. The composite condition on the left, involves, we have seen, a number of elementary conditions equal to $r-1-a+r-1-b$, or $2r-2-a-b$. This is also the number of elementary conditions for a line to meet a space $[a+b-r+1]$; also, as the number of conditions for a line to lie in an $[r-1]$ is 2, $2r-2-a-b$ is also the number of elementary conditions for a line to satisfy the composite condition represented by the second term on the right. The equation is thus equivalent also to saying that, the number of lines satisfying the condition (a, r) (b, r), which also satisfy further specified $a+b$ elementary conditions (thus satisfying in all the number, $2r-2$, of elementary conditions which suffice to determine a finite number of lines), is equal to the sum of the numbers of lines which satisfy these $a+b$ conditions and, also, satisfy one of the two alternative conditions represented by the terms on the right of the equation.

To obtain the equation, we argue as follows: In general, the two spaces $[a]$, $[b]$ meet in a space $[a+b-r]$ (or do not meet, if $a+b=r-1$), and are not both contained in a space $[s]$ with $s<r$; suppose, however, in particular, that they intersect in a space $[a+b-r+1]$, and are therefore both contained in a prime space, $[r-1]$, of the given space $[r]$. Then, the lines meeting both $[a]$ and $[b]$ evidently consist of, (1), the lines meeting their common $[a+b-r+1]$, without, necessarily, lying in the $[r-1]$, together with, (2), the lines lying in this $[r-1]$, which meet both $[a]$ and $[b]$, without, necessarily, meeting the $[a+b-r+1]$ intersection of these. There may also, (3), be lines which satisfy both the conditions (1) and (2), in the particular figure considered. But the number of elementary conditions for a line to do this is greater than for either (1) or (2) alone, and the aggregate of lines of this kind is of less freedom than in either of these cases; such lines will not enter therefore in the solution for the general figure, wherein all the alternative conditions, in terms of which the composite condition is to be expressed, are to have the same *power* as this. The conclusion is illustrated by the example discussed, of the lines of a congruence in [3] which meet two lines.

We have thus justified the equation put down when $a<r-1$, $b<r-1$, $a+b\geqslant r-1$. The equation evidently remains true when one, or both, of a, b is as great as $r-1$, provided the second term on the right be omitted, that is if we put $(r-1, r-1)=0$, and replace the unrestrictive factor $(r-1, r)$ by 1. But, retaining the conditions $a<r-1$, $b<r-1$, we can repeat the argument, and obtain

$$(a, \overline{r-1})(b, \overline{r-1})=(a+b-r+2, r-1)+(a, \overline{r-2})(b, \overline{r-2}),$$

and the process can be continued, provided a and b are not too large. If, with $a+b\geqslant r-1$, we have $a<b+1<r$, we can proceed as far as the equation

$$(a, \overline{b+2})(b, \overline{b+2})=(a-1, b+2)+(a, \overline{b+1})(b, \overline{b+1}),$$

where $(b, \overline{b+1})=1$. By addition of the equations we thus have

$$(a, r)(b, r)=(a+b-r+1, r)+(a+b-r+2, r-1)+\dots$$
$$+(a-1, b+2)+(a, b+1),$$

with $a+b \geqslant r-1$, $a < b+1 < r$; and this is conventionally true also if $a < b+1 \leqslant r$. Under these limitations this solves the problem of expressing the composite condition $(a, r)(b, r)$ as a sum of alternative conditions.

If, however, $a+b \leqslant r-1$, the spaces $[a]$, $[b]$, which do not intersect, lie both in the same space $[s]$, where $s = a+b+1$. In this case, a line which meets both $[a]$ and $[b]$ will lie in this space $[s]$, so that we have $(a, r)(b, r) = (a, s)(b, s)$. As we have $a+b \geqslant s-1$ (in fact $a+b = s-1$) and $b+1 \leqslant s$, it follows from the preceding case, provided $a < b+1$, that $(a, s)(b, s) = (0, s) + (1, s-1) + \ldots + (a, b+1)$. Reversing the order of the terms on the right, both results are included in the formula $(a, r)(b, r) = \sum\limits_{t=0}^{\lambda} (a-t, b+1+t)$, where $a \leqslant b$, and λ is the less of the two numbers $r-b-1$, a.

Now consider $(a, p)(b, q)$, with $a < p$, $b < q$, the space $\lfloor a \rfloor$ lying in the space $[p]$, and $[b]$ in $[q]$. A line which satisfies this condition must lie in both the spaces $[p]$ and $[q]$, and this cannot be unless these intersect in a line at least, or $p+q-r \geqslant 1$. When this is not so, the symbol has the value zero. When it is so, a line satisfying the condition, as lying in the space $[p+q-r]$ common to $[p]$ and $[q]$ and meeting the space $[a]$, must meet the common space of the $[p+q-r]$ and the $[a]$; as both these two spaces lie in the space $[p]$, this common space is a space $[a+q-r]$. In the same way, for the line to meet the common space of the $[p+q-r]$, and to meet $[b]$, both these lying in the space $[q]$, the line must meet a certain space $[b+p-r]$. Whence the condition $(a, p)(b, q)$ may be modified, writing $(a, p)(b, q) = (a+q-r, k)(b+p-r, k)$, where $k = p+q-r$. Thus the previously obtained equation, replacing a, b, r respectively by $a+q-r$, $b+p-r$, k, leads to

$$(a, p)(b, q) = \sum\limits_{t=0}^{\lambda} (a+q-r-t, b+p-r+1+t),$$

where, now, λ is the less of the two numbers $q-b-1$, $a+q-r$, and it is supposed, beside $p+q \geqslant r+1$, that $a+q \leqslant b+p$.

This result may be written in a form which is perhaps easier to grasp. Put $r-1-a = \alpha$, $r-p = \beta$, $r-1-b = \xi$, $r-q = \eta$, and use the condition symbols; then we have

$$\{\alpha, \beta\}\{\xi, \eta\} = \sum\limits_{t=0}^{\lambda} \{\alpha+\eta+t, \beta+\xi-t\},$$

where the distinction between the two symbols on the left is by the inequality $\alpha - \beta \geqslant \xi - \eta$ (this being the relation previously written as $a+q \leqslant b+p$); and the series on the right continues until the first index has reached the value $\alpha+\xi$, and the second index has fallen to $\beta+\eta$, unless the first index reaches $r-1$ before this (this being the fact that λ is the less of the two numbers $\xi - \eta$, $r - 1 - (\alpha + \eta)$), in which case the series stops at this term. The second index in this term would be $\alpha + \xi - [r - 1 - \eta - \beta]$. The value of $\{\alpha, \beta\}\{\xi, \eta\}$ is zero unless $r - 1 \geqslant \beta + \eta$ (this being the condition $p + q \geqslant r + 1$); this is involved, however, in the condition $r - 1 \geqslant \alpha + \eta$, which is obviously necessary in order that the term, for $t = 0$, on the right, may have a meaning. Thus, beside $\alpha - \beta \geqslant \xi - \eta$, which we may look upon as indicating the order in which to write the factors on the left of the equation, the only condition in the application of the formula is that involved in the rule for the last term on the right.

One of the simplest examples of the formula is that considered above for the number of lines of a congruence in ordinary space which meet two given lines; for this we obtained a result which may be written $\{1, 0\}\{1, 0\} = \{2, 0\} + \{1, 1\}$. We may give other simple examples, also for ordinary space: (a), It is clear that a line through an arbitrary point cannot lie in an arbitrary plane. Thus the product $(0, 3)(1, 2)$ should have the value zero. In fact this is $\{2, 0\}\{1, 1\}$, which is zero because $2 + 1 > 2$ $(\alpha + \eta > r - 1)$; (b), Again, the condition for a line to lie in two given planes is $(1, 2)^2$, which is $\{1, 1\}^2$; by the formula this is $\{2, 2\}$ or $(0, 1)$; namely, the line must coincide with a given line (the intersection of the planes); (c), Lastly, the condition for a line to meet four given lines is

$$(1, 3)^4 = [(1, 3)^2]^2 = [\{1, 0\}^2]^2 = [\{2, 0\} + \{1, 1\}]^2$$
$$= \{2, 0\}^2 + 2\{2, 0\}\{1, 1\} + \{1, 1\}^2;$$

of the terms on the right, $\{2, 0\}^2 = \{2, 2\} = (0, 1)$, while

$$\{1, 1\}^2 = \{2, 2\} = (0, 1);$$

and $\{2, 0\}\{1, 1\}$, or say $\{\alpha, \beta\}\{\xi, \eta\}$, is zero because $\alpha + \eta > r - 1$. Thus, as $(0, 1) = 1$, we have 2 as the number of lines meeting four given lines in ordinary space.

We also give some examples for lines in space of four dimensions $(r = 4)$: (a) It is known that there is one line meeting three given lines; this follows from

$$(1, 4)^3 = \{2, 0\}^3 = \{2, 0\}[\{2, 2\} + \{3, 1\}] = \{2, 0\}\{3, 1\} = \{3, 3\} = (0, 1) = 1,$$

the term $\{2, 0\}\{2, 2\}$, or, say, $\{\alpha, \beta\}\{\xi, \eta\}$, being zero because $\alpha + \eta > r - 1$. (b) There is one line through a given point to meet a given line and a given plane. And

$$(0, 4)(1, 4)(2, 4) = \{3, 0\}\{2, 0\}\{1, 0\} = \{3, 0\}[\{2, 1\} + \{3, 0\}]$$
$$= \{3, 0\}^2 = \{3, 3\} = (0, 1) = 1,$$

the term $\{3, 0\}\{2, 1\}$ being zero because $3 + 1 > 3$, as in the last example. (c) There are five lines meeting six given planes. For, first for four planes,

$$(2, 4)^4 = \{1, 0\}^4 = [\{2, 0\} + \{1, 1\}]^2 = \{2, 0\}^2 + 2\{2, 0\}\{1, 1\} + \{1, 1\}^2$$
$$= [\{2, 2\} + \{3, 1\}] + 2\{3, 1\} + \{2, 2\} = 3\{3, 1\} + 2\{2, 2\},$$

and hence for six planes,

$$(2, 4)^6 = [\{2, 0\} + \{1, 1\}][3\{3, 1\} + 2\{2, 2\}]$$
$$= 3\{2, 0\}\{3, 1\} + 2\{2, 0\}\{2, 2\} + 3\{3, 1\}\{1, 1\} + 2\{2, 2\}\{1, 1\}$$
$$= 3\{3, 3\} + 2[0] + 3[0] + 2\{3, 3\} = 5\{3, 3\} = 5(0, 1) = 5.$$

Another example of the formula is to prove, for any space $[r]$, the following formula, which we require below (though a direct proof is possible), namely $(p, q)(r-2, r) = (p, q-1) + (p-1, q)$. This supposes $p < q-1$; but, for $p = q-1$, we have, to express that a line lying in a $[q]$, and meeting a $[r-2]$, meets their $[q-2]$ intersection, the particular case $(q-1, q)(r-2, r) = (q-2, q)$. Suppose, then, $p \leqslant q-2$; the formula required is

$$\{r-1-p, r-q\}\{1, 0\} = \{r-1-p, r-q+1\} + \{r-p, r-q\},$$

and this is obtained from the formula by putting

$$\alpha = r-1-p, \quad \beta = r-q, \quad \xi = 1, \quad \eta = 0,$$

for which $\alpha - \beta \geqslant \xi - \eta$.

More generally, in space $[r]$, we can use the formula, above, for the product of two line symbols, to obtain a formula for the number of lines which meet s given spaces of dimension $r-2$; that is, to find $(r-2, r)^s$, or $\{1, 0\}^s$. The result is

$$\{1, 0\}^s = \{s, 0\} + c_1\{s-1, 1\} + c_2\{s-2, 2\} + \ldots + c_p\{s-p, p\},$$

where p is the greatest integer in $\frac{1}{2}s$, and the coefficient c_i is, with $\binom{s}{i}$ for the binomial coefficient,

$$c_i = \binom{s}{i}\frac{s-2i+1}{s-i+1}.$$

This can be proved by induction; for the initial values, $s = 1, 2, \ldots$, the formula can be proved directly; it is sufficient then to shew that, if the right side be multiplied by $\{1, 0\}$, the result can be arranged so as to give the same formula for $\{1, 0\}^{s+1}$. For this, denote the line symbol $\{m, s+1-m\}$ by λ_m, and notice that

$$\{s-k, k\}\{1, 0\} = \{s-k, k+1\} + \{s-k+1, k\} = \lambda_{s-k} + \lambda_{s-k+1};$$

thus, on multiplication by $\{1, 0\}$, the right side becomes

$$\lambda_s + \lambda_{s+1} + c_1(\lambda_{s-1} + \lambda_s) + c_2(\lambda_{s-2} + \lambda_{s-1}) + \ldots + c_p(\lambda_{s-p} + \lambda_{s-p+1}),$$

or $\lambda_{s+1} + (1+c_1)\lambda_s + (c_1+c_2)\lambda_{s-1} + \ldots + (c_{p-1}+c_p)\lambda_{s-p+1} + c_p\lambda_{s-p}$,

while, we see at once, $c_h + c_{h+1} = C_{h+1}$, where C_i is the value of c_i when s is changed to $s+1$. The new right side is thus

$$\lambda_{s+1} + C_1\lambda_s + C_2\lambda_{s-1} + \ldots + C_p\lambda_{s-p+1} + c_p\lambda_{s-p},$$

and the required result is obtained if we shew that the last effective term of this is

$$\binom{s+1}{q}\frac{s-2q+2}{s-q+2}\{s-q+1, q\},$$

where q is the greatest integer in $\frac{1}{2}(s+1)$. When s is even, we have $q=p$, and the last term $c_p\lambda_{s-p}$ which is written, is zero, since, in $\lambda_{s-p}=\{s-p, p+1\}$, we have $s-p=\frac{1}{2}s$, which is $<p+1$. When s is odd, say $2t+1$, we have $q=p+1$, $=t+1$, and it is to be verified that $c_p=C_{p+1}$; this is immediate. The formula for $\{1, 0\}^s$ is thus proved.

In particular, when $s=2r-2$, all the early terms $\{\alpha, \beta\}$, of the series, vanish because $\alpha>r-1$; the only surviving term is the last, for $p=r-1$. Thus, replacing $\{r-1, r-1\}$ by 1, the number of lines in space $[r]$ which meet $2r-2$ given spaces of dimension $r-2$, is $(2r-2)!/r!(r-1)!$, as we have seen above for $r=3$ and $r=4$.

Many other formulae have been found, for the solution of the general problem of expressing the product of a number of simultaneous conditions as a sum of alternative conditions. We can only give space for a few.

(1) For a line in space $[r]$, which depends on $2r-2$ parameters, to meet a space $[r-3]$, is two elementary conditions; there is thus a finite number of lines meeting $r-1$ given spaces $[r-3]$. This number is given by

$$(r-3, r)^{r-1}= \sum_{i=0} (-1)^i \binom{r-1}{i} \binom{2r-4-3i}{r-3},$$

where each term involves the product of two binomial coefficients.

(2) The number of spaces $[k]$, in space $[r]$, which satisfy two general conditions $(a_0, a_1, \ldots, a_k)$ and $(b_0, b_1, \ldots, b_k)$, and also satisfy the further e conditions of meeting e given spaces $[r-k-1]$, where e is such that the total number of elementary conditions imposed is equal to the number $(k+1)(r-k)$, of parameters on which a $[k]$ depends, is given by the formula

$$(a_0, a_1, \ldots, a_k)(b_0, b_1, \ldots, b_k)(r-k-1)^e=e!\,|c_{ij}|,$$

where $e!=1.2.3\ldots e$ and the determinant is of $k+1$ rows and columns $(i, j=0, 1, \ldots, k)$, and the element c_{ij} is $1/(a_i+b_{k-j}-r)!$, to be replaced by zero for every element in which $a_i+b_{k-j}<r$.

As an example, for the lines in [4] which pass through a given point, meet a given line, and meet a given plane, we have

$$(0, 4)(1, 4)(2)= \begin{vmatrix} [0!]^{-1}, & [(-3)!]^{-1} \\ [4!]^{-1}, & [1!]^{-1} \end{vmatrix} =1;$$

and for the lines in [4] which meet, in all, six planes, we have

$$(2, 4)(2, 4)(2)^4=4!\begin{vmatrix} [2!]^{-1}, & [0!]^{-1} \\ [4!]^{-1}, & [2!]^{-1} \end{vmatrix} =5.$$

(3) A particular case of the formula in (2) is when

$$(b_0, b_1, \ldots, b_k)=(r-k, r-k+1, \ldots, r)=1.$$

Then, with the appropriate value of e,

$$(a_0, a_1, \ldots, a_k)(r-k-1)^e = e! \,|\, [(a_i-j)!]^{-1} \,|.$$

In the determinant, the elements in the $(i+1)$-th row are

$$[a_i!]^{-1}, \, [(a_i-1)!]^{-1}, \, \ldots, \, [(a_i-k)!]^{-1},$$

and these, all multiplied by $a_i!$, are

$$1, \, a_i, \, a_i(a_i-1), \, \ldots, \, a_i(a_i-1)\ldots(a_i-k+1).$$

Thus the determinant, multiplied by $a_0! a_1! \ldots a_k!$, if the columns be suitably combined, becomes the determinant of which the general row consists of the elements $1, \, a_i, \, a_i^2, \, \ldots, \, a_i^k$. Wherefore

$$(a_0, a_1, \ldots, a_k)(r-k-1)^e = e! \prod_{r,s}^{0\ldots k} (a_r-a_s)/a_0! \, a_1! \ldots a_k!$$

(4) As a particular case of this last formula, we may take $(a_0, \ldots, a_k) = (r-k, \, r-k+1, \, \ldots, \, r) = 1$, with e equal to $(k+1)(r-k)$. Then we have the result that the number of spaces $[k]$ which meet $(k+1)(r-k)$ given spaces of dimension $r-k-1$ is

$$e! \, \frac{((k)) \, ((r-k-1))}{((r))},$$

or, say $e!((k))((k'))/((r))$, where $k'=r-k-1$ and $((m))$ denotes $1! \, 2! \ldots m!$, and $e=(r-k)(k+1), \, =(k+1)(k'+1)$. This is also the number of spaces $[r-k-1]$ which meet $(k+1)(r-k)$ given spaces $[k]$.

This is a fundamental number. Writing the coordinates of $k+1$ points, of the original space $[r]$, which suffice to determine a space $[k]$, as the $k+1$ rows of a matrix of $r+1$ columns, the $(r+1, \, k+1)$ determinants of order $k+1$, which can be formed from this matrix, are the homogeneous coordinates of the $[k]$; these can be used to represent the $[k]$ by a point of space $[R]$, where $R=(r+1, \, k+1)-1$. The aggregate of all spaces $[k]$, in $[r]$, is then represented, in this space $[R]$, by a manifold M_e, of dimension e. The order of this manifold is the number we have found. For, by definition, the order of a manifold, of dimension e, is the number of its points which satisfy e equations linear in the coordinates of the space in which it lies; namely, the number of its points which lie on e primes of the space $[R]$. The condition for a space $[k]$, in the original space $[r]$, to meet a given $[r-k-1]$ is, however, a linear condition for the coordinates of the $[k]$; the coefficients in this linear equation, which are the coordinates of the $[r-k-1]$, are not general; but the relations which connect them are quadratic, not linear. Wherefore, the order of the manifold in $[R]$ which represents the $[k]$ of $[r]$ is given by the number of spaces $[k]$ which meet e given spaces $[r-k-1]$. All this is an easy generalisation of the familiar fact that the quadric, in space $[5]$, which represents the lines of ordinary space, has the order 2, this being the number of lines in ordinary space [3] which meet four given lines. More generally, the lines of space [4], as also the dually corresponding planes, are representable, in a space of 9 dimensions, by a manifold of dimension 6, whose order is 5, say $M_6{}^5$; and, for example, the planes of space [5], are representable, in a space of 19 dimensions, by a manifold of dimension 9, whose order is 42.

(5) Another noticeable formula, with the same notation, is

$$(a_0, a_1, ..., a_k)(r-k-1) = (a_0-1, a_1, ..., a_k)$$
$$+ (a_0, a_1-1, a_2, ..., a_k) + ... + (a_0, a_1, ..., a_{k-1}, a_k-1),$$

which, in condition symbols, is

$$\{\alpha_0, \alpha_1, ..., \alpha_k\}\{1, 0^k\} = \sum_{i=0}^{k} \{\alpha_0, ..., \alpha_{i-1}, \alpha_i+1, \alpha_{i+1}, ..., \alpha_k\}.$$

From this formula we can build up, by steps, the formula for $(a_0, ..., a_k)(r-k-1)^m$; it is necessary at each stage to omit the symbols on the right which are unmeaning because the indices are not in ascending order.

A generalisation in another direction is the formula

$$(a_0, ..., a_k)(r-k-m) = \sum_j (a_0-j_0, a_1-j_1, ..., a_k-j_k),$$

where, for brevity, $(r-k-m)$ means the condition that a [k] meets a $[r-k-m]$, which is equivalent to m elementary conditions. Here the summation extends to all positive, or zero, integers $j_0, ..., j_k$ whose sum is m, subject to $j_i < a_i - a_{i-1}, j_0 < a_0 + 1$.

Part II. The problem of multiple tangents of a manifold. Consider a primal, of order n, in space [r], which we may denote by $M^n{}_{r-1}$, defined by a single perfectly general polynomial equation connecting the coordinates. In order that a line should touch the primal at one of its intersections with it, a single condition must be imposed upon the line; in order that, at this intersection, there should be i coincident points of the primal upon the line, $i-1$ conditions must be imposed. Lines with such a multiple intersection are thus ∞^k in aggregate, where

$$k = 2r-2-(i-1), = 2r-1-i;$$

and this is also the freedom of the points where such lines touch the manifold. If we impose, upon these multiple tangents, a further condition equal to $2r-2-i$ elementary conditions, or say of power $2r-2-i$, the lines satisfying all the conditions will be ∞^1, and will form a ruled surface; the locus of the points of multiple contact with the manifold will then be a curve. For a moment, let the order of this ruled surface be denoted by g, and the order of the curve of contact by b; each generator will have $(n-i)$ further intersections with the primal, and the curve which is the locus of these further points will have an order f given by $f = ng - ib$.

Now take a pencil of primes, that is, a series of primes all passing through a definite space $[r-2]$. Consider the correspondence between a prime, (α), of this pencil which contains a point of multiple contact, with the primal, of a generator of the ruled surface, and a prime, (β), of this pencil, which contains one of the simple intersections of the same generator. On each prime (β) there are f of the simple intersections spoken of, each associated with the multiple contact on the same generator; on each prime (α) there are b of the multiple contacts, each associated with the $n - i$ simple intersections on the same generator. Thus the correspondence between the primes (α), (β) is of indices f and $(n - i)\, b$. There are, therefore, $f + (n - i)\, b$ primes of the pencil which are coincidences of a pair of corresponding primes. Such coincidences of primes arise, (1), for lines of the ruled surface which have $(i + 1)$ coincident intersections with the primal; denote the number of these by ϵ; they arise also, (2), for generators of the ruled surface which meet the base space $[r-2]$ of the pencil of primes; there are g such generators, and each of these gives $n - i$ coincidences of pairs of corresponding primes, the multiple intersection of the generator being taken with all the $n - i$ simple intersections in turn. On the whole, then, we have the equation $\epsilon + (n - i)\, g = f + (n - i)\, b$; and, in virtue of $f = ng - ib$, this leads to $\epsilon = (n - 2i)\, b + ig$.

We modify this equation, in two stages, so that it may be a statement, not in regard to a point only, or a line only, but in regard both to a line and a point lying thereon; that the point lies in a space $[k]$ will be denoted by a symbol (k), that the line meets a space $[h]$ will be denoted by a symbol (h, r); further, that the point is to be thought of as the coincidence of j points on the line will be denoted by the symbol ϵ_j. Thus, as b is the number of i-pointic intersections of a line (the generator of the ruled surface) with the primal, which lie in a prime $[r-1]$, we replace b by $\epsilon_i\, (r-1)$; and as g is the number of the lines considered which meet a space $[r-2]$, we replace g by $\epsilon_i\, (r-2, r)$. The equation can then, first, be replaced by

$$\epsilon_{i+1} = (n - 2i)\, \epsilon_i\, (r-1) + i\epsilon_i\, (r-2, r),$$

this being an abbreviation of the complete statement, for the composite entity consisting of a line and a point thereon, in which there would be supplied, in the three terms respectively, the factors $(r)(r-1, r)$, $(r-1, r)$, (r). This is an equation connecting numbers, obtained on the hypothesis that $2r - 2 - i$ elementary conditions have been imposed on the line, in addition to the condition of multiple intersection with the primal; without these additional conditions the line would not describe a ruled surface. We now suppose that these $2r-2-i$ elementary conditions for the line may be built up of a certain number s, of conditions for the line, with a number, t, of conditions for the multiple intersection with the primal (where $s+t=2r-2-i$); therein making the assumption that t elementary conditions, of the kind to be imposed, for this distinguishable point of the line, are equivalent to t conditions for the line. The s conditions for the line are that it lie in a space $[q]$ and meet a $[p]$ contained therein, or (p, q); this is equivalent, we have seen, to $(r-1-p)+(r-q)$, or to $2r-1-p-q$ elementary conditions; the t conditions for the multiple intersection are that it lie in a space $[r-p-q+i+1]$, that is, in $p+q-i-1$ primes, which is equivalent to $p+q-i-1$ elementary conditions. We have

$$s+t=2r-1-p-q+p+q-i-1=2r-2-i.$$

It is supposed that $p+q \geqslant i+1$. Denoting the condition for the point by $(r-p-q+i+1)$, we can now write the equation above in the form

$$\epsilon_{i+1}(p, q)(r-p-q+i+1)=(n-2i)\,\epsilon_i(r-1)(p, q)(r-p-q+i+1)$$
$$+i\epsilon_i(r-2, r)(p, q)(r-p-q+i+1),$$

of which each term, counting ϵ_j as of power $j-1$, is of the power $2r-2$, equal to the number of parameters on which a line depends. This equation, however, may be simplified: First, a point which lies both on a $[r-1]$ and on a $[r-p-q+i+1]$, lies on the $[r-p-q+i]$ in which they meet, so that, in the equation, $(r-1)(r-p-q+i+1)$ may be replaced by $(r-p-q+i)$. Second, we may use the formula, remarked above (p. 83), always true (for $p<q$) if impossible terms on the right are omitted, $(p, q)(r-2, r)=(p, q-1)+(p-1, q)$, to modify the last term on the right, so obtaining

$$\epsilon_{i+1}(p, q)(r-p-q+i+1)=(n-2i)\,\epsilon_i\,(p, q)\,(r-p-q+i)$$
$$+i\epsilon_i\,(p, q-1)(r-p-q+i+1)+i\epsilon_i(p-1, q)(r-p-q+i+1).$$

If then we put
$$H_i(p, q)=\epsilon_i(p, q)(r-p-q+i), \qquad p+q \geqslant i+1,$$
we have

$$H_{i+1}(p, q)=(n-2i)\,H_i(p, q)+iH_i(p, q-1)+iH_i(p-1, q).$$

The finite number of lines, $H_i(p, q)$, satisfying the condition $H_i(p, q)$, which is of power $2r-2$, can be obtained by induction from this equation if we obtain $H_1(p, q)$. This, being $\epsilon_1(p, q)(r-p-q+1)$, is the number of lines, lying in a $[q]$, which meet a $[p]$ lying therein, whose intersection with the primal M^n_{r-1} lies on a space $[r-p-q+1]$. This last space meets M^n_{r-1} in a manifold of dimension $r-p-q$, say M^n_{r-p-q}; unless $p=0$ this manifold will not in general intersect the space $[q]$, and there will be no lines in $[q]$ whose intersections with

M^n_{r-1} lie on the M^n_{r-p-q}; but, when $p=0$, the space $[q]$ will meet the M^n_{r-p-q} in n points, each of which can be joined to the point $[p]$, which lies in the $[q]$. Thus $H_1(p, q)$ is 0 or n, according as $p > 0$ or $p = 0$.

Now let P_i denote the polynomial in t, of order $i-1$, given by

$$P_i = n[n-1+t][n-2+2t] \dots [n-(i-1)(1-t)],$$

and let $c_i(p, q)$ denote the coefficient of t^{-1} in the ascending expansion of $P_i(t^{-p-1}-t^{-q-1})$; in particular $c_1(p, q)$, the coefficient of t^{-1} in the ascending expansion of $n(t^{-p-1}-t^{-q-1})$, is 0 or n, according as $0 < p < q$ or $0 = p < q$. Thus $H_1(p, q) = c_1(p, q)$. While

$$(n-2i)\, c_i\, (p, q) + ic_i\, (p, q-1) + ic_i\, (p-1, q),$$

being the coefficient of t^{-1} in the ascending expansion of the product of P_i with

$$(n-2i)(t^{-p-1}-t^{-q-1}) + i(t^{-p-1}-t^{-q}) + i\,(t^{-p}-t^{-q-1}),$$

or

$$[n-i(1-t)][t^{-p-1}-t^{-q-1}],$$

is equal to $c_{i+1}(p, q)$. Thus in general we have $H_i(p, q) = c_i(p, q)$, or $\epsilon_i(p, q)(r-p-q+i)$ is equal to

$$[nn_1 n_2 \dots n_{i-1}(t^{-p-1}-t^{-q-1})]_{t^{-1}},$$

where n_s means

$$n-s(1-t), \quad \text{and} \quad p+q \geqslant i > 0, \quad 0 \leqslant p < q;$$

this gives the number of lines, in space $[r]$, subject to the condition (p, q), which have i-pointic intersection with a primal M^n_{r-1}, with the point of contact in a space $[r-p-q+i]$. This number is expressible only by i, p, q, without r; as P_i is of order $i-1$ in t, the number is zero if $i-1 < p$; or, we must suppose $i > p$. Further, we consider only those lines which lie in a $[q]$; thus, effectively, we consider only the M^n_{q-1} which is the section, by the $[q]$, of the original M^n_{r-1}. Thus, changing the notation by supposing $q = r$, the formula is, effectively,

$$\epsilon_i(p, r)(i-p) = [nn_1 n_2 \dots n_{i-1}(t^{-p-1}-t^{-r-1})]_{t^{-1}}$$

with the obvious conditions $0 \leqslant p < r$, $0 \leqslant i - p \leqslant r$.

Some examples may be given:
(1) If $p=0$, $r=2$, and $i=1$ or 2, the formula gives, respectively, the order n, and the class $n(n-1)$, of a plane curve whose point equation is quite general. If $p=1$, $r=2$, $i=3$, the formula for $\epsilon_3(1, 2)(2)$ gives $3n(n-2)$ for the number of inflexional tangents of the curve.
(2) If $p=0$, $r=3$, $i=3$, the number, $\epsilon_3(0, 3)(3)$, of inflexional tangents of a surface, of order n, in ordinary space, which pass through a given point, is found as $n(n-1)(n-2)$. The points of the surface at which the inflexional tangents meet an arbitrary line, lie on a curve of order $\epsilon_3(1, 3)(2)$, which is $n(3n-4)$.

(3) If $p=2$, $r=3$, $i=4$, the order of the curve locus of 4-pointic tangents of a surface in ordinary space, is

$$\epsilon_4(2,3)(2), \quad = \{n(n-1+t)(n-2+2t)(n-3+3t)(t^{-3}-t^{-4})\}_{t^{-1}},$$
$$= n[6(n-1)+3(n-2)+2(n-3)-6],$$

or $n(11n-24)$. In particular, for $n=3$, this gives the number, 27, of lines on a general cubic surface.

The fact that the points of 4-pointic contact, of tangents of a surface of order n in ordinary space, lie on another surface of order $11n-24$, was obtained by Salmon in 1849 (*Camb. and Dub. Math. J.* IV, 1849, p. 260). See also *Quart. J. of Math.* I, 1857, p. 336; *Phil. Trans.* A, CL, 1861, p. 229; and Clebsch, *Crelle*, LVIII, 1861, pp. 93, 106. Salmon, *Solid Geom.* 1882, p. 559, quotes from Cayley the remark that this surface is obtainable by eliminating m from the two equations

$$z_{11}+2mz_{12}+m^2z_{22}=0, \quad z_{111}+3mz_{112}+3m^2z_{122}+m^3z_{222}=0,$$

($z_{11}=\partial^2 z/\partial x^2$, $z_{12}=\partial^2 z/\partial x \partial y$, etc.); and proves (*loc. cit.* p. 567) that the surface touches the Hessian of the original surface. Salmon also gives (*ibid.*) the order, $2n(n-3)(3n-2)$, of the ruled surface generated by the 4-pointic tangents, which is obtainable as $\epsilon_4(1,3)(3)$.

(4) The number of 5-pointic tangents of a general surface of order n in ordinary space is $\epsilon_5(2,3)(3)$, $=5n(n-4)(7n-12)$, for $n \geqslant 4$.

(5) The simple case $\epsilon_i(i-1,r)(1)=n(i-1)!$ is in agreement with the fact that the i-pointic tangents, at a point of the primal M_{r-1}^n, form a cone of dimension $r-i+1$, and of order $(i-1)!$. For instance, the 2-pointic tangents lie in a prime. The 3-pointic tangents form a cone of dimension $r-2$ and order 2 (for $r=3$ they are 2 in number; for $r=4$, they form a conical sheet of order 2; for $r=5$ a quadric solid cone). The 4-pointic tangents form a cone of dimension $r-3$ and order 6 (for $r=4$ they are 6 in number).

(6) For a general value of r, $\epsilon_r(0,r)(r)$, $=n(n-1)...(n-r+1)$ is the number of r-pointic tangents which can be drawn through an arbitrary point.

(7) The number of tangents of the primal with $(2r-1)$-pointic contact is $\epsilon_{2r-1}(r-1,r)(r)$. For example, in space [4], the number of tangents with 7-pointic contact is $35n(n-6)(7n-12)(3n-10)$.

(8) The order of the ruled surface generated by 5-pointic tangents of a primal in [4] is $\epsilon_5(3,4)(2)$, which is $10n(5n-12)$.

(9) The order of the curve of contact of 6-pointic tangents of a primal in [4] is $\epsilon_6(3,4)(3)$, which is $5n(45n^2-274n+360)$. In particular, a primal of order 5 in space [4], contains $5^3.23$ or 2875 lines (cf. Schubert, *Math. Ann.* XII, 1877, p. 192). In general a M_{r-1}^{2r-3} in [r] contains $\epsilon_{2r-2}(r-1,r)(r-1)$ lines.

A more general problem than the one solved in the preceding investigation, is that of a line meeting a given primal with several points of coincident intersection, of assigned multiplicities, satisfying possibly other conditions also, for the line itself or the points of contact. This is an interesting problem, generalising the investigation of de Jonquières explained above (p. 39), of the variable curves in a plane, having coincidences of assigned multiplicities in their intersections with a given curve. There is here no space for a detailed account of Schubert's remarkable solution of this problem;

the reader who finds the introductory account of Schubert's notation which is given below (pp. 96 ff.) to be interesting, is referred to Schubert's chapter (*Kalkül d. abzähl. Geom.* 1879, pp. 228–44); and also to Zeuthen's *Lehrbuch d. abzähl. Methoden d. Geom.* 1914, pp. 372–76. Some particular examples, mostly given by Salmon, *Solid Geometry*, 1882 (or Ed. of 1915, pp. 277 ff.) may be quoted. See also Roth, *Proc. Camb. Phil. Soc.* XXVI, 1930, p. 43.

(1) If for a general surface of order n, in ordinary space, a line be inflexional at one point and ordinarily tangent at another point of the surface (thus satisfying $2+1$ elementary conditions in all), the locus of the points of inflexion is a curve of order $n(n-4)(3n^2+5n-24)$, while the locus of the points of ordinary contact is a curve of order $n(n-2)(n-4)(n^2+2n+12)$, and the lines generate a ruled surface of order $n(n-3)(n-4)(n^2+6n-4)$. From these it is easily found that the simple intersectiòns generate a curve of order

$$n(n-4)(n-5)(n^3+6n^2-n-24).$$

For these results, see Salmon, *loc. cit.* pp. 558, 568, 570.

(2) In the same case, a line with three ordinary contacts generates a ruled surface of order $\frac{1}{3}n(n-3)(n-4)(n-5)(n^2+3n-2)$, the contacts generating a curve of order $\frac{1}{2}n(n-2)(n-4)(n-5)(n^2+5n+12)$, and the simple intersections a curve of order

$$\frac{1}{3}n(n-4)(n-5)(n-6)(n^3+3n^2-2n-12).$$

(Salmon, *loc. cit.* p. 569.)

(3) The lines having four ordinary contacts are in number

$$\tfrac{1}{12}n(n-4)(n-5)(n-6)(n-7)(n^3+6n^2+7n-30).$$

(Salmon, *loc. cit.* p. 575, with reference to Schubert.)

(4) The number of lines with 4-pointic contact at one place, and an ordinary contact elsewhere, is $2n(n-4)(n-5)(3n-5)(n+6)$. (Salmon, *ibid.*)

(5) The number of lines which are inflexional at two places, is $\frac{1}{2}n(n-4)(n-5)(n^3+3n^2+29n-60)$. (Salmon, *ibid.*)

(6) The lines inflexional at one place and ordinarily tangent at two others are in number $\frac{1}{2}n(n-4)(n-5)(n-6)(n^3+9n^2+20n-60)$. (Salmon, *ibid.*)

(7) The number of points of the surface where both inflexional tangents have four coincident intersections is $5n(7n^2-28n+30)$. (Salmon, p. 572; Schubert, *Kalkül*, p. 246.) The surface has ∞^2 points, and the condition for each inflexional tangent is onefold. For a cubic surface the number is three times the number (45) of tritangent planes.

(8) The points where the two inflexional tangents of the surface coincide in a line having 4-pointic intersection are in number $2n(n-2)(11n-24)$. (Salmon, p. 573.)

(9) The inflexional tangents at the parabolic points of the surface, where the inflexional tangents coincide, generate a ruled surface of order $2n(n-2)(3n-4)$. (Schubert, p. 244.) The parabolic curve is the intersection of the surface with its Hessian, and the tangent planes of the surface at two coinciding points of the parabolic curve intersect in the inflexional line. Thus the number in question is the order of the developable surface enveloped by the tangent planes of the surface at the points of the parabolic curve; as such, it is given by Salmon (p. 546). Salmon also gives (p. 547) the order of the developable surface formed by the

tangent planes of the given surface, of order n, at the points of the curve
of intersection of this surface with another surface of order k, namely
$nk(3n+k-6)$. When $k=4(n-2)$, the second surface being the Hessian,
this number is $4n(n-2)(7n-14)$; and this exceeds the number
$2n(n-2)(3n-4)$ here given, by the number, $2n(n-2)(11n-24)$, given
in Ex. 8 preceding. In general, the number of tangents of the curve
(n, k) which are inflexional lines of the surface (n) is $nk(3n+2k-8)$; and
the inflexional lines of the surface (n), at the points of the curve (n, k),
form a ruled surface of order $nk(3n-4)$. (Salmon, p. 546.)

Part III. Correspondence of points of two manifolds. We
consider now the generalisation of the principle of correspondence,
as originally developed above for ∞^1 pairs of corresponding points,
to cases of multiple correspondence. The results in the simpler cases
were first enunciated by Salmon and Zeuthen (Salmon, *Solid Geom.*
1882, pp. 550, 598; Zeuthen, *Compt. rend.* LXXVIII, 1874, p. 1553).
We may, for instance, in a plane, have ∞^2 pairs of points (P, P'),
such that, to any position of P corresponds only a finite number
of positions of P'; and similarly, to any position of P', a finite number
of positions of P. This does not involve, however, necessarily, that
points P, or P', occur everywhere in the plane. We may denote the
number of positions of P' which correspond to an assigned position
of P by $(0)(2)'$, the first symbol referring to the assigned point P,
and the second symbol to the corresponding positions of P', anywhere in those parts of the plane where points P' occur; similarly,
the number of positions of P when P' is assigned may be denoted
by $(2)(0)'$. As P takes positions upon an arbitrary line, P' will take
positions upon a certain curve, and there will be a certain number
of points of this curve lying upon a second arbitrary line. We thus
consider a number, $(1)(1)'$, of pairs of corresponding points, in
which P and P' lie, respectively, on given arbitrary lines. The
formula we arrive at expresses the sum $(0)(2)'+(1)(1)'+(2)(0)'$
in terms of the number of coincidences of P and P'. But such coincidences are of two kinds: It may happen that, as the pair of corresponding points, P, P', approaches a point C at which they coincide, the line PP' likewise approaches a definite line through C,
which is independent of the manner in which P approaches to C;
or, it may happen that the line PP', by suitable approach of these
points to C, can be made to approach to any assigned line through
C. For the number of points of coincidence of the second kind we
may use the symbol $\epsilon\,(0, 2)(2)$, where the ϵ, after Schubert's use,
indicates that we are considering a coincidence; the $(0, 2)$ refers to
the line which is the ultimate position of PP' as these points approach C, and indicates that this line may be supposed to pass
through an arbitrary point of the plane; and the (2) refers to the
point of coincidence C, and indicates that we are considering all
such coincidences as occur anywhere in the plane. Points of coin-

cidence of the first kind, for which PP' ultimately takes up a definite position as P, P' approach C, are generally of infinite number, as we shall see immediately, and lie upon a curve, so that there is a definite number of such points lying upon an arbitrary line. The whole number of these points is therefore denoted by $\epsilon(1, 2)(1)$, where the symbol $(1, 2)$ has only a formal significance, as indicating that the ultimate line PP' cannot be supposed to pass through an arbitrary point of the plane, though it meets an arbitrary line, and the symbol (1) indicates that we consider all points of coincidence, C, of the kind in question, which lie upon an arbitrary line. Under the hypothesis that all the symbols have definite finite values, the equation we are to prove is

$$(0)(2)' + (1)(1)' + (2)(0)' = \epsilon(0, 2)(2) + \epsilon(1, 2)(1).$$

It must be understood that, when the contrary is not stated, the corresponding points P, P' are supposed to be so defined as to be distinguishable from one another. For instance, the theory applies when the P, P' are in a $(1, 1)$ correspondence defined by general linear equations. But, if this linear correspondence be involutory, two points, A, B, which are a pair of the involution, give two pairs of the correspondence; for P' is at B when P is at A, and also P' is at A when P is at B. The steps where the proofs of the fundamental formulae need such modification are briefly noticed below (p. 101).

Before proceeding to the proof of this equation, we consider the statements we have made as to the possible modes of coincidence. Denote the equations which express the correspondence, between the point (x, y) and the point (x', y'), for a limited region of the plane containing both these points, by

$$x' - f(x, y) = 0, \quad y' - \phi(x, y) = 0;$$

the functions f and ϕ, in the applications made, are generally algebraic; but we suppose them at least analytic and holomorphic (and single valued) in the regions in which they are used. The coincidences of the correspondences are then the points (x, y) which are the solutions of the equations $x - f(x, y) = 0, y - \phi(x, y) = 0$. If (a, b) be such a coincidence, and $(x, y), (x', y')$ be both in the neighbourhood of this, the equations of the correspondence lead, in virtue of $a - f(a, b) = 0, b - \phi(a, b) = 0$, to

$$x' - x = (x-a)[f_1(a, b) - 1] + (y-b)f_2(a, b) + U_2,$$
$$y' - y = (x-a)\phi_1(a, b) + (y-b)[\phi_2(a, b) - 1] + V_2,$$

where U_2, V_2 are series in $x-a, y-b$, involving squares and higher powers, and $f_1(x, y) = \partial f/\partial x$, etc. If then (a, b) do not lie on the curve $\Delta = 0$, where

$$\Delta = \begin{vmatrix} f_1(x, y) - 1, & f_2(x, y) \\ \phi_1(x, y), & \phi_2(x, y) - 1 \end{vmatrix},$$

the equations, by an appropriate limiting value of the ratio
$(y-b)/(x-a)$, lead to any assigned limiting value of $(y'-y)/(x'-x)$;
that is, the joining line of the coinciding points may be regarded as
capable of an arbitrary ultimate direction. Then we have a point
included in the formula $\epsilon(0, 2)(2)$. But if (a, b) lie on the curve
$\Delta = 0$, the limiting value of $(y'-y)/(x'-x)$ is definite, expressible
as $\phi_1(a, b)/[f_1(a, b)-1]$, or $[\phi_2(a, b)-1]/f_2(a, b)$; we have then a
point of coincidence included in the formula $\epsilon(1, 2)(1)$. We omit
the consideration of the cases in which neither of the alternative
forms for $(y'-y)/(x'-x)$ is definite in value.

Similarly, we may have a correspondence in which there are ∞^3
pairs of corresponding points P, P', which then we suppose to lie
in a space of three dimensions. Then we have the formula

$$(0)(3)' + (1)(2)' + (2)(1)' + (3)(0)' = \epsilon(0,3)(3) + \epsilon(1,3)(2) + \epsilon(2,3)(1);$$

here, on the left, the symbol $(i)(j)'$ means the number of pairs of
corresponding points P, P', in which P lies in an arbitrary space $[i]$,
while P' lies in another arbitrary space $[j]$; while, on the right, three
possible kinds of coincidence are contemplated; these are classified
by the freedom of the ultimate position of the joining line PP' as
P and P' approach to a point of coincidence C, and it is assumed
that this is correlated with the freedom of position of the associated
points C. In particular $\epsilon(0, 3)(3)$ is the number of coincidences of
P and P', of such character that the limiting position of the line
PP' may be taken arbitrarily through the point of coincidence;
$\epsilon(1, 3)(2)$ is the number of coincidences lying in an arbitrary plane,
for which the ultimate PP' meets an arbitrary line, while, $\epsilon(2, 3)(1)$
is the number of coincidences lying on an arbitrary line, for which
the ultimate PP' satisfies only the (nugatory) condition of meeting
an arbitrary plane.

The general case is that of ∞^r pairs of corresponding points, in
a space $[r]$. The appropriate equation, to be obtained, is immediately written down from the two preceding cases. On the right of
this equation the general term, one of r terms, has the form
$\epsilon(i, r)(r-i)$, for $i = 0, \ldots, (r-1)$; this denotes the number of points
of coincidence, lying in an arbitrary space $[r-i]$, of which the
ultimate line PP' meets an arbitrary space $[i]$. For a point to lie
in a $[r-i]$ requires i conditions, and for a line to meet an $[i]$ requires $r-1-i$ conditions. We must then explain the justification
for the evident assumption, in the formula, that the conjoint figure,
of a point of coincidence with the ultimate line PP' through this
point, can satisfy, in all, $r-1$ conditions; the simple remark that
there are ∞^r pairs of corresponding points, and the coincidence of
a pair, upon the line which joins them, requires one condition, may
not seem enough. On the left of the equation to be obtained, the

general term, $(i)(r-i)'$, one of $r+1$ terms, requires $r-i+i$, that is r elementary conditions.

As in the case $r=2$, using non-homogeneous coordinates $x_1, ..., x_r$, we may represent the correspondence, when both P and P' are in a suitably limited space, by the r, regular analytic, equations $f_i(x_1, ..., x_r) - x_i' = 0$, a coincidence $(a_1, ..., a_r)$ being a point for which the r equations $f_i(a_1, ..., a_r) - a_i = 0$ hold. If then the corresponding points (x) and (x') are in the neighbourhood (a), we have, to the first approximation, r equations such as

$$x_i' - x_i = (x_1 - a_1)f_{i1} + ... + (x_i - a_i)(f_{ii} - 1) + ... + (x_r - a_r)f_{ir}, \quad \text{(I)},$$

where f_{ik} is the value at (a) of $\partial f_i / \partial x_k$. If then

(1), the determinant, Δ, of the coefficients of $x_1 - a_1, ..., x_r - a_r$ on the right, be not zero, then the r equations $f_i(x_1, ..., x_r) - x_i = 0$ are independent in the neighbourhood of (a); in this case, by suitable limiting values of the ratios of $x_1 - a_1, ..., x_r - a_r$, that is, by a suitable direction of approach of (x) to (a), we can secure that the ultimate direction of PP', determined by the limits of the ratios of $x_1' - x_1, ..., x_r' - x_r$, is arbitrary. This corresponds to the hypothesis of discrete points of coincidence, and the term $\epsilon(0, r)(r)$. While

(2), if this determinant Δ be zero, but not every minor of $(r-1)$ rows and columns therein, then the equations $f_i(x_1, ..., x_r) - x_i = 0$ determine a curve of points of coincidence lying in the neighbourhood of (a). In this case, the equations (I), by elimination of $x_1 - a_1, ..., x_r - a_r$, imply a linear relation connecting $x_1' - x_1, ..., x_r' - x_r$; thus the ultimate line PP' may be, by suitable approach of (x) to (a), any line through (a) lying in a certain prime, and there is, in general, one such ultimate PP' meeting an arbitrary line of the space $[r]$. The equations (I) are only a first approximation; the more accurate statement is that the ultimate lines PP' lie on a cone, with vertex at (a), of dimension $r-1$, having a definite number of intersections with an arbitrary line of the space $[r]$. The symbol $\epsilon(1, r)(r-1)$ thus represents a definite number when the complete equations representing the correspondence are algebraic. Next

(3), if every first minor of the determinant Δ be zero, then the equations $f_i(x_1, ..., x_r) - x_i = 0$ determine an ∞^2 of points of coincidence in the neighbourhood of (a), say a surface; and the equations (I) imply two necessary linear relations connecting $x_1' - x_1, ..., x_r' - x_r$; in this case, the ultimate lines PP' are any lines through (a) in a space $[r-2]$. Or, more generally, considering the complete equations to which the equations (I) are an approximation, these lines PP' are the generators of a conical locus, of dimension $r-2$, having a finite number of intersections with an arbitrary plane of the space $[r]$; and the symbol $\epsilon(2, r)(r-2)$ has a definite value, when the correspondence is algebraic.

And so in general, the term $\epsilon(i, r)(r-i)$ arises from points of co-incidence lying on one or more manifolds, M_i, of dimension i, through any point of which the ultimate lines PP' are the generators of a cone of dimension $r-i$. For instance, when $r=3$, we have the cases, (1), of discrete coincidences, the lines PP' being all the lines through the point of coincidence; (2), of curves of points of coincidence, the ultimate lines PP' being the generators of a conical sheet, with vertex at the point of coincidence; and (3), of surfaces of points of coincidence, there being a finite number of ultimate positions PP' through a point of coincidence.

The reasonableness of the formula being thus made clear, we pass to a detailed proof. For the cases $r=2$ and $r=3$ this can be given without much difficulty so as to explain also Schubert's symbolism, as we first shew (Schubert, *Kalkül*, 1879, p. 45; see also *Math. Ann.* x, 1876, p. 54). But some readers may prefer to pass at once to the less symbolical general proof given later (pp. 99 ff.), which includes the cases $r=2, 3$.

Supposing the corrésponding points P, P' to lie in a space [3], associate these points, respectively, with the letters p and q; the primary meaning of these letters is in equations of condition, in which they signify the condition of lying in an arbitrary plane, imposed on P or P' respectively; but here p denotes the number of corresponding pairs, P, P' whereof P lies in an arbitrary plane, with a corresponding meaning for q. Similarly let g denote the condition that a specified line, g, should meet an arbitrary line; in our case the specified line is the join, PP', of the corresponding points P and P'. Further let ϵ denote a coincidence, in our case of the corresponding points P, P'. We then postulate the symbolic equation $\epsilon = p+q-g$.

This is justified by first supposing that limitations equivalent to ∞^2 elementary conditions are imposed, so as to reduce the aggregate of pairs of points P, P' to ∞^1; and then considering the correspondence of a pair of planes, of a pencil of planes, with a line g as axis, of which one plane contains P and the other P'. In each plane of the pencil, in accordance with the definition of P, lie p points P, each of which gives a point P', and hence a corresponding plane; while, reversely, in such a plane, lie q points P', each of which gives a point P, and hence a corresponding plane. There is thus a correspondence, in the planes of the pencil, of indices (q, p), and there are $q+p$ coincidences of pairs of corresponding planes. These arise, however, from ϵ coincidences of the points P, P'; and from the g cases in which the line PP' intersects the axis of the pencil. Thus $\epsilon + g = q + p$, which is the symbolic equation put down. It will be seen that this equation, modified by such multiplications, indicating further conditions, as render the terms capable of concrete interpretation, is the basis of Schubert's method. We also employ the symbolic equations of condition $g^2 = g_p + g_e$, $pg = p^2 + g_e$, which introduce two fresh symbols, g_p, the condition that a line g should contain an arbitrary point, and g_e, the condition that a line g should lie in an arbitrary plane. The former of these equations is one we have remarked upon above (p. 78), expressing that the condition, for a line g, of meeting two arbitrary lines, is compounded of the alternative conditions, of passing through an arbitrary point and lying in an arbitrary plane; it may be looked upon as a particular case of a relation (p. 83) for lines in space $[r]$ expressible by

$$(r-2, r)(p, q) = (p-1, q) + (p, q-1).$$

The second of these equations is an equation of condition for the composite entity consisting of a line (g) with a distinguishable point (p) thereon. The left side is the condition that the point lies on a given arbitrary plane, while the line meets a given arbitrary line; in a notation already employed (p. 93) it may be denoted by $(2)(1, 3)$. On the right side, the term p^2 is the condition for the point to lie in two arbitrary planes, that is, to lie on an arbitrary line; supplying a unitary factor $(2, 3)$ as condition for the line g, the condition may be denoted also by $(1)(2, 3)$. The term g_e is the condition for the line g to lie in a given plane, and may be denoted by $(1, 2)$, to which, if desired, may be supplied the unitary factor (3) for the point p. The whole equation may thus be denoted by $(2)(1, 3) = (1)(2, 3) + (1, 2)$. As the equation is to be applied to cases in which, by imposing complementary conditions, it becomes an equation in the numbers of the composite entity (g, p) which satisfy the conditions, it may be justified, as explained before (p. 78), by taking the case when the arbitrary plane, in which p is to lie, contains the arbitrary line which g is to meet. The composite condition can then obviously be satisfied in two ways; either, by the point p, of the line g, lying on the arbitrary line (g not necessarily being in the plane); this corresponds to the term p^2 on the right: or, by the line g lying in the arbitrary plane, in which case the point p lies in this plane; this corresponds to the term g_e on the right. It may be similarly proved that, for a line in space $[r]$, associated with a distinguishable point thereon, there is an equation

$$(a, r)(b) = (b, r)(a) + (a, b),$$

where $a < b$; which may be proved by considering the particular case when the space $[a]$ lies in the space $[b]$. Assuming then

$$\epsilon = p + q - g, \quad g^2 = g_p + g_e, \quad pg = p^2 + g_e, \text{ and hence } qg = q^2 + g_e,$$

the first gives, using the others,

$$\epsilon g = pg + qg - g^2 = (p^2 + g_e) + (q^2 + g_e) - (g_p + g_e) = p^2 + q^2 + g_e - g_p,$$

and also gives

$$\epsilon p = p^2 + pq - pg = p^2 + pq - (p^2 + g_e) = pq - g_e;$$

hence

$$\epsilon g + \epsilon p = p^2 + pq + q^2 - g_p, \text{ or } p^2 + pq + q^2 = (g_p + \epsilon g) + \epsilon p.$$

This equation leads to the equation for the coincidences of a correspondence of ∞^2 pairs of points P, P' in a plane ϖ, enunciated above. For first, p^2, the condition that P lies in two arbitrary planes, and therefore in their line of intersection, is the condition for P to be at the point where this line meets the plane ϖ; thus p^2 is what was before denoted by $(0)(2)'$. And similar remarks hold for the other two terms, pq which is $(1)(1)'$, and q^2 which is $(2)(0)'$. Again, g_p, expressing that the line PP', in the plane ϖ, passes through an arbitrary point, not necessarily in the plane ϖ, is zero; while ϵg, expressing a coincidence in which (the limiting position of) the line PP' meets an arbitrary line, that is, passes through the point in which this line meets the plane ϖ, will be the condition we have denoted by $\epsilon(0, 2)(2)$; and similarly ϵp will be $\epsilon(1, 2)(1)$.

But we can proceed further in the same way, and obtain the equation for the coincidences of a correspondence of ∞^3 pairs of points P, P' in space. It is convenient to introduce another symbol, g_s, the condition for a line g both to pass through a given point and to lie in a given plane (to be a line of a given flat pencil); evidently we have $gg_e = g_s = gg_p$. Then, from $pg = p^2 + g_e$, $g^2 = g_e + g_p$, we have

$$p^2g = p^3 + pg_e, \quad pg^2 = p^2g + g_e g_e, = p^2g + g_s, \quad pg^2 = pg_e + pg_p,$$

and, hence, $\quad p^3 + pg_e + g_s = p^2g + g_s = pg^2 = pg_p + pg_e,$

so that $pg_p = p^3 + g_s$, and $qg_p = q^3 + g_s$; but, we found above $\epsilon p = pq - g_e$, and we have $\epsilon = p + q - g$; thus

$$\epsilon g_p = pg_p + qg_p - gg_p = (p^3 + g_s) + (q^3 + g_s) - g_s = p^3 + q^3 + g_s,$$

as also

$$\epsilon g_e = pg_e + qg_e - gg_e, \quad \epsilon p^2 = (pq - g_e) p = p^2q - pg_e, \quad \epsilon q^2 = pq^2 - qg_e;$$

so that

$$\epsilon g_p + \epsilon g_e + \epsilon p^2 + \epsilon q^2 = (p^3 + q^3 + g_s) + (pg_e + qg_e - g_s)$$
$$+ (p^2q - pg_e) + (pq^2 - qg_e),$$

and hence

$$p^3 + p^2q + pq^2 + q^3 = \epsilon g_p + \epsilon p^2 + \epsilon(q^2 + g_e) = \epsilon g_p + \epsilon qg + \epsilon p^2,$$

if we recall that $qg = q^2 + g_e$. Here ϵg_p is $\epsilon(0, 3)(3)$, while ϵqg is $\epsilon(1, 3)(2)$, and ϵp^2 is $\epsilon(2, 3)(1)$, the factor $(2, 3)$ being supplied *pro forma*. The equation is therefore exactly that enunciated,

$$\sum_{i=0}^{3} (i)(3-i)' = \sum_{i=0}^{2} \epsilon(i, 3)(3-i).$$

We now give a less symbolic proof, by induction, of the equation for the coincidences in the general case of a correspondence of ∞^r pairs of corresponding points P, P' (Pieri, *Rend. Palermo*, v, 1891, p. 252; see also Pieri, *Rend. Ist. Lombardo*, xxvi, 1893; xxvii, 1894; xxviii, 1895). The proof is to be built up from the two formulae

$$\sum_{i=0}^{r-1}(i)(r-i)' - (0, r)(r-1) = \sum_{i=1}^{r-1}\epsilon(i, r)(r-i), \qquad (a),$$

$$(r)(0)' + (0, r)(r-1) = \epsilon(0, r)(r), \qquad (b),$$

which, by addition, give the desired formula

$$\sum_{i=0}^{r}(i)(r-i)' = \sum_{i=0}^{r-1}\epsilon(i, r)(r-i), \qquad (c).$$

The equation (c) holds for $r = 1$, becoming then Chasles' fundamental formula, $(0) + (0)' = \epsilon$, in the form $(0)(1)' + (1)(0)' = \epsilon(0, 1)(1)$. We thus assume (c) to hold when $r - 1$ is put for r; from this the equation (a) is deduced. The equation (b) is capable of direct proof, without induction, provided $r > 1$; for $r = 1$ it becomes the Chasles' formula $(0)' + (0) = \epsilon$.

We first give the direct proof of (b), with $r > 1$. By considering a pencil of primes in the space $[r]$, all through a base space $[r-2]$, and considering two primes which join this base to corresponding points P, P' as corresponding primes, we deduce the fundamental symbolic equation of condition $(r-1) + (r-1)' - (r-2, r) = \epsilon$, where $(r-1)$ is the condition for P to lie in a prime, $(r-1)'$ the same condition for P', $(r-2, r)$ is the condition for the line PP' to meet a space $[r-2]$, and ϵ is the condition for coincidence of P and P'. This is in fact Schubert's fundamental equation, discussed above (p. 96). From this equation we have

$$(0, r)(r-1) + (0, r)(r-1)' - (r-2, r)(0, r) = \epsilon(0, r)(r), \qquad (d).$$

We combine this with the equation

$$(0, r)(r-1)' - (0, r-1) = (r)(0)', \qquad (e),$$

which we now prove. Consider, of the ∞^r couples of corresponding points P, P', those for which the line PP' passes through an arbitrary point; this is an $(r-1)$-fold condition, and there will be ∞^1 couples P, P' subject to this condition, namely P' (as well as P) will describe a curve. The correspondence being, as is assumed here throughout, of algebraic character, this curve will be algebraic. Its order will be the number of points P' lying in an arbitrary prime when the line PP' passes through an arbitrary point, namely $(0, r)(r-1)'$. This curve locus of P' will pass through the arbitrary fixed point as many times as there are pairs, of the ∞^r original pairs P, P', for which P' is at this arbitrary point; namely the curve has

a multiple point of order $(r)(0)'$ at this arbitrary point. The order
of the curve is equal to this multiplicity together with the number
of intersections of the curve with an arbitrary prime through this
point, which is $(0, r-1)$. Thus we have

$$(0, r)(r-1)' = (r)(0)' + (0, r-1);$$

which is the equation (e). From this and (d), since

$$(0, r-1) = (r-2, r)(0, r),$$

equation (b) follows.

Assuming this, we can prove that the equation (a), as written,
follows from the hypothesis that this equation holds when $r-1$ is
put for r. Take a fixed point O, and a fixed prime Ω (not containing
O); and, from the ∞^r original pairs of corresponding points P, P',
consider the ∞^{r-1} pairs in which P lies in Ω; projecting each of the
points P', which correspond to such points P, from O, into points
P'' which lie in Ω, we have a correspondence of ∞^{r-1} pairs P, P''
in Ω. For this correspondence, from the hypothesis made for the
equation (a), we have

$$\sum_{j=0}^{r-1}(j)(r-1-j)'' = \sum_{j=0}^{r-2} \epsilon'(j, r-1)(r-1-j),$$

where $(r-1-j)''$ refers to a position of P'' in a space $[r-1-j]$ lying
in Ω, and ϵ' refers to a coincidence of P and P'', which lies in a
$[r-1-j]$ of Ω, the ultimate line PP'' meeting a $[j]$ of Ω. But, the
point P'' will lie in a $[r-1-j]$ of Ω, if and only if P' lie in the $[r-j]$
which joins the $[r-1-j]$ to O. The left side of this equation is
therefore the same as the terms $\sum_{i=0}^{r-1}(i)(r-i)'$ occurring on the left
side of the equation (a). The right side of this equation has, for first
term, $\epsilon'(0, r-1)(r-1)$, referring to coincidences of P and P'' in
which the ultimate line PP'' has an arbitrary direction in Ω. Such a
coincidence arises when the point P', corresponding to P, is itself
in Ω and coincides with P, in such a way that the ultimate line PP'
can be regarded as meeting an arbitrary line (through O); such a
coincidence also arises when P and P', not coinciding, are in a line
through O. We may thus replace the term $\epsilon'(0, r-1)(r-1)$ by
$\epsilon(1, r)(r-1) + (0, r)(r-1)$, referring to the original ∞^r correspondence. For $j > 0$, however, the term $\epsilon'(j, r-1)(r-1-j)$ of the right
side, or say the term $\epsilon'(i-1, r-1)(r-i)$, where $i = j+1$, so that
$i > 1$, is the same as the term $\epsilon(i, r)(r-i)$ applied to the original
correspondence in $[r]$. The right side of this equation is thus
$(0, r)(r-1) + \sum_{i=1}^{r-1} \epsilon(i, r)(r-i)$, applied to the original correspondence.
Thus the equation (a) is established.

Thus, finally, the desired formula (c) is obtained.

Ex. 1. For the composite entity consisting of a line with two distinguishable points thereon, we have used the unsymmetric equation (*b*). Prove also the symmetric equation

$$(0)(r)' + (r)(0)' + (0, r-1) = \epsilon(0, r)(r),$$

and the unsymmetric generalisation of (*b*)

$$(r-i)(i)' + (i, r)(r-i-1) - (i-1, r)(r-i) = \epsilon(i, r)(r-i).$$

Ex. 2. For a correspondence of ∞^2 pairs of corresponding points in a plane, prove that $(1)(1)'$ is equal to the number of pairs of corresponding, but not coinciding, points upon an arbitrary line, together with the number, $\epsilon(1, 2)(1)$, of pairs of coinciding corresponding points upon the line.

Ex. 3. We have remarked that, in the preceding theory, it is supposed that the corresponding points P, P' are distinguishable from one another, by their geometrical construction or otherwise. Consider now a correspondence of points P, P', in a plane, which are harmonic conjugates of one another, in regard to a fixed point O, and a fixed line l, to examine the application of the formulae given, to this involutory case. The points P, P' are in line with O, and l is a range of self-corresponding points at each of which the ultimate line PP' has a definite direction, passing through O. In this case, in the formula $\epsilon g = p^2 + q^2 + g_e - g_p$, of p. 97, we have $p^2 = q^2 = 1$, $g_e = g_p = 0$, and hence ϵg, which we denote by $\epsilon(0, 2)(2)$, becomes equal to 2. Similarly, the formula (*d*), of p. 99, for $r = 2$, namely $(0, 2)(1) + (0, 2)(1)' - (0, 1) = \epsilon(0, 2)(2)$, putting $(0, 1) = 0$, gives again $\epsilon(0, 2)(2) = 2$. The general formula of coincidence

$$(0)(2)' + (1)(1)' + (2)(0)' = \epsilon(0, 2)(2) + \epsilon(0, 1)(1)$$

is then valid, in the form $1 + 1 + 1 = 2 + 1$. For a general homographic transformation in a plane, it is valid, because $\epsilon(0, 2)(2) = 3$ and $\epsilon(0, 1)(1) = 0$.

Ex. 4. Another involutory correspondence in a plane is that of the pairs of points in which the plane is met by the two inflexional tangents at a point of a surface, say of order n. For the indices of the correspondence we have $(0)(2)' = (2)(0)' = n(n-1)(n-2)$, a point P of the plane, not on the surface, being on an inflexional tangent at a point X, of the surface, if X lie on the first and second polars of P. It has been remarked (p. 91) that the ruled surface formed by the inflexional tangents at the parabolic points of the surface is of order $2n(n-2)(3n-4)$. Hence, taking account of the involutory character of the correspondence, we infer $\epsilon(1, 2)(1) = 2n(n-2)(3n-4) + 2n$; also, the number of pairs of inflexional tangents which meet a line is $n(n-1)^2$, and each gives two pairs of corresponding points on this line. Hence, by Ex. 2 above, we have, for the correspondence in the plane,

$$(1)(1)' = 2n(n-1)^2 + 2n(n-2)(3n-4) + 2n, \quad = 4n(2n^2 - 6n + 5).$$

Ex. 5. As an application of Ex. 4, the number of points of the surface, for which the inflexional tangents meet two given arbitrary lines in a plane—either by one inflexional tangent of the pair meeting each of the two lines, or by both meeting the same—is

$$4n(2n^2 - 6n + 5) + n(n-1)^2 + n(n-1)^2, \quad = 2n(5n^2 - 14n + 11).$$

Again, the number of points of the surface for which the inflexional tangents both meet the same repeated line of the plane, is

$$2n(n-2)(3n-4) + 2n + 4n(n-1)^2, \quad = 2n(5n^2 - 14n + 11).$$

If we regard the pair of lines in the plane, or the repeated line, as alternative degenerations of the absolute conic of the plane, either of these formulae gives the number of umbilici of the general surface of order n. Cf. Schubert, *Kalkül*, p. 244; and Voss, *Math. Ann.* IX, 1876, p. 241.

Ex. 6. As an example of Schubert's symbolism, the following determination of the number of torsal lines of a ruled surface may be given (Schubert, *Kalkül*, p. 60, No. 22; see also p. 196, F. 1. And Chap. I above, p. 26). Of the ∞^1 generators consider a correspondence between two generators g and h; let g also denote the number of corresponding pairs in which g meets an arbitrary line, with a similar meaning for h; β denote the condition that g and h meet an arbitrary plane in points of a ray of an arbitrary flat pencil of rays in that plane, and ϵ the condition that g and h coincide, being "consecutive". Then, by considering a pencil of rays in a plane, we have the equation of condition $g + h = \epsilon + \beta$. If also σ be the condition that two generators g, h intersect without coinciding, we again have, considering an arbitrary plane, $\sigma + \epsilon = \beta$. Thus $\sigma = 2\beta - g - h$. We have however ϵg, $= \epsilon h$, $=$ order of the ruled surface; also $\epsilon \beta$ is the class of a plane section of the surface. Thus $\epsilon \sigma$, which is the number of torsal generators, being $2\epsilon\beta - \epsilon g - \epsilon h$, or $2\epsilon\beta - 2\epsilon g$, is equal to $2(2n + 2p - 2 - n)$, if n be the order of the surface, and p the genus of a plane section.

Ex. 7. The following example of the general formula gives what we may describe as the point-equivalence of an unusual intersection of two manifolds in space $[r]$, whose usual intersection consists of points. Let $M_s{}^m$ and $N_t{}^n$ denote two algebraic manifolds in space $[r]$, of orders m and n, whose dimensions s and t are complementary, namely $s + t = r$; suppose that these have, as part intersection, a manifold F_k, of dimension k, $(k > 0)$, at every point of which $M_s{}^m$ and $N_t{}^n$ have definite distinct tangent spaces, $[s]$ and $[t]$; but otherwise meet only in points. We prove that the number, ξ, of these points is

$$\xi = mn - \sum_{i=0}^{k} \epsilon(i, k+1, k+2, \ldots, r)(r-i), \text{ or } mn - \sum_{i=0}^{k} \epsilon\{k-i, 0^{r-k}\}(r-i),$$

where $\epsilon\{k-i, 0^{r-k}\}(r-i)$ denotes the number of points of the section of F_k by a space $[r-i]$, at which the space $[r-k]$, or $[s+t-k]$, containing the tangent spaces of M and N (which both pass through the tangent space $[k]$ of F), meets an arbitrary space $[i]$. In space $[r]$, this is equivalent to $r - (r-k) - i$, or $k - i$, elementary conditions, while the condition for the point to lie in a space $[r-i]$ is equivalent to i elementary conditions; there will then presumably be a finite number of such points existing on the F_k. The symbol $\epsilon\{k-i, 0^{r-k}\}(r-i)$, relating to the join, $[r-k]$, of the tangent spaces $[r]$, $[s]$, is different from the grade $\omega_{k-i} = \epsilon\{k-i, 0^k\}(r-i)$, defined above (p. 73), relating to the meet $[k]$, tangent to F_k, of these tangent spaces $[r]$, $[s]$.

The formula is a direct application of the equation

$$\Sigma(i)(r-i)' = \Sigma\epsilon(i, r)(r-i),$$

applied to the correspondence of points P, P', in which P is any point of $M_s{}^m$, and P' is any point of $N_t{}^n$; as, to any point P, of M, there corresponds any one of the ∞^s points P' of N, and to any point P' of N correspond ∞^r positions of P, anywhere on M, there are in all ∞^{s+t} or ∞^r couples of corresponding points. By the same condition, however, as there is no position of P at an arbitrary point of the space $[r]$, the term $(0)(r)'$ is zero; and so, likewise, for the same reason, are all the terms $(1)(r-1)'$, $(2)(r-2)'$, … which precede the term $(r-s)(r-t)'$ on the left side of the equation; this term is however equal to mn, there being m points P of the manifold $M_s{}^m$ lying on an arbitrary space $[r-s]$, and, corresponding to each, n points P' of the $N_t{}^n$ lying on an arbitrary space $[r-t]$; as spaces $[r-i]$, with $i>t$, do not intersect $N_t{}^n$, the succeeding terms, of the series on the left of the equation, all vanish. Now consider the right side of the equation. Let ξ be the number of points common to M and N which do not lie on F; such a point is a coincidence of the correspondence; and, we assume, the points P, P', of M and N, can be made to approach this intersection so that the ultimate line PP' is any line through this; there is thus such a line through an arbitrary point of the space $[r]$. Wherefore, these intersections furnish a contribution ξ to the first term, $\epsilon(0, r)(r)$, on the right of the equation. The other coincidences of the correspondence are the common points of M and N which lie on F; at such a point, we assume, the ultimate line PP' is any line, through the point, which lies in the space $[r-k]$ which is the join of the tangent spaces of M and N, at this point. Just as, for example, if P, P' be points of two curves which intersect at O, the line PP', when P and P' approach to O, is any line in the plane containing the tangent lines of the curves at O. The number of such ultimate lines PP' which intersect an arbitrary space $[i]$ is therefore the number of such joining spaces $[r-k]$ which intersect this space $[i]$. When $i>k$, the space $[i]$ being arbitrary, this number is zero; but for less values of i, the terms $\Sigma\epsilon(i, r)(r-i)$, of the correspondence equation, for coincidences lying on F, furnish the terms

$$\sum_{i=0}^{k} \epsilon\{k-i, 0^{r-k}\}(r-i).$$ These terms together then give the point-equivalence of the intersection F, of M and N, whose remaining intersections have the number ξ given by the formula.

Ex. 8. As an illustration of the formula $(a, r)(b) - (b, r)(a) = (a, b)$, for a line on which is a distinguishable point, which has been quoted (p. 97), the following may be taken. Suppose that we have an algebraic system of manifolds M_k, of dimension k, in space $[r]$, the system being of aggregate ∞^ρ. Further let a, b be such that $0 \leqslant a < b \leqslant r$, and $a + b = 2(r-k) - \rho$. Consider the aggregate of all possible tangent lines of all these manifolds; on each M_k are ∞^k points, at each of which (we assume) are ∞^{k-1} tangent lines; the aggregate of all these tangent lines is thus $\infty^{2k-1+\rho}$. The condition (a, b), or $\{r-1-a, r-b\}$, for a line is of power $2r-1-a-b$, and this, by the assumed relation, is $2k-1+\rho$. Of all the tangent lines, of the system of manifolds, there is thus a finite number, which satisfy the condition (a, b), of lying in an arbitrary $[b]$ and meeting an arbitrary $[a]$ therein. The number $(a, r)(b)$ is that of points, on the manifold M_k, where these are intersected by a space $[b]$, at which the tangent lines meet a space $[a]$; with a similar meaning for $(b, r)(a)$.

For instance, for $r=3$, $k=1$, $a=1$, $b=2$, taking an algebraic system of ∞^1 curves in ordinary space, the number (a, b) is that of these curves which touch an arbitrary plane; the number $(1, 3)(2)$ is the order of the ruled surface generated by the tangents of these curves at points where the curves meet an arbitrary plane; and the number $(2, 3)(1)$, in which the symbol $(2, 3)$ is unitary, is the number of curves of the system which meet an arbitrary line.

This particular case, for ordinary space, is remarked by Schubert (*Kalkül*, p. 27). The general formulation was made by Pieri.

Ex. 9. A simple application of the principles here developed is to the normals of a surface in ordinary space. Suppose this is of order n, of rank r, and of class n'. Let ϖ denote the plane of the Absolute conic; let P' be the pole, in regard to the Absolute conic, of the line in which the

plane ϖ is met by the tangent plane at a point P of the surface. We define
the normal at P as the line PP'. It may easily be seen that, as P describes
a plane section of the surface, P' describes a curve in ϖ of order r; and
that the points P, P' are then in $(1, 1)$ correspondence. Thus the normals
of the surface at the points of a plane section generate a ruled surface of
order $r+n$; and there are r normals of the surface in a general plane. But,
in particular, there are ∞ normals in the plane ϖ, enveloping a curve of
class $r+n$. To find how many normals of the surface pass through an
arbitrary point, O, of the space, join O to any point P of the surface, and
let the joining line meet the plane ϖ in Q; as before let P' be the point
of the plane ϖ determined from P. Consider the correspondence of ∞^2
pairs Q, P' in the plane ϖ; to these coincidences of O correspond
to the coincidences of this correspondence. In the general case, these
are finite in number, and given by the formula $(0)(2)'+(1)(1)'+(2)(0)'$,
where (0), (1), (2) refer to Q and $(2)'$, $(1)'$, $(0)'$ refer to P'. The terms herein
are easily seen to be n, r and n', respectively; so that the number of
normals of the surface from O is $n+r+n'$.

We may also find the number of lines normal to two given general surfaces. For the order and class of the congruences of normals of these two
surfaces, respectively, are given by $(n_1+r_1+n_1',r_1)$ and $(n_2+r_2+n_2',r_2)$;
while there is a curve of class n_1+r_1 in the plane ϖ, whose tangents
satisfy the definition of a normal of the first surface, and a similar curve
of class n_2+r_2 for the second surface. The number of normals common
to the two surfaces is, therefore,

$$(n_1+r_1+n_1')(n_2+r_2+n_2')+r_1r_2-(n_1+r_1)(n_2+r_2),$$
or
$$(n_1+r_1)n_2'+(n_2+r_2)n_1'+r_1r_2+n_1'n_2'.$$

Schubert obtains the number, r, of normals of a surface which lie in
an arbitrary plane, by supposing the Absolute conic, defined tangentially,
to degenerate into two pencils of lines, with centres I, J, together with
the joining line IJ, say g, counted doubly. Then the normals of the surface,
lying in an arbitrary plane, are assumed to be the same in number as
those in a plane which passes through the point I; and these consist of
the tangent lines of the surface, in this plane, which pass through I,
which are r in number. Again, the normals of the surface from an
arbitrary point are assumed to be the same in number as those passing
through the point I; and these consist, first, of n lines, all coinciding
with g, one at each intersection of the surface with this line g; then,
second, of r tangents to the surface from I, in an arbitrary plane through
I; and, then, lastly, of n', arising from points of contact of tangent planes
of the surface drawn through the line g. See *Kalkül*, p. 16, where the
reference for the results is to Sturm, *Math. Ann.* VII, 1873, p. 567.
Schubert also considers the Absolute conic, defined punctually, as
degenerating into two lines (*loc. cit.* p. 244).

Ex. 10. For two general primals, of orders n, n', in Euclidean space
of r dimensions, the number of common normals (defined, as above, by
an Absolute quadric in a particular prime space $[r-1]$) is given by Pieri
(*Rend. Palermo*, v, 1891, p. 323), with reference to Fouret (*Bull. d. l. Soc.
Math. d. France*, VI, 1878, p. 43) in the form

$$\sum_{k=1}^{r}\sum_{j=0}^{r-k}\rho_{k+j}\rho'_{r-j},$$

where ρ_i is the rank, defined on p. 73 above, given by

$$\epsilon\{1^i,\,0^{r-i}\}(i+1),\;=n(n-1)^{i-1}$$

in general, for $i = 1, \ldots, r$. For instance, when $r = 4$, the number is

$$(\rho_1\rho_4{}' + \rho_2\rho_3{}' + \rho_3\rho_2{}' + \rho_4\rho_1{}') + (\rho_2\rho_4{}' + \rho_3\rho_3{}' + \rho_4\rho_2{}') + (\rho_3\rho_4{}' + \rho_4\rho_3{}') + \rho_4\rho_4{}'.$$

Putting $\rho_1{}' = \rho_2{}' = \ldots = \rho'_{r-1} = 0$, $\rho_r{}' = 1$, the number of normals of a single primal which pass through an arbitrary point, is $\rho_1 + \rho_2 + \ldots + \rho_r$.

Part IV. Pairs of corresponding spaces $[k]$, in space $[r]$. If, in space $[r]$, we have ∞^e pairs of corresponding spaces $[k]$, where $e, = (r-k)(k+1)$, is the freedom of such spaces, then, in case the coincidences of two corresponding spaces are only finite in number, this number is $\Sigma(a_0, \ldots, a_k)(r - a_k, \ldots, r - a_0)'$, where the two Schubert symbols refer to the two spaces of a pair, respectively, and the summation extends over all sets of integers $a_0, \ldots, a_k$, subject to $0 \leqslant a_0 < a_1 < \ldots < a_k \leqslant r$.

This summation, as we have remarked (p. 72), contains, therefore, in general $(r+1, k+1)$ terms. For $k = 0$, the sum is that, $\Sigma(i)(r-i)$, obtained for ∞^r pairs of points. For ∞^2 pairs of lines in a plane, the number of coincidences, when finite, is

$$(0, 1)(1, 2)' + (0, 2)(0, 2)' + (1, 2)(0, 1)', \text{ or } l + cc' + l',$$

where l is the number of the first elements (lines) of a pair, which coincide with an arbitrary line, c is the number which pass through an arbitrary point, and l', c' have the same meanings for the second elements of a pair; a simple application is to the case when the ∞^2 lines are the tangents of two curves in the plane, and any tangent of one curve corresponds to every tangent of the other. For an ∞^4 of pairs of lines in ordinary space [3], the number of coincidences is

$$(0, 1)(2, 3)' + (0, 2)(1, 3)' + (0, 3)(0, 3)' + (1, 2)(1, 2)' + (1, 3)(0, 2)'$$
$$+ (2, 3)(0, 1)';$$

for instance, if we consider two congruences of lines, and any line of one of these congruences is regarded as corresponding to every line of the other, the formula reduces to Halphen's well-known result $(0, 3)(0, 3)' + (1, 2)(1, 2)'$, for the number of lines common to the congruences; or, as another instance, for the number of lines common to a ruled surface and a complex of lines, we have $(1, 3)(0, 2)'$, the product of the orders of the ruled surface and the complex. For an ∞^3 of pairs of corresponding planes in ordinary space, the formula gives

$$(0, 1, 2)(1, 2, 3)' + (0, 1, 3)(0, 2, 3)' + (0, 2, 3)(0, 1, 3)'$$
$$+ (1, 2, 3)(0, 1, 2)';$$

for instance, there are ∞^1 planes tangent to a cone, and ∞^2 planes passing through the tangent lines of a curve; the formula then gives the number of tangent lines of the curve which touch the cone; and, it reduces to $(0, 2, 3)(0, 1, 3)'$, the product of the class of the cone and the rank of the curve.

The general formula above was enunciated and proved by Severi,

Rend. Lincei, IX, 1900, p. 321. In what follows we shall illustrate this proof by carrying it out when $r=3$, for the case $k=1$. For the case of lines see also Pieri, *Atti...Torino*, xxv, 1890, p. 365. For the cases when the pairs of corresponding elements are those of a manifold ∞^s, each taken with those of a manifold ∞^t, with $s+t=r$ (as in all the examples given above), see Schubert, *Mittheil. d. math. Ges. Hamburg*, I, 1886, p. 134.

In general the proof of the formula is by induction; being supposed to be proved for space $[r-1]$, it is deduced for space $[r]$. As we have proved it to hold for points in space $[2]$, and can hence deduce it, by duality, for lines in space $[2]$, this is sufficient. And in proceeding now to prove the theorem for lines in space $[3]$, we can assume it true in space $[2]$.

Consider, in ordinary space $[3]$, an aggregate, Γ, of ∞^4 pairs of corresponding lines. From this aggregate we select an aggregate, γ, of ∞^1 pairs of corresponding lines, so chosen that all the coinciding pairs of the aggregate Γ are equally coincidences in γ; to find the number of coincidences in Γ, it is thus sufficient to find the coincidences in γ. The deduction of γ from Γ is as follows: Take an arbitrary point P, and an arbitrary plane ϖ; for a pair of lines of Γ to intersect is a one-fold condition; for the point of intersection to lie on the plane ϖ is a one-fold condition; for the plane of the lines to pass through P is a further one-fold condition. The aggregate γ consists of the ∞^1 pairs from Γ which satisfy these three conditions. Evidently any pair of coinciding lines of Γ is a pair of coinciding lines of γ. In the aggregate γ, the point of intersection of a pair of the lines describes a curve in the plane ϖ, whose order we denote by m; and the common plane of a pair envelops a cone, of vertex P, whose class we denote by μ. The two lines of a pair, in the aggregate γ, are (as in all the cases discussed in this chapter) supposed to be distinguishable; we denote them, respectively, by l and l'. Now we take a general plane α, and therein a point A, and determine a correspondence of the rays of the flat pencil (α, A); namely, by the condition that two such rays correspond when the first ray meets a line l of the aggregate γ, and the second ray meets a corresponding line l'. As it is a one-fold condition for a line to meet a given line, there will, of the aggregate γ, be a finite number of pairs for which the line l meets a given ray of the pencil (α, A), say y pairs; and as it is a two-fold condition for a line to lie in a given plane, the lines l' of these y pairs will meet the plane α in points, and so determine y rays of the pencil (α, A) corresponding to the first given ray. Similarly, a second given ray of the pencil is met by a definite number, say y', of lines l' of the aggregate γ, and the corresponding lines l determine y' rays of the pencil, of the first kind. There is thus a (y', y) correspondence of rays of the pencil;

and therefore there are $y + y'$ rays of coincidence in the pencil, each met by both the corresponding lines l, l' of a pair of the aggregate γ. These $y + y'$ coincidences in the pencil (α, A) may arise in three ways: (1), through a coincidence of the lines l, l' of γ; (2), because the point of intersection of l and l' is on the plane α—that is, at one of the m points where the plane α meets the curve, of the plane ϖ, which is the locus of such intersections; (3), because the pair l, l' meet the plane α in points lying in line with A, that is, because the plane containing l and l' passes through A—namely, when this plane is one of the μ tangent planes from A to the cone of vertex P enveloped by the planes (l, l'). The number of coincidences of lines of a pair l, l', of the aggregate γ, is thus $y + y' - m - \mu$.

Now the number y was that of the lines l, of a pair (l, l') of γ, which meet an arbitrary line. By considering the case when this line lies in the plane ϖ, whereby in general y is unaltered, we can find an expression for $y - m$. First, consider the original aggregate Γ of ∞^4 pairs of lines l, l'; and then, of these, the ∞^2 pairs which satisfy the two-fold condition that the line l lies in the plane ϖ; project the lines l', of these ∞^2 pairs, from P, obtaining thence lines l'' in the plane ϖ. We thus have ∞^2 pairs of corresponding lines (l, l''), in the plane ϖ, of which every l'' arises from a line l', of the aggregate Γ, of which the corresponding line l lies in the plane ϖ. If, in the plane ϖ, the line l'' satisfy the Schubert condition represented by (b_0, b_1), the line l', from which it arises, satisfies, in the space [3], a condition $(b_0 + 1, b_1 + 1)$, wherein, however, $0 \leqslant b_0 < b_1 \leqslant 2$. Assuming the formula under discussion to be proved for a plane, the number of coincidences of lines l, l'' in the plane ϖ, is given by $\Sigma(a_0, a_1)(2 - a_1, 2 - a_0)'$, with $0 \leqslant a_0 < a_1 \leqslant 2$. Wherefore, the number of pairs of lines l, l', of the aggregate Γ, which satisfy the conditions that l lies in the plane ϖ and its corresponding line l' lies with l in a plane through P, is given by $\Sigma(a_0, a_1)(3 - a_1, 3 - a_0)'$, with $0 \leqslant a_0 < a_1 \leqslant 2$. All these pairs (l, l') evidently belong to the aggregate γ, and are all the pairs of the aggregate γ of which l lies in ϖ. This being so, consider, as suggested, the number y, of lines l, of a pair (l, l') of the aggregate γ, which meet an arbitrary line, p, of the plane ϖ. These consist of, (1), those lying in the plane ϖ; their number has just been expressed; (2), those, not in this plane, whose intersection with the corresponding line l' is at one of the m points in which the line p meets the curve in the plane ϖ, which is the locus of intersections of lines l, l' of the aggregate γ. Thus we have $y = m + \Sigma(a_0, a_1)(3 - a_1, 3 - a_0)'$. A like argument applied to the lines l' of the aggregate γ leads to the equation

$$y' = m + \Sigma(a_0, a_1)'(3 - a_1, 3 - a_0),$$

with the same limitations for the values of a_0 and a_1.

We can now prove further that the order m of the curve in the plane ϖ, and the class μ of the cone with vertex at P, which arise from the definition of the aggregate γ, are such that

$$m - \mu = (0, 3)(0, 3)' - (1, 2)(1, 2)',$$

where the line symbols refer to the lines l, l' of the aggregate Γ. For first, m, by definition, is the number of points of intersection, of two corresponding lines l, l' of Γ lying in a plane through P, which lie in the line p of the plane ϖ. Consider now how this number is made up when P lies on this line p: the point, H, where l, l' meet, may lie on p at a point which does not coincide with P; then the plane of l and l', which contains P, will contain the line p. Let u be the number of such cases. But, the intersection H may fall at P; then the plane of l, l' need not contain the line p. As P is, essentially, an arbitrary point, the number of such cases is that denoted, for the lines l, l' of the aggregate Γ, by the symbol $(\bar{0}, 3)(\bar{0}, 3)'$, the notation $\bar{0}$ denoting that the same arbitrary point is used in both symbols. Thus we have $m = u + (\bar{0}, 3)(\bar{0}, 3)'$.

By a dual argument, in which the use of the plane ϖ and the point P is interchanged, we may similarly prove $\mu = u + (1, \bar{2})(1, \bar{2})'$. The cases for which, in considering m, we obtained the number u, are evidently self-dual in the sense indicated; and the dual of the possibility of both l and l' passing through P is that in which they both lie in ϖ.

Finally, then, the number $y + y' - m - \mu$, of coincidences of a corresponding pair (l, l') in the aggregate Γ, is expressed by

$$\Sigma(a_0, a_1)(3 - a_1, 3 - a_0)' + \Sigma(3 - a_1, 3 - a_0)(a_0, a_1)'$$
$$+ (\bar{0}, 3)(\bar{0}, 3)' - (1, \bar{2})(1, \bar{2})',$$

subject to $0 \leqslant a_0 < a_1 \leqslant 2$. If we suppose that when the line l, of a corresponding pair (l, l'), passes through an arbitrary point, the condition for l' to pass through the same point, remains two-fold, as for an arbitrary point, we may replace $(\bar{0}, 3)(\bar{0}, 3)'$ by $(0, 3)(0, 3)'$. With a similar hypothesis we may replace $(1, \bar{2})(1, \bar{2})'$ by $(1, 2)(1, 2)'$. It is then easy to see that the whole formula reduces to

$$\Sigma(a_0, a_1)(3 - a_1, 3 - a_0)', \qquad 0 \leqslant a_0 < a_1 \leqslant 3,$$

which we desired to obtain.

For the proof of the general case, for the ∞^e aggregate Γ of corresponding spaces (l, l'), of dimension k, where $e = (k+1)(r-k)$, in space $[r]$, we can similarly select, from Γ, an aggregate ∞^1, γ, constituted by the pairs (l, l') which meet in a space $[k-1]$, lying in a given prime ϖ, whose containing space $[k+1]$ passes through a given point P. And then the proof proceeds as here.

Appendix. Some enumerative formulae. Consider a matrix of $(p+1)$ rows and $(q+1)$ columns $(p \leqslant q)$, of which the element in the (i,j)th place is homogeneous in the $(r+1)$ coordinates of a space $[r]$, of order $r_i + c_j$. We consider the manifold given by the simultaneous vanishing of all determinants of the matrix which have $(c+1)$ rows and columns. For suitable c such a manifold exists, and is of dimension $r-(p-c+1)(q-c+1)$. To express the order of this manifold, let

$$(1+c_1 t) \ldots (1+c_{q+1}t) = \sum_{i=0}^{q+1} \gamma_i t^i, \quad (1-r_1 t) \ldots (1-r_{p+1}t) = \sum_{i=0}^{p+1} \rho_i t^i,$$

so that $\rho_1, \rho_2, \ldots$ are alternately negative and positive; form the determinant of $p+q-2c+2$ rows and columns

$$\begin{vmatrix} 1, & \rho_1, & \rho_2, & \ldots, & \rho_{p+q-2c+1} \\ 0, & 1, & \rho_1, & \ldots, & \rho_{p+q-2c} \\ \multicolumn{5}{c}{\cdots\cdots\cdots\cdots\cdots\cdots\cdots\cdots} \\ 0, & \ldots, 0, & \ldots, 1, & \ldots, & \rho_{p-c+1} \\ 1, & \gamma_1, & \gamma_2, & \ldots, & \gamma_{p+q-2c+1} \\ 0, & 1, & \gamma_1, & \ldots, & \gamma_{p+q-2c} \\ \multicolumn{5}{c}{\cdots\cdots\cdots\cdots\cdots\cdots\cdots\cdots} \\ 0, & \ldots, 0, & \ldots, 1, & \ldots, & \gamma_{q-c+1} \end{vmatrix},$$

in which there are $q+1-c$ rows containing symbols ρ, and $p+1-c$ rows containing symbols γ. The determinant is of a form familiar in considering the resultant of two polynomials (Vol. v, Chap. viii, p. 202). The order of the manifold in question is given by the value of this determinant.

In particular, if $c=p$, this is a determinant of $q-p+2$ rows and columns, with only one row containing elements γ, the last of the rows containing elements ρ consisting of the two elements $1, \rho_1$. In this case the determinant is equal to the coefficient of t^{q-p+1} in the ascending expansion of

$$(1+\gamma_1 t + \ldots + \gamma_{q+1} t^{q+1})/(1+\rho_1 t + \ldots + \rho_{p+1} t^{p+1}).$$

In terms of the numbers r and c, associated with the columns and rows of the original matrix to specify the orders of its elements, this order of the manifold spoken of may be remembered in the abbreviated form $[(c)/(-r)]_t q-p+1$.

If, as another particular case, we have $r_1 = r_2 = \ldots = r_{p+1} = 0$, the order of an element of the original matrix depending then only on the column in which the element occurs, then the determinant giving the order of the manifold in question reduces to

$$\begin{vmatrix} \gamma_{q+1-c}, & \ldots, & \gamma_{q+1-c+p-c} \\ \gamma_{q-c}, & \ldots, & \gamma_{q+1-c+p-c-1} \\ \multicolumn{3}{c}{\cdots\cdots\cdots\cdots\cdots\cdots} \\ \gamma_{q-p+1}, & \ldots, & \gamma_{q-c+1} \end{vmatrix},$$

with $p-c+1$ rows and columns. If, still more particularly, we have also $c_1=c_2=\ldots=c_{q+1}=\mu$, so that every element of the original matrix is of order μ, then $\gamma_i=\mu^i(q+1, i)$, where $(q+1, i)$ is a binomial coefficient. In this case, putting m for $q+1$, and transposing rows and columns, the determinant which gives the order may be written

$$\mu^{(p-c+1)(q-c+1)}\begin{vmatrix} (m, c) & , (m, c+1) & , \ldots, (m, p) \\ (m, c-1) & , (m, c) & , \ldots, (m, p-1) \\ \cdot & \cdot & \ldots, & \cdot \\ (m, 2c-p), & (m, 2c-p+1), & \ldots, (m, c) \end{vmatrix},$$

where each element is a binomial coefficient, to be replaced by zero when its second index is negative. It can however be proved that

$$\begin{vmatrix} (m,c) & , (m,c+1) & , \ldots, (m,c+d) \\ (m,c-1), & (m,c) & , \ldots, (m,c+d-1) \\ \cdot & \cdot & \ldots, & \cdot \\ (m,c-d), & (m,c-d+1), & \ldots, (m,c) \end{vmatrix} = (m,c)\frac{\{m+d, d\}}{\{c+d, d\}\{m-c+d, d\}}$$

in, which, if $((k))$ mean $1!2!\ldots k!$, the notation $\{n, d\}$ stands for $((n))/((d))\,((n-d))$; it is understood that $d\geqslant 0$ and $c+d<m$. For the case under consideration we have $d=p-c$; thus the required order of the manifold, given by the vanishing of all determinants of order $c+1$, in a matrix of type $p+1, q+1$, wherein each element is of order μ in the coordinates, is

$$\mu^{(p-c+1)(q-c+1)}(q+1, c)\frac{\{q+p+1-c, p-c\}}{\{p, p-c\}\{q+p+1-2c, p-c\}}.$$

This agrees with a result found by Segre, *Rend. Lincei*, IX, 1900, p. 253.

The preceding results assume the original matrix to be of general form. If this be square $(p=q)$, and be either symmetrical, or skew symmetrical, general formulae can also be given. In the symmetrical case, if the order of the (i, j)th element of the matrix, supposed to be of $p+1$ rows and columns, be r_i+r_j, wherein either $r_1, \ldots, r_{p+1}$ are all integers, or else $r_1-\frac{1}{2}, \ldots, r_{p+1}-\frac{1}{2}$ are all integers, the dimension of the manifold upon which all determinants, in the matrix, of $c+1$ rows and columns, vanish, is $r-\frac{1}{2}(p-c+1)(p-c+2)$; and if we define $h_1, h_2, \ldots$ by

$$(1-r_1 t)\ldots(1-r_{p+1}t)=\sum_{i=0} h_i t^i,$$

then the order of this manifold is

$$2^{p-c+1}\begin{vmatrix} h_1 & , h_0, 0, 0, 0, & \ldots\ldots \\ h_3 & , h_2, h_1, h_0, 0, & \ldots\ldots \\ \ldots\ldots\ldots\ldots\ldots\ldots\ldots\ldots\ldots\ldots \\ h_{2p-2c+1}, & \ldots\ldots\ldots\ldots, & h_{p-c+1} \end{vmatrix},$$

where the determinant has $p-c+1$ rows and columns. In particular, if every element of the original matrix be of order μ, and $r_1=r_2=\ldots=\frac{1}{2}\mu$, we have $h_s=2^{-s}\mu^s(p+s,p)$, where $(p+s,p)$ is the binomial coefficient. In this case the order of the manifold is

$$\mu^e \prod_{i=0}^{p-c} (2p-c+1, 2i+1)/(2p-c+1, i),$$

with $e=(p-c+2, 2)$. When the original matrix is skew symmetric, but p is odd and c is odd, precisely the same formulae hold with the one change, that the two given values for the order of the manifold must be divided by 2^{p-c+1}.

These results are due to Giambelli (Segre, *Enzyk. Math. Wiss.* III, C 7, pp. 828, 856, 930). The following references may be given: Salmon, *Higher Algebra*, 1866, Lesson XVIII, p. 212; S. Roberts, *Crelle*, LXVII, 1867 (with reference to de Jonquières, *Crelle*, LXVI); Schubert, *Jahresber. Deut. math. Ver.* IV, 1894–5; Vahlen, *Crelle*, CXIII, 1894; Pieri, *Rend. Palermo*, XI, 1896; Segre, *Rend. Lincei*, IX, 1900; Palatini, *Rend. Lincei*, XI, 1902; Nanson, *Mess. Math.* XXXIII, 1903; Kohn, *Archiv d. Math. u. Phys.* IV, 1903; Giambelli, *Rend. Lincei*, XII, 1903; *ibid. Atti...Torino*, XXXVIII, 1903; *ibid. Mem. Ist. Lombardo*, X, 1903 and XI, 1904; *ibid. Atti...Torino*, XLI, 1905.

CHAPTER III

TRANSFORMATIONS AND INVOLUTIONS FOR THE MOST PART IN A PLANE

THE present chapter is designed to indicate some preliminary theorems, for the most part dealing with plane geometry, which it is desirable for the reader to have in mind in dealing with the subsequent theory of surfaces. The treatment is only introductory, but can be supplemented by the wide literature, extending over a long time, which treats of the questions referred to.

Cremona transformations in a plane. Take a set of distinct points in a plane; and consider the plane curves, of given order, say n, which have multiple points of assigned order at these; suppose that the assigned orders are such that the general curve satisfying the conditions has an equation of the form $\lambda u + \mu v + \nu w = 0$, wherein $u = 0$, $v = 0$, $w = 0$ are particular curves of the system, and λ, μ, ν are arbitrary parameters; suppose further that any two curves of the system have only one intersection, outside the assigned points common to all. Then the equations $x'/u = y'/v = z'/w$, corresponding to any point, (x, y, z), whose coordinates enter in u, v, w, define another point, (x', y', z'), of the plane. Conversely, if (x', y', z') be any given point of general position, these equations define a point (x, y, z), which is the common point, outside the given base points, of the two curves $x'v - y'u = 0$, $x'w - z'u = 0$. Thus the equations are capable of solution in the form $x/u' = y/v' = z/w'$, where u', v', w' are homogeneous polynomials of the same order in x', y', z'. There is thus established a $(1, 1)$ birational correspondence of points (x, y, z), (x', y', z') of the plane.

We may denote the prescribed multiple points by a scheme $(i_1^{p_1}, i_2^{p_2}, \ldots)$, meaning that the curves are to have p_1 distinct i_1-ple points (with tangents all unassigned), p_2 i_2-ple points (with unassigned tangents), and so on. Suppose that these numbers are such that

$$p_1 i_1 + p_2 i_2 + \ldots = 3n - 3, \quad p_1 i_1^2 + p_2 i_2^2 + \ldots = n^2 - 1,$$

which we may denote by $\Sigma i = 3n - 3$, $\Sigma i^2 = n^2 - 1$. The second equation secures that two curves of the system which are such as have a finite total number (n^2) of common points, have one intersection not at the base points; and, if the prescribed conditions at the base points lead to linearly independent conditions for the coefficients in the general curve which is subject thereto, the number of such curves which are linearly independent will be

$$\tfrac{1}{2}(n+1)(n+2) - \tfrac{1}{2}\Sigma i(i+1), \; = \tfrac{1}{2}(n^2 + 3n + 2) - \tfrac{1}{2}(n^2 - 1 + 3n - 3), \; = 3,$$

as we have supposed above. But the numerical conditions, even when independent, do not ensure that the resulting curves are without a common part, which is a necessary condition for the proof that any two curves of the system have only one free intersection. For instance, quintic curves with the prescribed base system $(3^2, 1^6)$ break up into the line joining the two triple points, and quartic curves with the base $(2^2, 1^6)$, of which any two have 2 intersections outside this base.

Suppose now that the conditions at the base are independent, and that the resulting curves have no common part, so that, conversely, the numerical conditions are satisfied; and the reverse expressions for x, y, z in terms of x', y', z' follow as above. Then the curve $\lambda u + \mu v + \nu w = 0$ has a genus given by

$$\tfrac{1}{2}(n-1)(n-2) - \tfrac{1}{2}\Sigma i(i-1),$$

that is $\tfrac{1}{2}(n^2 - 3n + 2) - \tfrac{1}{2}(n^2 - 1 - 3n + 3)$, or zero, the same as that of the line $\lambda x' + \mu y' + \nu z' = 0$, to which it corresponds. The curves of the plane (x', y', z') which correspond to the lines $\lambda x + \mu y + \nu z = 0$, given by $\lambda u' + \mu v' + \nu w' = 0$, will have an order equal to the number of common solutions, for general values of λ, μ, ν and p, q, r, of $\lambda u' + \mu v' + \nu w' = 0$ and $p x' + q y' + r z' = 0$; this is the same as for $\lambda x + \mu y + \nu z = 0$ and $p u + q v + r w = 0$; thus the curves

$$\lambda u' + \mu v' + \nu w' = 0$$

have also the order n. From the $(1, 1)$ correspondence between the two planes, they will then be of genus zero, and their base system will satisfy equations which we can represent by

$$\Sigma j = 3n - 3, \quad \Sigma j^2 = n^2 - 1,$$

provided their base points be distinct. We consider only cases when this is so.

To any point, (x, y, z), in the neighbourhood of one of the base points, O, of the curves $u = 0$, $v = 0$, $w = 0$, will correspond a point (x', y', z') which, in general, will assume a definite position as (x, y, z) approaches O along a definite line. The points (x', y', z') so obtained, which we may speak of as corresponding to the neighbourhood of O, will describe a curve, of which the points correspond severally to the rays of the pencil of lines through O. This curve will then be rational. We can obtain the equation of this curve by supposing u, v, w expressed in non-homogeneous coordinates ξ, η, both vanishing at O; and, then, in $x'/u = y'/v = z'/w$, retaining only terms of the lowest order in ξ, η, we have the ratios of x', y', z' expressed by polynomials, in ξ/η, whose order, i, is the common multiplicity of $u = 0$, $v = 0$, $w = 0$ at the point O. In general terms, it appears that the base points in the plane (x, y, z) correspond severally to curves in the plane (x', y', z'), which we may call *exceptional* (or fundamental) curves. Similarly, there are exceptional curves in the plane (x, y, z), which correspond to the base points in the plane (x', y', z'). Such a curve has the property that the substitution, in u, v, w, of the coordinates of a point of this curve, gives, ultimately, for (x', y', z'), the same values whatever be the point, namely the coordinates of the corresponding base point in the plane (x', y', z'). Using u_1 for $\partial u/\partial x$, etc., it follows that, if (x, y, z) be a point of an exceptional curve of the plane (x, y, z), there are ratios of dx, dy, dz for which the three equations

$$u_1 dx + u_2 dy + u_3 dz = 0, \quad v_1 dx + v_2 dy + v_3 dz = 0,$$
$$w_1 dx + w_2 dy + w_3 dz = 0,$$

are all satisfied. The exceptional curves are thus part of the curve expressed by the vanishing of the Jacobian of u, v, w, or, say, $\partial(u, v, w)/\partial(x, y, z)$. The order of this Jacobian is $3n - 3$; we have seen that, in general, an exceptional curve, in the plane (x, y, z), has the order, j, of the base point, in the plane (x', y', z'), to which it corresponds, and that $\Sigma j = 3n - 3$. Thus we infer that the Jacobian curve referred to is made up of the exceptional curves; and likewise, in the general case under consideration, the Jacobian curve $\partial(u', v', w')/\partial(x', y', z') = 0$ is made up of the exceptional curves in the plane (x', y', z').

After the case of linear transformations, the simplest application of the preceding theory is for $n = 2$. The curves $\lambda u + \mu v + vw = 0$ are then conics, with three distinct simple base points. We may thus take $x' = yz$, $y' = zx$, $z' = xy$, with the reverse formulae $x = y'z'$, $y = z'x'$, $z = x'y'$. The exceptional curves, in either plane, are now the joining lines of the base points; and in fact the Jacobian of yz, zx, xy is $2xyz$. In this simple case, however, we can see directly that the restriction we have imposed, of distinctness in the base points of the curves $u = 0, v = 0, w = 0$, is not necessary for obtaining a $(1, 1)$ relation of the two planes. We can use a system of conics subject to the three conditions of passing through two points but with a given tangent at one of these; that is, with three base points of which the third is "infinitely close" to the second. In fact the equations $x' = x^2$, $y' = xy$, $z' = yz$ have the unique reverse $x = x'y'$, $y = y'^2$, $z = x'z'$, and the base points in the plane (x', y', z') are of similar character to those in the plane (x, y, z). The exceptional curves, in either plane, consist then of the line joining the two distinct base points, taken twice over, together with the line joining the coincident base points. As the reader is aware, the quadratic transformations have been used to establish an analysis of multiple points of a plane curve, when these consist of limits of simple distinct multiplicities lying infinitely close. With this analysis it is possible to attach a definite meaning to a prescription for a curve, that it should have multiple points with other multiplicities infinitely close thereto, all of assigned orders. The results obtained are simple; but the justification, to be complete, is long, and we do not give space to it here. The reader may consult the references given below.

A very general (1, 1) transformation of the plane, which includes the quadratic transformation just referred to, is that which is effected by curves, of order n, having one general $(n-1)$-fold multiple point, and $(2n-2)$ simple base points, supposed distinct from one another, and from the $(n-1)$-fold point. The numbers of the appropriate specification, $[(n-1)^1, 1^{2n-2}]$, evidently satisfy the equations $\Sigma i = 3n - 3$, $\Sigma i^2 = n^2 - 1$, namely $(n-1) + (2n-2) = 3n - 3$, and $(n-1)^2 + 2n - 2 = n^2 - 1$. If the coordinates (x, y, z) be taken with the $(n-1)$-fold point at $x = 0 = y$, the equations for the transformation are of the forms

$$x' = zu_{n-1} + u_n, \quad y' = zv_{n-1} + v_n, \quad z' = zw_{n-1} + w_n,$$

where, for $i = n - 1$, $i = n$, the symbols u_i, v_i, w_i denote homogeneous polynomials in x, y, of order i. But we can conveniently modify the coordinates x', y', z'. For there exists, in general, a definite curve of order $(n-1)$, with a $(n-2)$-fold point at $x = 0 = y$, passing through the $(2n-2)$ assigned base points, these conditions being equivalent to $\frac{1}{2}(n-2)(n-1) + 2n - 2$, or $\frac{1}{2}(n-1)(n+2)$ conditions. If this particular curve have the equation $za_{n-2} + a_{n-1} = 0$, where a_{n-2}, a_{n-1} are definite homogeneous polynomials in x, y, then we can take the coordinates x', y', z', so that

$$x' = x(za_{n-2} + a_{n-1}), \quad y' = y(za_{n-2} + a_{n-1}), \quad z' = zb_{n-1} + b_n,$$

where again b_{n-1} and b_n are homogeneous in x and y. Writing, momentarily, $Y = y/z$, $Z = z/x$, the equations of transformation are then $Y' = Y$, $Z' = (Zxb_{n-1} + b_n)/(Zx^2a_{n-2} + xa_{n-1})$; if then a_i', b_i' be the same polynomials as a_i, b_i, with x', y' put for x, y, these equations lead to $Z = (-Z'x'a'_{n-1} + b_n')/(Z'x'^2a'_{n-2} - x'b'_{n-1})$. Thus the reverse transformations are

$$x = x'(z'a'_{n-2} - b'_{n-1}), \quad y = y'(z'a'_{n-2} - b'_{n-1}), \quad z = (-z'a'_{n-1} + b'_n);$$

these are obtained from the direct transformations by replacing a_{n-1}, b_{n-1} respectively by $-b'_{n-1}$, $-a'_{n-1}$, and a_{n-2}, b_n respectively by a'_{n-2}, b'_n. This reverse transformation is effected by certain curves of order n in (x', y', z'), likewise with an $(n-1)$-fold point at $x' = 0 = y'$, and $(2n-2)$ other simple base points.

Consider now the exceptional curves in the plane (x, y, z). The rational curve $za_{n-2} + a_{n-1} = 0$ is the exceptional curve which corresponds to the neighbourhood of the multiple base point at $x' = 0 = y'$ in the second plane. For, in the equations which express x, y, z in terms of x', y', z', if we retain only the lowest powers of x', y', we find that x, y, z are in the ratios of $x'a'_{n-2}$, $y'a'_{n-2}$, $-a'_{n-1}$; so that we have $za_{n-2}+a_{n-1}=0$. Similarly, a line joining the multiple base point $(0, 0, 1)$, in the plane (x, y, z), to one of the $2n-2$ points, other than $(0, 0, 1)$, which are common to the two curves

$$za_{n-2}+a_{n-1}=0, \quad zb_{n-1}+b_n=0,$$

is an exceptional curve corresponding to one of the $2n-2$ simple intersections of the two curves $z'a'_{n-2}-b'_{n-1}=0$, $z'a'_{n-1}-b'_n=0$. The aggregate of these lines is given by $a_{n-1}b_{n-1}-a_{n-2}b_n=0$; and if the transformation be denoted by $x'/u=y'/v=z'/w$, the Jacobian of u, v, w is $n(za_{n-2}+a_{n-1})(a_{n-1}b_{n-1}-a_{n-2}b_n)$. The exceptional curves of the plane (x', y', z') are similar.

It is clear that the quadratic transformation considered above is a particular case of the present transformation ($n=2$). More generally, the present transformation may be regarded as the result of a succession of a finite number of quadratic and linear transformations. For if, momentarily, we put $\xi=x/y$, $\eta=z/y$, with $\xi'=x'/y'$, $\eta'=z'/y'$, the transformation is $\xi'=\xi$, $\eta'=(\eta\alpha+\beta)/(\eta\gamma+\delta)$, where α, β, γ, δ are polynomials in ξ, of respective orders $n-1$, n, $n-2$, $n-1$, for which it is supposed that $\alpha\delta-\beta\gamma$ is not identically zero. Provided neither α, nor γ, nor δ, is identically zero, this transformation is made up by the succession of the following four, in which ξ remains unaltered, but η is changed by

$$\eta_1=\frac{\delta}{\gamma}\left(\eta\frac{\gamma}{\delta}+1\right), \quad \eta_2=\frac{1}{\eta_1}, \quad \eta_3=\frac{\beta\gamma-\alpha\delta}{\gamma^2}\eta_2, \quad \eta'=\frac{\alpha}{\gamma}\left(\eta_3\frac{\gamma}{\alpha}+1\right);$$

to prove the statement made, in the general case contemplated, it is thus sufficient to shew that each of the three transformations following, in which m denotes a rational function of ξ, namely

$$(\xi'=\xi, \ \eta'=\eta+1), \quad (\xi'=\xi, \ \eta'=1/\eta), \quad (\xi'=\xi, \ \eta'=m\eta),$$

is obtainable by a finite number of linear and quadratic transformations. Of these three, however, the first is the linear transformation $x'=x$, $y'=y$, $z'=z+y$; and the second is the quadratic transformation $x'=xz$, $y'=yz$, $z'=y^2$; the third, considering in detail the possible forms for the rational function m, is obtainable by combining a transformation of one of the two forms

$$\xi'=\xi, \ \eta'=(a\xi+b)\,\eta, \quad \text{or} \quad \xi'=\xi, \ \eta'=(a\xi+b)^{-1}\eta,$$

in which a, b are constants, with a transformation of the form $\xi'=\xi$, $\eta'=\mu\eta$, in which μ is rational in ξ, but with either its denominator or its numerator of one less order than those of m. The two particular transformations are the quadratic transformations in which x', y', z' are respectively proportional to xy, y^2, $z(ax+by)$ and $x(ax+by)$, $y(ax+by)$, yz. By continued reduction of the rational functions m, μ, ... thus arising, the general theorem is therefore proved by induction. If one of α, γ, δ be zero the corresponding theorem is similarly obtained. With $\alpha\delta - \beta\gamma \neq 0$, the theorem is therefore true.

We have supposed the $(n-1)$-fold point, and the $(2n-2)$ points, of the transforming curves, to be distinct. This is not essential. An extreme case is the transformation $x' = xy^{n-1}$, $y' = y^n$, $z' = zy^{n-1} + b_n$, for which $a_{n-2} = 0$ and $a_{n-1} = b_{n-1} = y^{n-1}$. In this case the exceptional curves consist of $y^{n-1} = 0$ (replacing $za_{n-2} + a_{n-1} = 0$), and $y^{2n-2} = 0$ (replacing the $2n-2$ exceptional lines). The composition of this transformation, by linear and quadratic transformations, is easily obtained.

In such a case, when the transforming curves (of order n) have coincident base points, the necessary conditions for a $(1, 1)$ transformation are that the base points imposed are equivalent to $\frac{1}{2}(n+1)(n+2)-3$ linear conditions, that they absorb n^2-1 intersections of any two of the curves, and are such that the curves are irreducible; though, when the analysis of complex multiple points into a succession of ordinary multiple points following one another in infinitely near neighbourhood, has been carried out, the conditions can be stated as in the case of distinct base points. Utilising the description of the multiple points obtained by such an analysis, it can be shewn that any $(1, 1)$ birational transformation of the plane can be obtained by compounding transformations of the character just considered (with base points $(n-1)^1$, 1^{2n-2}). This was proved by Castelnuovo, *Atti...Torino*, xxxvi, 1900–1, p. 861 (cf. Alexander, *Trans. Amer. Math. Soc.* xvii, 1916, p. 295). As we have shewn that such a transformation is compounded of quadratic transformations, it follows that the general $(1, 1)$ transformation may be compounded of quadratic transformations (for a recent direct proof of this see Enriques-Chisini, *Teoria Geom.* iii, 1924, p. 170).

The theory of the resolution of a complex singularity by a succession of quadratic transformations was enunciated by Noether (*Math. Ann.* ix, 1875; *Math. Ann.* xxiii, 1883). The theorem of the decomposition of a $(1, 1)$ transformation into quadratic transformations was given, almost at the same time, by Noether, *Math. Ann.* iii, 1870, p. 167; by Rosanes, *Crelle*, lxxiii, 1870; and by Clifford (see Cayley, *Papers*, vii, p. 223, and Clifford, *Papers*, p. 542); but the completion of the proof (for cases of complex base points) is due to a remark made by Segre (*Atti...Torino*, xxxvi, 1900–1901, p. 645). A recent exposition of Noether's analysis of a singular point, is given in Enriques-Chisini, *Teoria Geom.* ii, pp. 327–535. A detailed account of $(1,1)$ birational transformations, in the plane, and in ordinary space, with an exhaustive bibliography, is found in Hudson, *Cremona Transformations* (Cambridge, 1927). See also Clebsch-Lindemann-Benoist, *Leçons sur la Géométrie*, ii, 1880, pp. 188–219; Brill-Noether, *Die Entwicklung der Theorie der algebraischen Functionen*, 1894, vi. Abschnitt; and Coble, *Bulletin of Nat. Research Council, Washington, D.C.*, No. 63, 1928, Chaps. iv and viii.

The transformation of specification $(n-2)^1$, 1^{2n-2}, which we have discussed, is called after its discoverer, de Jonquières (*Battaglini's Gior.* XXIII, Naples, 1885, dating from 1859). If in the formulae

$$x' = x(za_{n-2} + a_{n-1}), \quad y' = y(za_{n-2} + a_{n-1}), \quad z' = zb_{n-1} + b_n,$$

the polynomial b_{n-1} is the negative of the polynomial a_{n-1}, then the reverse formulae are of precisely the same form. In this case, if we suppose the planes, and coordinates, (x, y, z) and (x', y', z'), to be identified, the formulae determine, in the plane (x, y, z), an *involution* of pairs of points, in which any point of the plane belongs to one such pair. This *de Jonquières involution* may be simply defined as consisting of pairs of points, of which the points of a pair have coordinates (x, y, z) and (x, y, z') and lie upon a line through the point $(0, 0, 1)$, the values of z, z' being connected by the relation $zz'a_{n-2} + (z+z')a_{n-1} - b_n = 0$; the points of a pair are thus harmonic conjugates of one another in regard to the two points (other than $(0, 0, 1)$ in which their joining line meets the curve whose equation is $z^2a_{n-2} + 2za_{n-1} - b_n = 0$. In general terms, the curve is any curve of order n with a $(n-2)$-ple point; it is of genus $n-2$, and of hyperelliptic character; and, it can in fact be conversely shewn that any hyperelliptic plane curve can be birationally reduced to this equation. The curve in this case is the locus of self-corresponding points of the involution, there being two such points upon any line through $(0, 0, 1)$. For this involution, the exceptional curve, corresponding to the multiple point, given as in the general case by the equation $za_{n-2} + a_{n-1} = 0$, is the first polar of $(0, 0, 1)$ in regard to the curve locus, Ω, of self-corresponding points; the curves $(\lambda x + \mu y)(za_{n-2} + a_{n-1}) + \nu(zb_{n-1} + b_n) = 0$, which determine the involution $(b_{n-1} = -a_{n-1})$, have in common the points of contact of the $2n-2$ tangents from $(0, 0, 1)$ to this curve Ω, and these tangents, given by $a_{n-2}b_n + a^2_{n-1} = 0$, are the exceptional curves corresponding thereto.

It is a general problem, to represent the ∞^2 pairs of a plane involution, each by a point, say (ξ, η, ζ), of a plane, so that conversely to every point (ξ, η, ζ) corresponds one such pair; it is a theorem of importance that this can always be done, as we shall see. In the case of the de Jonquières involution, a simple solution of this problem is to take, corresponding to the pair (x, y, z), (x, y, z') of the involution, the point $\xi : \eta : \zeta = x : y : \frac{1}{2}(z+z')$; then, when (ξ, η, ζ) is given, the pair are determined by taking $x : y = \xi : \eta$, the values of z/x and z'/x being given by the equation

$$(z^2\xi - 2zx\zeta)\,\xi\alpha_{n-2} - x^2(2\zeta\alpha_{n-1} - \beta_n) = 0,$$

where α_{n-2}, α_{n-1}, β_n are the same polynomials in ξ, η as are a_{n-2}, a_{n-1}, b_n in x, y. This is easy to verify. In a less simple, but more representative manner, following Bertini (*Rend. Ist. Lombardo,*

XXII, 1889, p. 773), we may solve this problem as follows: It can be shewn, as below, that a system of curves exists in the plane with the property that all curves of the system, which pass through an arbitrary point, pass also through the point corresponding thereto in the involution (having in general no other common points, outside the base points of the system). Let such a system of curves have the equation $\lambda_0\phi_0 + \ldots + \lambda_r\phi_r = 0$, with $r \geqslant 2$; then consider the point $(\xi_0, \ldots, \xi_r)$, in a space $[r]$, for which $\xi_0, \ldots, \xi_r$ are in the ratios of $\phi_0, \ldots, \phi_r$. As the point with coordinates x, y, z, which occur in $\phi_0, \ldots, \phi_r$, varies in the plane, the point $(\xi_0, \ldots, \xi_r)$ describes a surface; conversely, to any point (ξ) of this surface, correspond the pair of points in the plane common to all the curves $\xi_i\phi_j - \xi_j\phi_i = 0$; and these, by hypothesis, are a pair of the involution. Thus, if the surface obtained be rational, its points being in $(1, 1)$ birational correspondence with the points of a plane, then, it is clear, the points of this plane form a representation of the couples of the involution. To find, for the de Jonquières involution, a system of curves, of freedom at least 2, with the property that all curves of the system described through an arbitrary point of the plane pass also through the point which corresponds thereto in the involution, denote the curve, Ω, considered above, which is the locus of coincident pairs of the involution, by the equation $z^2u + 2zv + w = 0$, where u, v, w are homogeneous polynomials in x and y, respectively of orders $n-2$, $n-1$ and n. On any line through the point $(0, 0, 1)$, a pair of points which are harmonic in regard to the points in which this line meets the curve Ω, are a pair of the involution; these lines form a system of curves of freedom 1; but a system of greater freedom is obtainable by curves Ω', given by an equation of the form $z^2u' + 2zv' + w' = 0$, in which u', v', w' are homogeneous polynomials in x, y of respective orders $n'-2$, $n'-1$, n'; we require only that the curve Ω' should meet any line through $(0, 0, 1)$ in points harmonic in regard to the intersections of this line with the curve Ω. The condition for this relation is $u'w - 2v'v + w'u = 0$, which leads to $n + n' - 1$ homogeneous conditions for the $3n'$ unspecified homogeneous coefficients in the polynomials u', v', w'; thus the curves Ω' are of freedom at least $2n' - n$. At a point of intersection of a curve Ω' with the curve Ω, other than the $(n-2)(n'-2)$ intersections which fall at $(0, 0, 1)$, the tangent line of one of the two curves passes through $(0, 0, 1)$; this is in accordance with

$$nn' = (n-2)(n'-2) + 2(n-1) + 2(n'-1).$$

If the general curve Ω' have an equation $\lambda_0\phi_0 + \ldots + \lambda_{2n'-n}\phi_{2n'-n} = 0$, the equations $\xi_0/\phi_0 = \ldots = \xi_{2n'-n}/\phi_{2n'-n}$ determine, in space $[2n'-n]$, a surface whose points represent the pairs of the plane involution; and, as two curves Ω', beside the fixed intersections at $(0, 0, 1)$ and at the points of contact of tangents drawn from this point to Ω, have $n'^2 - (n'-2)^2 - 2(n-1)$, or $2(2n'-n-1)$ intersections—which will consist of pairs of points corresponding in the involution—the surface will be of order $2n'-n-1$. Here n' is arbitrary (subject to $2n'-n \geqslant 2$); when n is odd, we may take $n' = \frac{1}{2}(n+3)$, and obtain a quadric surface in ordinary space. It may be shewn that the points of the generators of one system, of this surface, represent the pairs of the plane involution lying on lines through the point $(0, 0, 1)$; while the generators of the other system correspond to the ∞^1 curves Ω' for which $n' = \frac{1}{2}(n+1)$. When n is even, if we take $n' = \frac{1}{2}n + 2$, we obtain a cubic surface in space of four dimensions. This is known to be rational, the coordinates of a point being expressible, in terms of two parameters θ, ϕ, by θ^2, θ, 1, $\phi\theta$, ϕ; such a surface is obtainable, if we take a rational quartic curve in the space [4], and, thereon, any involution of pairs of points, as the locus of the chords of the curve which join the pairs of this involution. In general, however, it is known that (irreducible) surfaces of order $r-1$ in space $[r]$, as in this case, with $r = 2n'-n$, are always rational, being, unless $r = 5$, ruled surfaces, normal in this space $[r]$, that is, unobtainable as projections of surfaces of the same order lying in higher space (Del Pezzo, *Rend....Napoli*, 1885, 1886; Bertini, *Geom. d. iperspazi*, 1907, p. 319). That, in our case, the representative surfaces are rational, follows also, because, as the curves Ω' are hyperelliptic, the prime sections are hyperelliptic curves (Castelnuovo, *Rend....Palermo*, IV, 1890, p. 73).

Ex. 1. A particular property of the related curves Ω, Ω' considered here, suggested by a note of Bertini's (*Atti...Lincei*, Transunti (3ª), I, 1877, p. 92; cf. Caporali, *Rend....Napoli*, XXI, 1882, p. 227, or *Mem. d. Geom.* 1888, p. 164), may be noted. Let the curves Ω, Ω' be given, as above, by equations $z^2u + 2zv + w = 0$ and $z^2u' + 2zv' + w' = 0$, the relation $u'w - 2v'v + w'u = 0$ being assumed. Denote the point $(0, 0, 1)$ by O; the $2(n-1)$ points of contact of tangents from O to Ω, that is, the points, other than O, common to Ω and the curve $zu + v = 0$, be denoted by A; and the $(n-2)$ points of intersection, other than O, of the tangent lines at O, with the curve (one on each of the tangents), be denoted by T. Use a similar notation for the curve Ω'. Then, neglecting intersections at O, it may be proved:

(1), That the intersections of Ω, Ω' are $A + A'$, as has been remarked above; (2), that the first polar of O, with respect to Ω, meets Ω', partly in the set A, and partly in the set T', the "tangentials" of O on Ω'; or, say, $(\Omega', zu + v) = A + T'$. This is in accordance with

$$n'(n-1) - (n-2)(n'-2) = (2n-2) + (n'-2).$$

Similarly we have $(\Omega, zu' + v') = A' + T$. For the relation connecting Ω and Ω' involves, we easily see, the two identities

$$w\Omega' + w'\Omega = 2(vz+w)(v'z+w'), \quad (zu'+v')(zu+2v) - u'\Omega = u(zv'+w'),$$

from which the results can be read off.

For instance, let Ω be a quintic curve, with triple point at O. Let the tangents at O meet the curve again in P, Q, R; let any conic through

O, P, Q, R meet the quintic curve again in the four points A, B, C, D. There are eight points of contact of tangents drawn from O to the quintic curve; these lie, with the five points O, A, B, C, D, upon a cubic curve, touching OA, OB, OC, OD at A, B, C, D. The conic $OABCD$ is the first polar of O in regard to this cubic curve. Moreover, the tangential point of O, upon this cubic curve, lies on the first polar of O with respect to the quintic curve, being the other intersection of this first polar with the cubic.

Ex. 2. Suppose that the three polynomials, $uv' - u'v$, $uw' - u'w$, $vw' - v'w$, arising for the curves Ω, Ω' have no common factor; denote them, respectively, by u'', $2v''$, w'', so that, beside the harmonic relation $uw' - 2vv' + wu' = 0$, for u, v, w and u', v', w', we have also the corresponding relations for both the couples $(u, v, w; u'', v'', w'')$, $(u', v', w'; u'', v'', w'')$. Consider a curve $z^2u'' + 2zv'' + w'' = 0$, or, say, Ω''. By any one of the three curves Ω, Ω', Ω'' a plane de Jonquières involution may be determined. Shew that any two of these involutions carried out in succession, in either order, leads to the third.

Ex. 3. Shew that the plane quadratic transformation expressed by $x' = y(z - x)$, $y' = z(z - x - y)$, $z' = z(z - x)$, or, in non-homogeneous variables, by $x' = y$, $y' = (1 - x - y)/(1 - x)$, or, say, $(x', y') = T(x, y)$, is such that $T^5 = 1$. This gives rise therefore to sets of five points in the plane, of which any set is determined indifferently by any one of its points, say to an involution of sets of five points. Consider also the quadratic involution, U, expressed by $x' = -x/(1 - x)$, $y' = y/(1 - x)$, which is a harmonic inversion, with $(2, 0)$ as centre, and $x = 0$ as axis. It can be shewn that the products of U with the powers of T, in all possible ways, give rise to a finite group of plane transformations having precisely the structure of the symmetric group of permutations of five symbols (Burnside, *Groups*, 1911, p. 469; Moore and Slaught, *Amer. J. Math.* xxii, xxiii, 1900 and 1901). This example is indicative of a wide range of possible investigations. (Cf. Wiman, *Math. Ann.* xlviii, 1897, p. 195; also p. 278, and p. 299 below.)

Bertini's four types of involution in a plane. If a transformation in a plane expressed by the rational equations $x' = \phi(x, y)$, $y' = \psi(x, y)$, where (x, y), (x', y') are the non-homogeneous coordinates of two points, be such that $\phi(x', y') = x$ and $\psi(x', y') = y$, then the two points are the pairs of an involution, of sets of two points. By any $(1, 1)$ Cremona transformation of the plane, whereby (x, y) is placed in $(1, 1)$ relation with a point (ξ, η) and (x', y') is placed in the same relation with a point (ξ', η'), we can hence deduce an involution of pairs of points (ξ, η) and (ξ', η'). The question then arises whether there is a finite number of involutions, sets of two points, from which all others are so derivable, by application of Cremona transformations. This question was answered, in the affirmative, by Bertini (*Ann. d. Mat.* viii, 1877; see the further references below); and his results have been confirmed from other points of view. By Cremona transformation, all plane involutions, of sets of two points, can be obtained from *four* types: namely (i), the familiar harmonic inversion, expressible by the linear equations $x' = x$, $y' = -y$, $z' = z$; (ii), the de Jonquières involution, considered above, expressible by equations $x' = x$, $y' = y$ with

$zz'u+(z+z')\,v+w=0$, where u, v, w are homogeneous polynomials
in x, y, of respective orders $n-2$, $n-1$ and n; (iii), the so-called
Geiser involution, of which the pair of a set is formed by the residual intersections of two cubic curves described through seven
given points; and (iv), the appropriately called Bertini involution.
This may be defined by sextic curves prescribed to have eight given
double points; two such curves have four further intersections; but,
all sextic curves with the eight given double points which pass
through another arbitrary point P, have a single further point P' in
common; and the pair (P, P') are a pair of the involution in question.

We do not prove this theorem of the reduction to four types of
involution; but it is desirable to have a familiarity with these four
involutions. Of the de Jonquières involution we have already
treated, explaining a method by which the pairs can each be represented by a single point of a plane. We can pass over the case of the
harmonic inversion with only the notice of such representations.
An obvious one is to represent the pair (x, y, z), $(x, -y, z)$, of
the involution, by the point (ξ, η, ζ), where $\xi=x^2$, $\eta=y^2$, $\zeta=xz$.
A second more geometrical representation may be given by a con-
struction in space; the points P, P', of a pair of the involution, may
be regarded as arising by projection from two points Q, Q' of a
quadric surface, whose joining line passes through a fixed point O
(the centre of the harmonic inversion in the plane); the involution
is represented by the points of intersection of the lines QQ' with the
plane which is the polar plane of O in regard to the quadric surface.
The point from which P, P' are projected to Q, Q' lies on the section
of the quadric surface by this polar plane.

Pass now to the Geiser involution (Geiser, *Crelle*, LXVII, 1867,
p. 78; Milinowsky, *Crelle*, LXXVII, 1874, p. 268). We have defined it
as consisting of the pairs of variable intersections of two plane cubic
curves of a system having seven given fixed common points. We
obtain an interesting representation of the involution in space; we
shew that the points of the plane, at which a pair of the involution
are coincident points, describe a sextic curve with double points at
the seven base points; we shew that the order of the functions by
which a point P', of a pair, is expressed in terms of P, is eight, or,
more precisely, that, as P describes an arbitrary line, P' describes
an octavic curve with triple points at the seven base points, determining the exceptional curves of the transformation from P to P';
and we shew how to represent the pairs of the involution by the
points of a plane. From the seven base points, choose out six; and
consider the cubic surface, which, as is well known, is obtained, as
a representation of the plane, from the cubic curves which pass
through these six points. These curves are then represented by the
plane sections of the surface. And the lines of the plane are repre-

sented by rational cubic curves on the surface, which have no intersections with the six lines of the surface which represent the base
points in the plane. The cubic curves of the plane, through the
original seven base points, are thus represented by plane sections
of the cubic surface which all have a single common fixed point on
the surface, say O. Hence the pairs of the involution in the plane
are represented by points Q, Q' of the cubic surface, whose joining
line passes through the fixed point O. Incidentally, the *pairs* of the
involution are represented by the points in which lines through O
meet an arbitrary fixed plane. The coincidence points, of pairs on
the cubic surface, are then the points of contact of tangent lines to
the surface drawn from O; the locus of these points is known to be
a sextic curve, of genus 3, having a double point at O. This curve,
meeting an arbitrary plane section in six points, and meeting, as
we know, every line on the cubic surface in two points, represents
a curve in the plane with a double point at each of the *seven* base
points; as such a curve has twelve fixed intersections, with each of
the fundamental cubic curves of the plane, at the base points of
these, it is of order six, having six other intersections with each of
the plane cubic curves. The order of the functions which express
the plane involution is the number of pairs P, P' in which each of
P and P' lies on an arbitrary line in the plane. This is then the
number of pairs Q, Q', on the cubic surface, collinear with O, in
which each of Q and Q' lies on an arbitrary rational cubic curve of
the surface, both belonging to the system of cubic curves which do
not intersect the lines of the surface representing the six base points
in the plane. Two such cubic curves have a single intersection,
corresponding to the intersection of the lines in the plane which the
curves represent; the line from O to this intersection is one of the
nine common generators of the cones, with vertex at O, which stand
on these curves. The other eight common generators meet the curves
in such pairs Q, Q' as we seek. Thus 8 is the order of the functions
representing the plane involution. Consider one of the six lines of
the cubic surface representing the six base points of the cubic curves
in the plane; the plane joining O to this line is met by a rational
cubic curve of the surface which does not meet this line, in three
points; and the line joining O to one of these intersects the line;
there are therefore three pairs Q, Q', of the involution, in which Q
lies on the cubic curve and Q' lies on the line. Thus the octavic curve
in the plane which is the locus of P' when the other point P, of a
pair of the involution, moves on an arbitrary line, has a triple point
O at each of the six base points of the system of plane cubic curves.
The curve has also a triple point at the seventh point of the plane,
represented by the point O of the surface; for, the tangent plane of
the surface at O meets a cubic curve on the surface in three points,

and as a point Q describes this curve, the corresponding point Q' will pass three times through O. A plane octavic curve with 7 triple points has genus $\frac{1}{2}7.6-21$, or zero; this is the genus of the curve described by P' when P describes a line. The exceptional curves, in the plane transformation by which we pass from a point P to the other point P' of the involutory pair, are the curves described by P' when P is in the neighbourhood of one of the seven triple base points of the octavic curves. It may be proved that these curves are cubics, passing through these seven points, each with a double point at one of them; in fact, on the cubic surface, the tangent plane at O meets the surface in a curve with a double point at O, and a plane drawn through a line of the surface meets the surface otherwise in a conic having two intersections with the line. Recurring to the introductory remarks in regard to Cremona transformations made above, we notice that the equations $\Sigma i = 3n - 3$, $\Sigma i^2 = n^2 - 1$ are satisfied here, with $n = 8$ and $i = 3$ (seven cases); the Jacobian of the three octavic curves which can be constructed with seven triple points is the product of the seven exceptional curves, the multiplicity of the Jacobian, at each of the seven s-ple points of the octavics ($s = 3$), being $3s - 1$.

Ex. There is similarly an involution in ordinary space of three dimensions, determined by quadric surfaces having six given fixed points, of which the pairs are the residual intersections of three quadrics through the base points. It is well known (e.g. *Principles of Geometry*, iv, p. 156) that the three-fold space can be represented on a rational three-fold locus, of order 3, say $M_3{}^3[4]$, in space of four dimensions, by means of quadric surfaces having five base points in the original space. The involution is then represented by pairs of points Q, Q' on this manifold, whose joining line passes through a fixed point O thereof. The locus of coincident points Q, Q' in this involution is then a surface which projects from O (on to space of three dimensions) into a quartic surface with sixteen nodes (the Kummer surface). The reader may determine the surface in the original space which is the locus of coincident points; and also the order of the functions representing the involution. As the prime section of the $M_3{}^3[4]$ is a cubic surface, the present involution leads to the Geiser involution. The present involution is considered in Hudson's *Cremona Transformations*, 1927, p. 323.

We now consider the Bertini involution. Defining the *characteristic series*, for a linear system of curves with assigned points, as the series determined on any curve of the system, outside the base points, by the other curves of the system, we may assume the theorem that the redundancy of the assigned conditions at the base points, is equal to the index of specialness of the characteristic series on any of the curves—and, in particular, is zero when the number of points in a set of the characteristic series exceeds the number of points in a canonical set on any curve of the system. Consider now sextic curves in the plane with 8 assigned double points (O). Such curves are hyperelliptic, of genus 2, and have a characteristic series of sets of 4 points; thus the base points are independent, and the curves, linearly independent of one another, are 4 in number. Let P be an arbitrary point of the plane, and let $\phi = 0$ be the cubic curve through the eight points (O), and through P, while $\psi = 0$ is a particular independent curve through the points (O) only. The curves $\phi = 0$, $\psi = 0$ will have common a further point, say K, which is common to all cubic curves through the base points (O), and independent of P. The general sextic curve, with double points at (O), which passes through P, is thus given by an equation $\phi(\lambda\phi + \mu\psi) + f = 0$, wherein λ, μ are arbitrary parameters, and $f = 0$ is a particular sextic curve of this character. The curve $\phi = 0$ is met by $f = 0$ in another point P', beside P; and all the sextic curves $\phi(\lambda\phi + \mu\psi) + f = 0$, having the points ($O$) as double points, which pass through P, pass through P' also. Conversely, when P' is assigned, an unique position for P is determined by the same construction. Thus the involution (P, P') is defined. It will appear that the curves which enter into the expression of P' in terms of P, are of order 17, with each of the points (O) as six-fold points (so that the involution may be said to be of order 17); and also that the curve of united points, where P and P' coincide, is of order 9, with the points (O) as triple points. Further, as also follows from this, the transformation from P to P' has, as exceptional curves, sextic curves, each with a double point at seven of the points (O), and a triple point at the remaining point.

We consider three representations of the involution. For the first, let $\phi = 0$, $\psi = 0$ be any two independent cubic curves through the base points (O). The general sextic curve with these as double points is representable linearly by means of the four curves $\phi^2 = 0$, $\phi\psi = 0$, $\psi^2 = 0$, $f = 0$, where $f = 0$ is a particular sextic curve having the points (O) for double points. Now take a point (ξ, η, ζ, τ) in three-fold space, such that ξ, η, ζ, τ have the ratios of ϕ^2, $\phi\psi$, ψ^2 and f. The points of the original plane are then represented by the points of a quadric cone, with equation $\zeta\xi = \eta^2$, whose vertex arises from the point, K, common to all cubic curves through the eight points (O); any generator of this cone represents a particular cubic curve $\phi + \nu\psi = 0$. If a point, P, of the plane, determine the point (ξ, η, ζ, τ) of the cone, then, as all the sextic curves of the plane, with the base (O), which pass through P, also pass through another point, P', it follows that the ratios of ϕ^2, $\phi\psi$, ψ^2, f at P' are the same as at P; and thus the point (ξ, η, ζ, τ) represents the two points P, P', forming a pair of the plane involution. The four intersections of two sextic curves with base (O), outside this base, are then represented by the two points in which the cone is met by a line. We have seen that a pair P, P' of the involution lie on a cubic curve with simple points at (O); let this be

$\phi = 0$; then the united points of the involution, which lie on $\phi = 0$, are the double points of the ∞^1 series of sets of two points, determined on $\phi = 0$ by the sextics with double points at (O), that is, by the curves with equation of the form $C\psi^2 + Df = 0$, where C, D are parameters. Such double points are 4 in number, but one of these is the point K, where $\phi = 0$, $\psi = 0$ intersect; and there are three others. As we may take, in place of $\phi = 0$, any cubic curve $\phi + \lambda\psi = 0$, we see that the curve of united points in the plane is represented, on the cone, by a curve meeting each generator in three points, other than the vertex. This curve is then the sextic intersection of the cone with a cubic surface not passing through the vertex; for clearness denote this sextic curve on the cone by γ. As in general, the neighbourhood of any one of the double base points (O), is represented on the cone by a curve of order 2, a conic; the plane of this, determined by the form of the general sextic curve, $A\phi^2 + B\phi\psi + C\psi^2 + Df = 0$, near this base point, has six intersections with the sextic curve γ. Thus the curve of united points in the plane has a triple point at each of the points (O); and the order, say ν, of this curve, is such that $6\nu - 8.3.2$ is the order, 6, of the curve γ, or $\nu = 9$. This curve of united points is then of genus 4, and the original sextic curves, with base (O), are the adjoint curves of order 6 of this nonic curve. To determine the order of the curves expressing the plane involution, we are to find the number of pairs of which one point lies on one arbitrary line, while its corresponding point lies on another such line. Either line is represented on the quadric cone by a sextic curve, given, we suppose, by the intersection of the cone with a cubic surface. The number of pairs required is then only the number of intersections of two such sextic curves, since each point of such a curve represents a pair of corresponding points in the plane. The number of intersections of two such sextics is 18 (being the number of common points of the cone and two cubic surfaces); but from these must be omitted the point arising from the intersection of the two lines in the plane. There remain then 17 pairs; so that the curves representing the involution in the plane are of order 17. If these curves pass x times through each of the points (O), there will remain $17.6 - 8.2.x$ intersections, of any one of these curves, with the general sextic curve of base (O), in the plane; as the curve on the cone which represents a line of the plane meets an arbitrary plane section of the cone in 6 points, we must thus have $17.6 - 16x = 6$, or $x = 6$. This deduction is confirmed by the equations $\Sigma i = 3n - 3$, $\Sigma i^2 = n^2 - 1$, with $n = 17$, $i = 6$. The exceptional curves, corresponding to the sextuple points of these representative curves of order 17, can be similarly found; as above stated, they are rational sextic curves, each with 7 double points and one triple point, at points of the set (O).

The Bertini involution may be obtained also by considering a certain involution in space of three dimensions. In this space consider seven arbitrary fixed points, which we denote by (H); all quadric surfaces through (H) have another common point, which we denote by K. An elliptic quartic curve, of the kind which is determined by the intersection of two quadric surfaces, can be put through 8 independent points. There passes then such a curve through the points (H) and any arbitrary point, P, of the space, other than K. Upon this curve, another point, P', may be determined, by describing a *quartic* surface, say $F=0$, to have a double point at each of the seven points (H), and to pass through P; the seven double points with P constitute 15 of the 16 intersections of the surface with the quartic curve. This point P' is, in fact, the same, whatever be the quartic surface subject to the conditions. For, let $\phi_1=0$, $\phi_2=0$, $\phi_3=0$ be three independent quadric surfaces passing simply through the points (H), of which $\phi_1=0$ and $\phi_2=0$ are two which also contain the point P; thus $\phi_1=0$, $\phi_2=0$ contain the quartic curve on which P was taken; then, if a, b, c, d, e be arbitrary parameters, a quartic surface having double points at the points (H), and passing through P, is given by the equation $a\phi_1^2+b\phi_2^2+c\phi_1\phi_2+\phi_3(d\phi_1+e\phi_2)+F=0$; this is then the general quartic surface of the description, because the equation of a quartic surface contains 35 terms, and to have a double point at a given point involves 4 conditions, and $35-7.4-1=6$. All these surfaces meet the quartic curve $\phi_1=0$, $\phi_2=0$ in the same point P'. Thus we reach the conclusion that all quartic surfaces described to have double points at the seven points (H), which pass through an arbitrary general point P, pass through another point P'; and this point P' similarly determines P. The pairs P, P' form the involution in question. From this involution, we may determine the Bertini involution in a plane, by first deducing an involution of pairs of points on a quadric surface, and then projecting this on to a plane. For consider, of the space involution just described, those pairs, (P, P'), of which P lies on a definite quadric surface, say $\phi_1=0$, put through the seven points (H); by what we have seen, the corresponding point P' equally lies on this quadric surface; it is given in fact as the point common to the octavic curves on $\phi_1=0$, having double points at the points (H), and passing through P, which are defined by quartic surfaces of the form $b\phi_2^2+e\phi_2\phi_3+F=0$. Each of these quartic surfaces meets the two generators, of the quadric surface $\phi_1=0$, which pass through any particular point H_1, of the set (H), in two points not at H_1; the octavic curves thus also meet these two generators each in two points not at H_1. If we now project the quadric surface $\phi_1=0$, from the point H_1, on to a plane, the octavic curves project into sextic curves, with double points at the

two points of the plane lying on the generators at H_1, and with six
other double points, projections of the points (H) other than H_1.
The sextic curves of this system through an arbitrary point Q of
the plane (the projection of P) all pass through another point Q'
(the projection of P')—which is Bertini's result. The involution in
space is dealt with, in this way, by Hudson, *Cremona Transformations*, Cambridge, 1927, pp. 97, 326, 127.

Ex. The pairs of points of the involution in three dimensions, which
we have considered, can be represented by the points of a three-fold locus,
in a manner analogous to that followed for the pairs of points of the
Bertini involution in the plane. For, if $\phi_1 = 0$, $\phi_2 = 0$, $\phi_3 = 0$ be three
independent quadric surfaces passing through the seven fixed points (H),
the general quartic surface having these points as double points has an
equation with seven parameters entering linearly, containing the six
terms ϕ_i^2, $\phi_i\phi_j$ and another term F, where $F = 0$ is a particular quartic
surface of the kind in question. By taking homogeneous coordinates
$\xi_0, \ldots, \xi_6$, of a space [6], respectively in the ratios of ϕ_1^2, ϕ_2^2, ϕ_3^2, $\phi_2\phi_3$,
$\phi_3\phi_1, \phi_1\phi_2, F$, the three-fold space is represented upon a point cone in this
space [6], with vertex at $(0, 0, 0, 0, 0, 0, 1)$, of which a generator,
characterised by definite ratios of ϕ_1, ϕ_2, ϕ_3, represents an elliptic quartic
curve in the space [3], passing through the points (H). The cone meets
the prime $x_6 = 0$ in the so-called Veronese surface, of order 4, upon which
$\xi_0, \ldots, \xi_5$ have the ratios of ϕ_i^2, $\phi_i\phi_j$; and consists of the lines joining its
vertex to the points of this surface. Any point of this cone represents
a pair of points in the involution in space [3]. Like the Veronese surface,
the cone is rational; but—it may be remarked—it is not to be assumed,
in general, that the three-fold locus which represents the pairs of an
involution in three-fold space, is necessarily rational (see below, in this
chapter). The characteristics of the involution may be determined in
connexion with the cone. In particular, it may be shewn that the united
points are represented by a surface of order 12 lying thereon, the intersection of the cone with a cubic primal of the space [6].

A third way of considering the Bertini involution is by representing the plane upon a cubic surface so that six of the eight double
points of the defining sextic curves become lines of the surface. The
sextic curves have then six intersections, not at the base points,
with the cubic curves, through the base points, which are represented by the plane sections of the surface. Thus the sextic curves
are represented by sextic curves on the cubic surface, having two
intersections with each of the lines of the surface which represent
the six base points in the plane; the curves on the surface have,
moreover, two double points, at the points, say A and B, which
represent the remaining two of the original eight double points of
the defining sextic curves in the plane. The sextic curves on the
cubic surface, by which the corresponding involution thereon is to
be determined, are thus given by quadric surfaces subject only to
the condition of touching the surface at the points A, B. Let C be
the third fixed point in which the line AB meets the cubic surface.
If P be an arbitrary point of the surface, let T be the tangential

point of C on the cubic curve in which the plane PAB meets the surface; thus T is on the section of the surface by the tangent plane at the fixed point C. The plane, TAB, say γ, meets a quadric surface drawn to touch the cubic surface at A and B, in a conic touching the section of the surface, by the plane γ, in A and B; and this conic contains P if the quadric surface contains P; the conic has then another intersection with the section γ. By the ordinary theory of residuation on the cubic curve γ, this remaining intersection is the third point, say P, in which the line TP meets the surface. Thus P' is uniquely determined when P is given, and the relation of these points is mutual. Moreover P' is common to all the sextic curves on the cubic surface representing the original defining sextic curves in the plane, which pass through P. This constitutes then another proof of the existence of the Bertini involution in the plane. Denoting the point where the line PP' meets the fixed line AB by U, a pair of the involution is given by taking any point T on the section of the cubic surface by the tangent plane at C, and joining this to any point U of the line AB; then P, P' are the two points, other than T, in which this line meets the surface. As every plane through the line AB meets the section of the surface by the tangent plane at C, in only one point T beside C, there is one line such as TU through an arbitrary point of space; thus an incidental consequence is that the pairs of the involution may be represented by the points in which an arbitrary fixed plane is met by the lines TU. That, as we have otherwise seen, the order of the curves representing the plane involution is 17, follows if we shew that there are 17 lines which meet the four curves: (1), the line AB; (2) and (3), two cubic curves on the cubic surface representing lines in the plane; (4) the section by the tangent plane at C. This is the number given by the formula of Ex. 10 d, p. 30, of Chapter I, namely

$$2n_1 n_2 n_3 n_4 - \Sigma n_p n_q i_{rs} + \Sigma i_{pq} i_{rs},$$

if we put $n_1 = 1$, $n_2 = 3$, $n_3 = 3$, $n_4 = 3$, with $i_{12} = 0$, $i_{13} = 0$, $i_{14} = 2$, $i_{23} = 1$, $i_{24} = 3$, $i_{34} = 3$. To find the curve of united points of the involution on the cubic surface, we are to consider the lines TU which touch the surface. These it is easy to see form a ruled surface having the line AB as six-fold line, and the rational tangent section, by the tangent plane at C, as a triple line; this ruled surface is then of order 9. The curve of united points is then represented on any plane by the curve in which this plane meets the ruled surface. This is a curve of order 9, with one six-fold point, on the line AB, and three triple points, these being on the line in which the tangent plane at C is met by the plane of the nonic curve. A curve of this character, we can easily see, has an equation which is capable of the form

$$x^3 z^3 (x+mz)^3 + yx^2 z^2 (x+mz)^2 u_2 + y^2 xz (x+mz) u_4 + y^3 u_6 = 0,$$

where u_i is homogeneous in x and z, of order i. By putting x^{-1}, y^{-1}, z^{-1} for x, y, z; then $z+mx=z_1$; and then $z_1=1$, this equation reduces to $y^3+y^2v_2+yv_4+v_6=0$, where v_i is a polynomial in x, of order i. This is a sextic curve having the singularity which is described, in Noether's phraseology, as two consecutive triple points. That the curve of united points is equivalent to such a curve, follows from the preceding representation of the pairs of a Bertini involution by the points of a quadric cone; we saw that the curve of united points is representable by the intersection of this cone with a cubic surface. The projection of this intersection on to a plane, from an arbitrary point of the cone, gives such a sextic curve. We see also, from this point of view, that the curve is equivalent to a quintic curve having one point of self-contact. (For this curve, the reader may compare Noether, *Math. Ann.* XXXIII, 1889, p. 543; and also Brill u. Noether, *Deut. math. Ver.* III, p. 512; Schottky, *Crelle*, CIII, 1888, p. 185.) But we may consider the curve of united points on the cubic surface on which the Bertini involution is represented; it is the locus of points of contact of quadric surfaces having beside two given contacts with the cubic surface, at A and B. This curve has triple points at A and B, and can be shewn (as was remarked by Cremona, *Ann. d. Mat.* VIII, 1877, p. 273 f.n.) to be the intersection of the given cubic surface U with another cubic surface V. The surface V touches U at A and B; the tangent planes of U at these points each meet V in three lines, one of which is the intersection of these planes; and the two lines, lying on V, which intersect in A, are the inflexional tangents of U at A (with a similar relation at B) (Hilton, *Journ. Lond. Math. Soc.* I, 1926, p. 2; and W. L. Edge).

The four involutions and rational double planes. The fact, due to Bertini, that all involutions of pairs of points, in a plane, are reducible by Cremona transformations of the plane, to one of the four types we have considered: (i), Harmonic inversion; (ii), a de Jonquières involution; (iii), a Geiser involution; (iv), a Bertini involution, is intimately related with a theorem due to Noether (*Sitzungsber. Phys.-med. Soc.* Erlangen, X, 1878, p. 81); this theorem states that, of the so-called double planes, namely the surfaces with an equation of the form $t^2z^m = \phi(x, y, z)$, where ϕ is a homogeneous polynomial in x, y, z, those which are rational are reducible, by a Cremona transformation of the plane (x, y, z), to one of the three in which the curve $\phi(x, y, z)=0$ is (i), an ordinary quartic curve; (ii), a sextic curve with two consecutive triple points, considered above; (iii), a curve of order $2n$ with a $(2n-2)$-fold point. That the surfaces should be rational means that there exist two functions, ξ and η, both rational, and homogeneously of zero order, in the four coordinates x, y, z, t, which are such that, in virtue of the equation of the surface, the ratios of x, y, z, t are expressible rationally in ξ and η. For a summary proof that the surfaces are rational in the three cases named, the reader may consult *Proc. Lond. Math. Soc.* XII, 1913, p. 35. It is clear that the surface $t^2 z^m = \phi(x, y, z)$, contains the pairs (x, y, z, t), $(x, y, z, -t)$, of an involution existing thereon; the curve of united points $\phi = 0$, in the three cases named, is such as arises, respectively, for a Geiser involution, a Bertini involution, and a de Jonquières involution. In this connexion, the reader may consult Bertini, *Ann. d. Mat.* VIII, 1877; Bertini, *Rend. Ist. Lombardo*, XIII, 1880; Bertini, *ibid.* XXII, 1889; also Lüroth, *Rationale Flächen und involutorische Transformationen*, Freiburg, 1889; beside the fundamental paper of Noether (Erlangen, 1878), referred to above. Also Vol. V, p. 110, Ex. 3.

Involutions of sets of more than two points. The involutions of sets of two points which we have considered may be defined by rational transformations of the coordinates which have the property that, on repetition, they lead to the identical transformation. We may similarly consider transformations which lead to the identical transformation after $k(>2)$ repetitions. Such is the transformation $(x', y') = T(x, y)$, remarked above (Ex. 3, p. 121), given, in non-homogeneous coordinates, by $x' = y$, $y' = (1 - x - y)/(1 - x)$, for which $T^5 = 1$. In this case there are sets of 5 points in the plane, of which any set is determined, by the repetition of the same rational formula, from any point of the set. More generally we may consider the possibility of sets of points, in any space, of which a set, and only one set, is determinable algebraically to contain an arbitrary point of the space; thus the set is equally determinable from every point of the set. Such a system of sets is called an involution; and is of *index* 1. We may in fact have involutions with the property that there are i sets containing an arbitrary point of the space; such an involution is said to be of index i. Involutions arising by repetitions of a single transformation were considered by Kantor, *Compt. rend.* fév. 1885; see *Atti...Napoli*, IV, 1892.

An interesting question in regard to an involution is whether its *sets* can be represented by the points of a (linear) space, so that every set of the involution gives one point of this space (whose coordinates will then be rational symmetric functions of the coordinates of the points of a set of the involution), and conversely any point of this space gives one set of the involution (the rational symmetric functions of the coordinates of the points forming a set of the involution being expressible rationally by the coordinates of the representative point). When this is so, the involution is said to be rational. We have illustrated the question, by shewing that the involutions of sets of two points in a plane are rational; it is true that all involutions in a plane are rational, however many be the

points in a set; we give some account of the proof of this, due to
Castelnuovo, in what follows. It is believed, however, that in higher
space there exist involutions which are not rational.

The question of the rationality of an involution arises naturally
in connexion with the theory of rational curves, or surfaces, or
manifolds of higher dimension. Consider, first, a curve. Suppose
that it is found that the coordinates of a point of the curve are expressible rationally in terms of a parameter, the equations which
determine the curve being satisfied identically when the appropriate expressions are substituted for the coordinates. If we interpret this parameter as a coordinate on a line, to any point, P, of
the line, corresponds, by rational equations, a single point, Q, of
the curve. But it may happen that the same point Q of the curve
arises, by the same formula, from several different points, P, of the
line; this will be so if the parameter is not expressible as a single-
valued (rational) function of the coordinates of the point of the
curve. Then, to any point P, of the line, will correspond, first, a
single point, Q, of the curve, and then, the set $P, P', \ldots$, of all the
points of the line from which this point Q may equally be determined. The sets $(P, P', \ldots)$ evidently form an involution on the line.
In this case it was proved by Lüroth (*Math. Ann.* IX, 1876, p. 163;
see *Principles of Geometry*, II, p. 136) that a new parameter can be
chosen, a rational function of the original one, of which the coordinates of the point of the curve are rational functions, which is
itself conversely expressible as a rational function of the coordinates
of the point of the curve. This new parameter represents then the
sets of the involution on the line, which is therefore said to be
rational. Similarly, in the case of a surface, it may be that the coordinates of a point, Q, of the surface are known to be rational
functions of two parameters, which we may interpret as the coordinates of a point, P, of a plane; and the functions may be such
that to the point Q correspond reversely, not only P, but also other
points, $P', \ldots$, of the plane; the sets $(P, P', \ldots)$ then constitute an
involution in the plane. In this case also, as has been stated, the
sets of the involution can be represented uniquely, both directly
and reversely, by the points R of another plane; thus the points Q
of the surface are in unique correspondence with the points R of this
other plane; and the surface is rational in this sense. But, contrary
to what may seem natural, the same procedure is not possible for
manifolds of more than two dimensions. For instance, consider, in
space of four dimensions, the general cubic primal, that is the locus,
of three dimensions, which is characterised by a single homogeneous
equation of order 3 in the coordinates; it is easy, as we shall see, to
express the coordinates of a point of the primal by rational functions of three parameters. But, interpreting these as coordinates

of a point in space of three dimensions, two points of this space, at least, give rise to the same point of the primal. The process which has been followed, in order to shew that any involution in a plane is rational, consists of two steps, which take the ideas we have explained in reverse order: (1), It can be shewn that any involution in a plane can be represented by an algebraic surface. This means (under the hypothesis, adopted throughout, that the points $(P, P', ...)$ of a set of the involution, are determined uniquely from any one of these, P, by algebraic equations), that we can find a surface, of points Q, of which the coordinates are rationally determinable from those of the points $P, P', ...$, of any set of the involution; and are, therefore, rational in the coordinates of any one point, P, of the set; while, conversely, any rational symmetric function of the coordinates of $P, P', ...$ is expressible rationally by the coordinates of Q. This theorem, which holds equally, with obvious changes, for an involution in space of any number of dimensions, is often assumed as obvious. But we shall formulate a proof. We may speak of a surface of this character as *representing* the involution; it will have properties, as regards the curves which lie upon it, which arise from the existence of the involution. The next step, (2), consists in deducing from these properties, with the help of general properties of surfaces, that this representative surface is rational. Here we must be content merely to indicate what is necessary for the step (2) to be made.

By what has been said, an involution in a plane consists of sets of, say, n, points, with, say, non-homogeneous coordinates $(x, y), (x_1, y_1), ..., (x_{n-1}, y_{n-1})$, such that any rational algebraic symmetrical function, of the coordinates of these n points, is expressible rationally in terms of the coordinates of any one of them, say of (x, y). We consider then, first, the whole aggregate of the functions which are rational algebraic symmetric functions of $(x, y), (x_1, y_1), ..., (x_{n-1}, y_{n-1})$; every function of this aggregate is a rational algebraic function of (x, y); and any rational function of two, or more, of the functions of this aggregate, is itself a function of the aggregate. Denote the aggregate of such functions by K. It may happen that there are two functions, belonging to the aggregate, in terms of which all other functions of the aggregate are expressible uniquely, and therefore rationally; this is the case when the involution is rational. In detail, however, let $u(x, y), v(x, y)$ be two functions of the aggregate K; and consider, for general assigned values of ξ and η, the solutions, for x and y, of the two equations $u(x, y) = \xi, v(x, y) = \eta$, excluding: (1), the solutions which do not vary with ξ, η; (2), the solutions which give indeterminate values for $u(x, y)$ or $v(x, y)$. By the nature of the aggregate K, if (x, y) be a solution of these two equations, so also are the other pairs

$(x_1, y_1), \ldots, (x_{n-1}, y_{n-1})$ defining, with (x, y), a set of the involution; and if these are all, for a given pair ξ, η, then the variable point (ξ, η) represents a set of the involution, which is therefore rational. But, generally, the solutions of these two equations may be taken to consist of the points of several sets of the involution. Now, let $w(x, y)$ be any other function of the aggregate K; thus $w(x, y)$ has the same value at all the points of any single set of the involution. If these values of $w(x, y)$ are different for all the different sets of the involution which are given by the equations $u(x, y) = \xi$, $v(x, y) = \eta$, then the assignment of the value of the function $w(x, y)$, say ζ, together with the values ξ, η of $u(x, y)$ and $v(x, y)$, will suffice to identify just one set of the involution. Suppose the equations $u(x, y) = \xi$, $v(x, y) = \eta$ are satisfied by μ sets of the involution, and consider the μ values ζ of $w(x, y)$ which correspond thereto; any rational symmetric function, of these values of ζ, will be uniquely determinate when ξ, η are given, and will therefore be expressible rationally in terms of ξ and η. Thus there exists an equation $F(\zeta, \xi, \eta) = 0$, rational in all of ζ, ξ, η, which becomes an identity when ξ, η, ζ are replaced by $u(x, y)$, $v(x, y)$, $w(x, y)$; and the order of this equation, in ζ, is μ when the μ sets in question lead to different values of ζ; but may be less than μ if $w(x, y)$ takes the same value for at least two of the sets of the involution which satisfy $u(x, y) = \xi$, $v(x, y) = \eta$. For simplicity, we prove that there certainly exist functions $w(x, y)$, in the aggregate K, for which the μ values in question are different. Let $\rho(x, y)$ be any rational function of (x, y); denote the μ sets (x, y) which satisfy the equations $u(x, y) = \xi$, $v(x, y) = \eta$, for a definite pair of values of ξ and η, by $(x, y), \ldots, (x_{n-1}, y_{n-1})$ and $(x^{(i)}, y^{(i)}), \ldots, (x_{n-1}^{(i)}, y_{n-1}^{(i)})$, with $i = 1, 2, \ldots, (\mu - 1)$. The function $w(x, y)$ which is the product of $\rho(x, y), \rho(x_1, y_1), \ldots, \rho(x_{n-1}, y_{n-1})$ is a function of the aggregate K. Consider the μ sets, of each n equations,

$$\rho(x, y) = k, \ldots, \rho(x_{n-1}, y_{n-1}) = k_{n-1};$$

$$\rho(x^{(i)}, y^{(i)}) = k^{(i)}, \ldots, \rho(x_{n-1}^{(i)}, y_{n-1}^{(i)}) = k_{n-1}^{(i)},$$

for $i = 1, \ldots, (\mu - 1)$; we assume that the arbitrary rational function $\rho(x, y)$ can be chosen so that these $n\mu$ equations are all satisfied for such values of the numbers k on the right that the μ products $kk_1 \ldots k_{n-1}, k^{(i)} k_1^{(i)} \ldots k_{n-1}^{(i)}$ are all different. The function $w(x, y)$, defined in terms of $\rho(x, y)$ as we have stated, has then different values for the chosen values of ξ and η; its values will therefore not be repeated for general values of ξ and η. We may thus assume that the equation $F(\zeta, \xi, \eta) = 0$, is of order μ in ζ. Hence we can deduce that any function, say $t(x, y)$, of the aggregate K, is rationally expressible by $u(x, y)$, $v(x, y)$ and $w(x, y)$, when $w(x, y)$ is chosen

as we have proved to be possible. The function $t(x, y)$ is in fact
determinate when a set of the involution is given, and such a set
is identified, we have remarked, when the values of $u(x, y)$, $v(x, y)$,
$w(x, y)$ are given; the necessarily algebraical expression of $t(x, y)$,
in terms of these three functions, is therefore rational. But we
can give a rule for the computation of $t(x, y)$, as a rational function.
Denote by $b_1, \ldots, b_\mu$ the values of $w(x, y)$ for the μ sets of the involution which satisfy $u(x, y) = \xi$, $v(x, y) = \eta$; the values of $t(x, y)$
for these sets are not necessarily all different; denote by $c_1, c_2, \ldots$ the
various *different* values of $t(x, y)$. Let θ be a number which is different
from every one of the finite number of fractions $(b_i - b_j)/(c_p - c_q)$.
Then the values of $w(x, y) - \theta t(x, y)$ cannot be the same for any
two of the μ sets in question, since $b_i - \theta c_p = b_j - \theta c_q$ is impossible
(even when $c_p = c_q$). It follows then, by the argument above, that
the function $w(x, y) - \theta t(x, y)$, or say $\zeta - \theta\tau$, satisfies an equation
$F_1(\zeta - \theta\tau, \xi, \eta, \theta) = 0$, rational in ξ, η, θ and $\zeta - \theta\tau$, and of order μ in
the last. We can then suppose that $F(\zeta, \xi, \eta) = F_1(\zeta, \xi, \eta, 0)$. As
the equation $F_1 = 0$ is unaffected when θ varies, it can be differ-
entiated with regard to θ; this gives $-\tau \partial F_1/\partial \zeta + \partial F_1/\partial \theta = 0$, which,
putting $\theta = 0$, gives the rational expression of τ in terms of ζ, ξ, η.
In its application to the involution we may sum up this result by
saying: *It is possible to find a surface of which the coordinates of any
point are rational symmetric functions of the coordinates of the n points
of a set of the plane involution; and any such function is expressible
rationally by the coordinates of the representative point of this surface.
The surface is thus an adequate reversible representation of the sets of
the involution.*

The functions of the aggregate K, defined as rational symmetric
functions of the coordinates of the n points of a set $(x, y), \ldots,$
(x_{n-1}, y_{n-1}) of the involution, are all rational functions of x and y
only, with the property of having the same value at all the points
of such a set. Conversely, any rational function of x, y, say $u(x, y)$,
which has this property, can be explicitly expressed as symmetric
in the n points, in the form $[u(x, y) + \ldots + u(x_{n-1}, y_{n-1})]/n$. We may
thus speak of the aggregate K as consisting of the (rational) *invariants* of the involution. As the aggregate consists of an infinite
number of functions, it may be of interest to exhibit a practical
method of computing a finite number of such invariants, from
which the functions $u(x, y)$, $v(x, y)$, $w(x, y)$ of the preceding theory
may be selected. Let θ be an undetermined number; consider the
product of the n functions $t - (x + \theta y), \ldots, t - (x_{n-1} + \theta y_{n-1})$. This is
of the form

$$\sum_{k=0}^{n} t^k [p + p_1\theta + \ldots + p_{n-k}\theta^{n-k}],$$

wherein, by the definition of the involution, $p, p_1, \ldots, p_{n-k}$ are expressible as rational functions of x and y; the product contains $2+3+\ldots+(n+1)$, or $\frac{1}{2}n(n+3)$, of these rational functions of x, y, of which, however, some may identically be constants. By the theory of symmetric functions, every rational symmetric function of (x, y), ..., $(\tilde{x}_{n-1}, y_{n-1})$ is expressible rationally by these $\frac{1}{2}n(n+3)$ functions (cf. Netto, *Algebra*, II, 1900, who refers to Poisson, *J. d. l'école polyt.* XI. Cah. (An x), p. 199). Thus every invariant of the involution is expressible rationally by these $\frac{1}{2}n(n+3)$ functions. The functions $u(x, y)$, $v(x, y)$, of our theory, were any two invariants of the involution; for these then we may take any two of the $\frac{1}{2}n(n+3)$ functions p_i which are not constants; the function $w(x, y)$ was an invariant, of sufficient generality, and is a rational function of the $\frac{1}{2}n(n+3)$ functions. All invariants of the involution are thus expressible rationally by any two, not constant, functions chosen from the $\frac{1}{2}n(n+3)$ functions p_i, and a further function, itself rationally expressible by these $\frac{1}{2}n(n+3)$ functions. The three chosen functions are connected by a rational equation. The necessary and sufficient condition for the involution to be rational is that all the invariants of the involution should be rationally expressible in terms of two properly chosen invariants; or, that all rational functions of the functions p_i should be rationally expressible in terms of two properly chosen rational functions of these; or, the condition is that the surface expressed by $F(\zeta, \xi, \eta)=0$ should be rational. The condition is satisfied when all the functions p_i are rational functions of two of these functions.

Ex. 1. Consider the simple involution of which the pairs of points, in non-homogeneous coordinates, are (x, y) and (x^{-1}, y^{-1}). In this case, two invariants of the involution are $u(x, y)$, $v(x, y)$ given by

$$u(x, y)=(x-1)^2(x+1)^{-2}, \text{ and } v(x, y)=(y-1)^2(y+1)^{-2}.$$

But the equations $u(x, y)=\xi$, $v(x, y)=\eta$ are satisfied by the two pairs of the involution $[(x, y), (x^{-1}, y^{-1})]$, $[(x^{-1}, y), (x, y^{-1})]$, or the number μ, of the theory above, is 2. If we take, however,

$$w(x, y)=(x-1)(y-1)(x+1)^{-1}(y+1)^{-1},$$

this is an invariant with different values for the two sets. Calling this last ζ, we have $\zeta^2=\xi\eta$, and every invariant is a rational function of ξ, η, ζ; in fact, therefore, is a rational function of ξ and ζ, the involution being rational. In this case, the product of $t-x-\theta y$ and $t-x^{-1}-\theta y^{-1}$ is $t^2-2tP+Q$, where P, Q, expressed in terms of ξ, η, ζ, are given, respectively, by

$$P=(1+\xi)(1-\xi)^{-1}+\theta(1+\eta)(1-\eta)^{-1},$$

$$Q=1+\theta^2+2\theta[(1+\xi)(1+\eta)-4\zeta](1-\xi)^{-1}(1-\eta)^{-1},$$

so that an inspection of P and Q might suggest the choice of ξ, η and ζ. Another set of three characteristic functions, for the involution, is given by ξ_1, η_1, ζ_1, where $\xi_1=\frac{1}{2}(x+x^{-1})$, $\eta_1=\frac{1}{2}(y+y^{-1})$, $\zeta_1=\frac{1}{2}(xy^{-1}+yx^{-1})$, which are connected by $\xi_1{}^2+\eta_1{}^2+\zeta_1{}^2-2\xi_1\eta_1\zeta_1=1$, likewise representing a rational surface.

Ex. 2. Consider the involution, in space of three dimensions, with non-homogeneous coordinates x, y, z, of which a pair is (x, y, z), (y, x, mz^{-1}), in which m is a given rational and symmetric function of x, y, only. Four invariants of this involution are ξ, η, ζ, τ, given by

$$\xi = \tfrac{1}{2}(mz^{-1} + z), \qquad \eta = \tfrac{1}{2}(mz^{-1} - z)(x - y)(x + y)^{-1},$$
$$\zeta = -(x + y)^2(x - y)^{-2}, \qquad \tau = x + y,$$

and these are a characteristic set, only one pair of the involution arising when the values of these four are given. For we have $xy = \tfrac{1}{4}\tau^2(1 + \zeta^{-1})$, and m is thus a given rational function of ζ and τ. Whence, when ζ and τ are given, the couple (x, y) is given, save for ambiguity between this and (y, x); and m is given. Thus if ξ and η be also given, $mz^{-1} + z$ is given, and $mz^{-1} - z$ is given save for sign, the signs being interchanged by interchange of x and y; whence, the only set satisfying the equations when ξ, η, ζ, τ are given, beside (x, y, z), is (y, x, mz^{-1}). Thus, by what we have proved, all invariants of the involution are rationally expressible by ξ, η, ζ, τ, which are connected by the equation $\xi^2 + \zeta\eta^2 = m$, where m is a given rational function of ζ and τ. For the involution to be rational, all invariants must be rationally expressible in terms of three properly chosen invariants; thus, in particular, ξ, η, ζ, τ must be rationally expressible in terms of three properly chosen rational functions of these; in other words, the manifold expressed, in the four-dimensional space in which ξ, η, ζ, τ are coordinates, by the equation $\xi^2 + \zeta\eta^2 = m$, must be rational. Conversely, if this manifold be rational, three rational functions of ξ, η, ζ, τ exist—which therefore are three invariants of the involution—in terms of which ξ, η, ζ, τ, and therefore all other invariants, are rationally expressible, and the involution is rational. In this case, with p, q, r undetermined, the product of $t + px + qy + rz$ and $t + py + qx + rmz^{-1}$ is $t^2 + tP + Q$, where $P = (p + q)\tau + 2r\xi$, and

$$Q = \tfrac{1}{4}(p^2 + q^2)\tau^2(1 + \zeta^{-1}) + \tfrac{1}{2}pq\tau^2(1 - \zeta^{-1}) + r^2m + pr(\xi + \eta)\tau + qr(\xi - \eta)\tau.$$

There are many cases for which the manifold $\xi^2 + \zeta\eta^2 = R(\zeta, \tau)$, where R is a given rational function of the arguments ζ, τ, is rational. For instance, if $R = \tau^2 + \zeta Z$, where Z is a given rational function of ζ only, the equation is $\xi^2 - \tau^2 = \zeta(Z - \eta^2)$, and if we take $\xi + \tau = \theta\zeta$, $\xi - \tau = \theta^{-1}(Z - \eta^2)$, then ξ and τ are rational in η, ζ and θ. But it seems impossible that the manifold should be rational for all forms of the rational function R. There is thus an *a priori* presumption that not all involutions, except on a line or in a plane, are rational.

The consideration of the preceding involution is suggested by the case of the cubic primal, represented by the vanishing of a single polynomial of the third order in the coordinates, in space of four dimensions. This primal contains lines (in fact ∞^2); take one of these, and take two points on this line. The tangent primes of the primal at these two points intersect in a plane; this plane contains the line, and meets the primal, beside, in a conic. Conversely, an arbitrary point of the primal may be joined to the line by a plane. The coordinates of a point of the primal are thus expressible rationally by three parameters: one to determine a point of the conic, the other two to determine the points of intersection of the conic with the line. But these parameters are not rational functions of the coordinates of the point of the primal, there being no rational discrimination between the parameters of the two points of the line (cf. Richmond, *Proc. Camb. Phil. Soc.* xv, 1910, p. 116; Enriques, *Rend....Lincei*, xxi, 1912, p. 81, who ascribes the preceding remark to Noether, and quotes, as another example, the three-fold intersection of a quadric and a cubic primal, in space of five dimensions, with reference to Fano, *Atti...Torino*, 1908).

The surface representing a plane involution cannot contain an irrational pencil of curves. Having thus dealt in detail with the representation of a plane involution by a surface (or, indeed, of any involution, existing in a linear space, by an appropriate manifold) we now approach the question whether such a surface is necessarily rational. We can prove without difficulty that the surface cannot be an irrational ruled surface, or indeed any surface upon which there exists an irrational pencil of curves. By a *pencil* of curves on a surface is meant a system of which there is just one curve of the system passing through an arbitrary point of the surface, so that the aggregate of the curves of the system is ∞^1. Such a system may be *linear*, being given by the intersection of the surface with a, singly infinite, linear family of surfaces, with equation of the form $\phi + \lambda\psi = 0$; when the surface under consideration is in higher space, the linear pencil is given by a family of *primals*, of equation $\phi + \lambda\psi = 0$. But the pencil may not be linear: it may be that there exist two definite rational functions of the coordinates, ϕ and ψ, not necessarily polynomials, whose values on the surface are connected by a definite irreducible rational equation $F(\phi, \psi) = 0$, satisfied identically in virtue of the equations which determine the surface; then, if ξ, η be numbers connected by the equation $F(\xi, \eta) = 0$, part (at least) of the intersection of the given surface with the primal whose equation is $\phi = \xi$, is a curve lying also on the primal $\psi = \eta$. In other words, if $F(\xi, \eta) = 0$ be regarded as the equation of a plane curve, to every point (ξ, η), of this curve, corresponds a curve on the surface, upon which both $\phi = \xi$ and $\psi = \eta$; conversely, in virtue of the identity $F(\phi, \psi) = 0$ on the surface, the values which $\xi = \phi$ and $\eta = \psi$ take, at an arbitrary point of the surface, correspond to a point of the curve $F(\xi, \eta) = 0$, and lead to a curve of the system passing through this point. The system of curves is thus ∞^1, and one curve of the system, lying on the surface, passes through every point of the surface. When the curve $F(\xi, \eta) = 0$ is rational, there is a rational function of ξ, η, say $\theta = r(\xi, \eta)$, in terms of which ξ, η have such rational expressions that the condition $F(\xi, \eta) = 0$ is identical, irrespective of the equations which determine the surface; in this case, to any point of the curve $F(\xi, \eta) = 0$, characterised by the value of θ, there is a curve on the surface upon which $\theta = r(\phi, \psi)$; and the equations $\phi = \xi, \psi = \eta$ are a consequence of this. The system of curves then becomes the linear ∞^1 system given, for variable θ, by $\theta = r(\phi, \psi)$; this is the first case considered. If, however, $F(\xi, \eta) = 0$ be not a rational curve, we have an *irrational pencil* of curves on the surface. The most familiar example of the theory is the case of a ruled surface in ordinary space, for which $F(\xi, \eta) = 0$ determines the plane section, and $\phi = \xi, \psi = \eta$ determine a generator. Conversely it will subsequently (see Note, at the end of the Chapter) be noticed that a surface upon which there exists a pencil of curves of which every one is rational, is birationally reducible to a ruled surface (or is a rational surface, when the pencil of curves is rational, or linear).

After this explanation of an irrational pencil of curves on a surface, we prove now that a surface on which such a pencil exists cannot be the representative surface for a plane involution. For clearness, let the coordinates in the plane be X, Y. The coordinates of a point of the surface will be rational in X, Y, and unaltered when (X, Y) is replaced by any other point of the set of the involution determined by (X, Y). A curve of an irrational pencil, supposed to exist on the surface, given by $\xi = \phi$, $\eta = \psi$, will thus give rise, on the plane, to a curve which satisfies equations of the forms

$$\xi = \Phi(X, Y), \quad \eta = \Psi(X, Y),$$

wherein Φ, Ψ are rational in their arguments; and there will exist an identical equation $F[\Phi(X, Y), \Psi(X, Y)] = 0$, in which X, Y are arbitrary. This is impossible unless the curve $F(\xi, \eta) = 0$ is a rational curve; it can conceivably hold for a rational curve for a particular numerical value of one of X, Y. The representative surface of the involution cannot, therefore, contain an irrational pencil of curves.

The rationality of the representative surface of a plane involution. The original very remarkable proof given by Castelnuovo in 1893 (*Math. Ann.* XLIV, pp. 145–55) that the representative surface of a plane involution is rational, has in effect been modified by subsequent work of the same writer, and of Enriques; first, by an investigation, given by Castelnuovo, of the conditions in general for a surface to be rational (*Mem. d. Soc. Ital. d. Scienze*, x, 1896, pp. 103–23); and then, by the results given in a paper by Castelnuovo and Enriques (*Ann. d. Mat.* VI, 1901). For any linear system of curves in a plane, without common base points, the number of points in a canonical set on any one of the curves is less than the number of intersections of this curve with any other curve of the system; and, for a linear system with assigned base points a "completed" canonical set likewise contains less points than the "completed" number of intersections of the curve with another curve of the system; here the completed canonical set consists of the whole number of intersections of the curve with the adjoint curve defining the set, including those at the base points, and the completed number of intersections of two curves includes their intersections at the base points. In the last paper referred to it is proved, on the basis of several particular theorems already known, that, if a surface, which has no multiple points, contain a linear system of curves for which the completed canonical sets contain a number of points less than the completed number of intersections of two curves of the system, then such a surface either contains an irrational pencil of curves, or is a rational surface. It had been proved in Castelnuovo's original paper of 1893, that the representative surface of a plane involution necessarily contains a linear system of curves for which the inequality above explained is fulfilled. We have proved that the surface does not contain an irrational pencil of curves. We assume that a representative surface can be found which has no multiple points. The conclusion, then, is that the surface is rational; and, therefore, as explained above, the plane involution is also rational. The general condition given above may be enunciated also thus: consider a surface, in space of any dimensions, not having multiple points; let the surface contain a linear system of curves with assigned base points, of typical multiplicity t; let these curves be of genus p, and n be the grade of the system, or the number of intersections of two curves of the system outside the base points; define the completed genus, π, of a curve, and the completed grade, ν, by $\pi = p + \frac{1}{2}\Sigma t(t-1)$, $\nu = n + \Sigma t^2$, and call $2\pi - 2 - \nu$ the *canonical number* of a curve of the system (see p. 222, below). We find $2\pi - 2 - \nu \leqslant 2p - 2 - n$. Then, *if there be such a linear system for which the canonical number is negative, the surface may be birationally transformed to a ruled surface.* The theorem is true when the linear system is of zero freedom; that is, consists of a single isolated curve, provided $\pi > 0$. The less detailed form of the result, which we have stated above, namely that a surface with the specified linear system of curves must contain an irrational pencil of curves, or be rational, is sufficient for our present purpose. We cannot here give the proof of this theorem. But we may give, in barest outline, the proof that the representative surface of a plane involution contains a linear system of curves of negative canonical number. For brevity, let two points of the plane which belong to the same set of the involution be called *conjugate*. To any curve on the representative surface will correspond a curve of the plane, with the property that all the conjugate points, of any point of the curve, equally lie on this curve; conversely any curve of the plane with this property corresponds to a curve of the surface; we may describe such a curve of the plane as "belonging to" the involution. A particular curve of this character is obtainable by taking an arbitrary line u, and then the curve ϕ which is the locus of the sets of points conjugate to all the points of the line; and then considering the composite curve $u + \phi$. Let k be the order of the curve of united points of the involution in the plane, the locus of points of which each coincides with one of its conjugate points; let j be the number of pairs of conjugate points which lie on an arbitrary line of the plane (called by Caporali the *class* of the involution; *Rend....Napoli*, XVIII, 1879, p. 212). The curve ϕ meets the line u in its k intersections with the locus of united points, and also in the j pairs of conjugate points lying on u. Thus ϕ is of order $k+2j$ and $u+\phi$ is of order $k+2j+1$. The irreducible curve $u+\phi$ defines a linear system of curves in the plane, all belonging to the involution (having base points at the exceptional points of the plane of which the conjugate points are not a finite set but all the points of an exceptional curve). From this system there arises a linear system of curves on the representative surface. *This is the system of which we can prove the canonical number to be negative.* The number of intersections of two curves of this system, on the surface, is equal to the number of sets of the involution, in the plane, which are common to the degenerate curve $u+\phi$, and to a general curve, of order $k+2j+1$, of the linear system defined thereby; from the nature of $u+\phi$, any set lying thereon has one point on u; the number of sets required is thus the number of intersections, with u, of a curve whose order is $k+2j+1$. The genus of the curves on the surface which correspond to the plane curves defined by $u+\phi$ is more difficult to find; it is in fact equal to j. As $2j-2-(k+2j+1)$, equal to $-(k+3)$, is negative, it follows that the canonical number in question is certainly negative.

Ex. 1. A very simple example is the involution before taken (p. 133) of which, in homogeneous coordinates, the sets are the pairs (x, y, z), (x^{-1}, y^{-1}, z^{-1}). The points conjugate to the points of a line u, of equation $lx+my+nz=0$, then describe the conic ϕ given by $lx^{-1}+my^{-1}+nz^{-1}=0$. The system defined by the curve $u+\phi$ then consists of the cubic curves given by $A(xz^{-1}+x^{-1}z)+B(yz^{-1}+y^{-1}z)+C(xy^{-1}+x^{-1}y)+D=0$; these have 3 base points, and any two of them meet beside in three pairs of the involution. With the invariants before used, $\xi=(x-z)^2(x+z)^{-2}$, $\eta=(y-z)^2(y+z)^{-2}$, $\zeta=(x-z)(y-z)/(x+z)(y+z)$, we have

$$xz^{-1}+x^{-1}z=2(1+\xi)(1-\xi)^{-1},$$

$$xy^{-1}+x^{-1}y=2[(1+\xi)(1+\eta)-4\zeta](1-\xi)^{-1}(1-\eta)^{-1},$$

and the system of cubic curves corresponds to the intersection of the cone $\xi\eta-\zeta^2=0$ with the quadric surfaces

$$P(\xi-\zeta)+Q(\eta-\zeta)+R(\xi\eta-\zeta)+H(1-\zeta)=0,$$

that is to a system of elliptic quartic curves. These curves contain the point $(1, 1, 1)$ and have, beside, two fixed points of contact on the cone. For this involution $k=0$, the united points being the four points $(\pm 1, \pm 1, 1)$; further, on an arbitrary line there lies one pair of conjugate points, or $j=1$. Thus j is the genus of the curves in which the cone is met by the system of quadric surfaces, and $2j+1$ is the number of variable intersections of two of these curves.

Ex. 2. For the involution of sets of five points obtainable from one point (x, y) of the set by repetitions of the transformation

$$x'=y, \quad y'=(1-x-y)/(1-x),$$

(Ex. 3, p. 121), obtain a representative surface, shewing that this is rational, and contains a linear system of curves of negative canonical number.

Number of cyclical sets in a correspondence. We have noticed that particular involutions are obtainable by repetitions of a (single valued, algebraic) transformation which is periodic (as in Ex. 3, p. 121). More generally, now, consider a correspondence between pairs of points, of such character that the number of coincidences is finite; we can find a formula for the number of cyclical sets, to be defined, contained therein.

In the first place, for definiteness, consider such a correspondence as we have already considered (Chap. II, p. 92), of ∞^r pairs of points in space [r], in which, to any point P, corresponds a finite number of points Q, and, reversely, to Q, corresponds a finite number of points P. Denote the correspondence by T; assume the correspondence to be such that P can be at any point of the space; then, when P takes the position of any one of the points Q which correspond to the original P, there will correspond points R, which we may denote by $T(Q)$; as already explained (p. 6), the aggregate of all such points R, for all the points Q, or $T(P)$, may be denoted by $T^2(P)$, and regarded as arising from P by the forward correspondence T^2. Assume then, further, that the correspondence is such that the numbers of coincidences in all the successive correspondences $T, T^2, T^3, \ldots$ are finite. We have proved, then (p. 99), that the number of coincidences for the correspondence T is $g(T)$, where $g(T)=(0)(r)'+(1)(r-1)'+\ldots+(r-1)(1)'+(r)(0)'$, in which $(0)(r)'$ denotes the number of points Q which correspond to an arbitrary point P, similarly $(r)(0)'$ denotes the number of points P which correspond to an arbitrary point Q, in the reverse correspondence; and, more generally, $(i)(j)'$ means the number of pairs of corresponding points P, Q, of which P lies in an arbitrary space of i dimensions, $[i]$, while Q lies in an independent arbitrary space $[j]$, with $i+j=r$. The correspondence being supposed to be expressed by algebraic equations, the number $(i)(j)'$ is the order of the algebraic manifold described by all the points Q when P describes the linear space $[i]$. As a simple example, consider the correspondence in a plane, between points (x, y) and (ξ, η), which is determined by equations $x^p = \xi^r, y^q = \eta^s$, wherein p, q, r, s are positive integers; then, as (x, y), or P, describes an arbitrary line $lx+my+n=0$, the rs points (ξ, η), or Q, which correspond thereto, describe a curve whose equation is obtained by rationalising the equation $l\xi^f+m\eta^g+n=0$, where $f=rp^{-1}$ and $g=sq^{-1}$. Suppose now, with a slight change of notation, that the points Q arising from P, are denoted by $Q^{(1)}$; we may write $(Q^{(1)})=T(P)$; next that all the points $T(Q^{(1)})$, arising from the aggregate of points $Q^{(1)}$, are denoted by $Q^{(2)}$, so that we write $(Q^{(2)})=T^2(P)$, and so on. Finally, that among the points $(Q^{(\nu)})$ given by $T^\nu(P)$, there is one particular point $Q^{(\nu)}$ which coincides with P; this is then a coincidence of the correspondence T^ν, and there is a succession of $\nu+1$ particular points $P, Q^{(1)}, \ldots, Q^{(\nu)}$ such that $Q^{(1)}=T(P)$, $Q^{(2)}=T(Q^{(1)})$, $\ldots$, $Q^{(\nu)}=T(Q^{(\nu-1)})$; moreover, the ν first points of this set will be such that, starting from *any one* of them, we can come back to it by ν successive applications of the correspondence T. In saying this, it is supposed that no one of the points $Q^{(1)}, Q^{(2)}, \ldots, Q^{(\nu-1)}$ coincides with P. Such a set as $P, Q^{(1)}, \ldots, Q^{(\nu-1)}$ is then called here a cyclical set. The result to which we proceed gives the number of *cyclical* sets which exist for a given value of ν, supposing the numbers of co-incidences $g(T)$, $g(T^2)$, $\ldots$, for the various powers of T, to be known. For example, if T be a Cremona transformation in a plane, of order n, that is, a $(1, 1)$ correspondence in which the point Q describes a curve of order n when P describes a line, then, in the most general case, the correspondence T^2 is of order n^2; and, by the formula $(0)(2)'+(1)(1)'+(2)(0)'$, the number of coincidences for this correspondence T^2 is $1+n^2+1$. This, however, includes the $1+n+1$ co-incidences for T itself. Therefore the number of coincidences *proper* to T^2, is n^2-n; and hence the number of cyclical sets of 2 points belonging to T is $\frac{1}{2}(n^2-n)$.

In general, among the $g(T^\nu)$ coincidences of T^ν will be found. not only the coincidences of T itself, but also the points of any existing cyclical set of d points, of which d is a divisor of ν; in particular the points of a cyclical set of ν points. Let γ_i denote the number of cyclical sets of i points; any such set gives i coincidences. Hence, if we put $p(T^i)$ for $i\gamma_i$, we have the equation

$$g(T^\nu)=p(T)+\Sigma p(T^d)+p(T^\nu),$$

wherein d denotes in turn every divisor of ν other than unity and ν itself.

The equations of this form, for successive values of ν, enable us to express $p(T^\nu)$ in terms of functions $g(T^i)$. We have in fact the theorem (*Proc. Lond. Math. Soc.* XXI, 1891, p. 30): If $f(a_1, \ldots, a_p)$ be any number uniquely determinate from the numbers $a_1, \ldots, a_p$, and symmetrically dependent thereon, and we put

$$F(a_1, \ldots, a_m)=f(a_1, \ldots, a_m)+\underset{\mu}{\Sigma\Sigma} f(a_i, a_j, \ldots, a_k)+\Sigma f(a_s)+f(0),$$

where $a_i, a_j, \ldots, a_k$ are in turn every $m-\mu$ of $a_1, \ldots, a_m$, and $f(0)$ is arbitrary, then, conversely,

$$F(a_1, \ldots, a_n)=F(a_1, \ldots, a_n)+\underset{\mu}{\Sigma\Sigma}(-1)^\mu F(a_p, a_q, \ldots, a_r)$$
$$+(-1)^{n-1}\Sigma F(a_s)+(-1)^n F(0),$$

where $F(0)=f(0)$, and $p, q, \ldots, r$ are in turn every selection of $n-\mu$ of $a_1, \ldots, a_n$. As a consequence of this: let $f(n)$, $F(n)$ be two determinate functions of the positive integer n, such that

$$F(n)=f(1)+\Sigma f(d)+f(n),$$

where d denotes in turn all the divisors of n, other than unity and n itself; also, let the expression of a positive integer $\nu(>1)$, in terms of its prime factors (>1), be $\nu=p_1{}^{\alpha_1}p_2{}^{\alpha_2}\dots p_s{}^{\alpha_s}$; then we have

$$f(\nu)=F(\nu)-\sum_{i=1}^{s}F(\nu/p_i)+\sum_{i\neq j}F(\nu/p_ip_j)+\dots+(-1)^sF(\nu/p_1\dots p_s);$$

cf. Dedekind, *Crelle*, LIV, 1857, p. 25; Moebius, *Werke*, IV, 1887, p. 591. Hence, we have the result

$$p(T^\nu)=g(T^\nu)-\Sigma g(T^{\nu_i})+\Sigma g(T^{\nu_{i,j}})-\dots+(-1)^sg(T^{\nu_{1,2,\dots,s}}),$$

where ν_i has all the values ν/p_i, $\nu_{i,j}$ has all the values ν/p_ip_j, ..., $\nu_{1,2,\dots,s}$ has all the values $\nu/p_1\dots p_s$.

We make application of this to the case of the correspondence of ∞^r pairs of points in space $[r]$, referred to, under the hypotheses we have stated. On the assumption, already made, that the number of coincidences in the correspondence T^ν is finite, this number, $g(T^\nu)$ is, in fact, given by

$$g(T^\nu)=[(0)(r)']^\nu+[(1)(r-1)']^\nu+\dots+[(r-1)(1)']^\nu+[(r)(0)']^\nu.$$

For, if the locus of points Q, corresponding to P in the correspondence T, when P describes a linear space $[i]$, be of order k, namely $(i)(r-1)'$, then, when P describes a manifold, $M_i{}^m$, also of dimension i, but of order m, the points Q will describe a locus of order mk. Whence, considering the sequence of corresponding points, P, $Q^{(1)}$, $Q^{(2)}$, ..., before described, when P describes a space $[i]$, the point $Q^{(1)}$ will describe a locus of order k, thence $Q^{(2)}$ will describe a locus of order k^2, and so on, and finally $Q^{(\nu)}$ will describe a locus of order k^ν. We can then use this form for $g(T^\nu)$ in the formula for $p(T^\nu)$.

A particular case is that of a general Cremona transformation in a plane, for which $(0)(2)'=(2)(0)'=1$, and $(1)(1)'$ is the order of the transformation, say n. Then $g(T^\nu)=n^\nu+2$. But, when $\nu>1$, the addition to $g(T^\nu)$ of a number independent of ν does not affect the value of $p(T^\nu)$, since, for $s>0$, the sum, for $i=0,\dots,s$, of the terms $(-1)^i(s,i)$, in which (s,i) is the binomial coefficient, is equal only to $(1-1)^s$ or zero. Hence, in general, for $\nu>1$,

$$p(T^\nu)=n^\nu-\Sigma n^{\nu_i}+\Sigma n^{\nu_{i,j}}-\dots+(-1)^sn^{\nu_{1,\dots,s}},$$

where $\nu_i=\nu/p_i$, $\nu_{i,j}=\nu/p_ip_j$, ..., $\nu_{1,\dots,s}=\nu/p_1\dots p_s$. And the number of cyclical sets is $p(T^\nu)/\nu$. Thus the number of cyclical sets of two points is $\frac{1}{2}(n^2-n)$; and, for example, is 1 for a quadratic Cremona transformation. If this be expressed by $ux'=vy'=wz'$, where u,v,w are linear in x,y,z, such a cyclical pair must also satisfy $u'x=v'y=w'z$; the elimination of x',y',z', from these, leads to two cubic curves having in common the three exceptional points $v=0=w$, $w=0=u$, $u=0=v$, and the four coincidences of the original transformation given by $ux = vy = wz$; the two remaining intersections constitute the cyclical pair. For cyclical sets of three points the number is $\frac{1}{3}(n^3 - n)$, which is 2 for a general quadratic transformation. More generally, in space $[r]$, with coordinates $(x_0, \ldots, x_r)$, let $y_0, \ldots, y_r$ denote general (homogeneous) linear functions of $x_0, \ldots, x_r$, and $y_0', \ldots, y_r'$ the same linear functions of $x_0', \ldots, x_r'$; we may consider the correspondence between $(x_0, \ldots, x_r)$ and $(x_0', \ldots, x_r')$ given by the equations

$$y_0' = x_0^{-1}, \; y_1' = x_1^{-1}, \; \ldots, \; y_r' = x_r^{-1}.$$

When $(x_0', \ldots, x_r')$, and therefore also a point $(y_0', \ldots, y_r')$, describes a linear space of dimension $r - s$, the point $(x_0, \ldots, x_r)$ will describe a manifold of $r - s$ dimensions given as the intersection of s primals each given by an equation of the form

$$a_{i0}x_0^{-1} + a_{i1}x_1^{-1} + \ldots + a_{ir}x_r^{-1} = 0, \qquad (i = 1, \ldots, s).$$

It can be shewn that this manifold is of order (r, s), that is $r!/s!(r-s)!$, having the vertices of the fundamental simplex as multiple points of order $(r-1, s)$, the edges of this simplex as multiple lines of order $(r-2, s)$, and so on, in general the bounding spaces $[k]$ of the simplex multiple of order $(r-1-k, s)$, the bounding spaces $[r-s-1]$ being simple thereon. Thus, in this case we have $g(T^\nu) = 1 + (r, 1)^\nu + (r, 2)^\nu + \ldots + (r, 1)^\nu + 1$. In particular, in ordinary space, we have $g(T^\nu) = 2 + 2 \cdot 3^\nu$, and the number of cyclic sets, of ν points, is $2^{\nu-1}[3^\nu - \Sigma \cdot 3^{\nu_i} + \Sigma \cdot 3^{\nu_{i,j}} - \ldots + (-1)^s 3^{\nu_{1,\ldots,s}}]$. This result was given by S. Kantor, *Compt. rend.* xc, 1880, pp. 1156–8, with proof. The result he states without proof for $r > 3$ is, however, not obtained by the formula given above.

Another application of the ideas developed here is to a *general* correspondence of two points on a curve of genus p, for which the indices are α, β and the valency is γ. In this case

$$g(T^\nu) = \alpha^\nu + \beta^\nu - 2p(-\gamma)^\nu$$

(Severi, *Mem....Torino*, LI, 1902, p. 88), under the same hypothesis as before, that the number of coincidences of T^ν is finite; an exception arises, for example, if, among the points corresponding to Q, where $Q = T(P)$, the point P occurs for every P, in which case the number of cyclical sets of two points is infinite.

Note in regard to surfaces containing a pencil of rational curves. In connexion with the proof that the surface representing a plane involution is rational, it is interesting to bear in mind the theorem that a surface having a pencil of rational curves is either rational, when the pencil is rational, or is birationally equivalent to a ruled surface whose genus is the same as that of the pencil of curves. We remark upon the proof of this theorem.

Lemma. Upon a rational curve, in space $[r]$, of order n, there exists a linear series of sets of $n-2$ points, of freedom $n-2$, a g_{n-2}^{n-2}, simple and without fixed point, which is determinate without any irrationality, when the curve is given. For a rational plane curve of order n, with multiple points, typically i-fold, supposed resolved in Noether's manner, we may assume, (a), that

$$\Sigma i(i-1) = (n-1)(n-2);$$

(b), that the number of independent conditions to be satisfied at the multiple points by an adjoint curve is not greater than $\frac{1}{2}\Sigma i(i-1)$; (c), that an adjoint curve, as involving symmetrically the whole set of multiple points, is determinable rationally. Thence, considering adjoint curves of order $n-2$, these determine a series of sets of each $n(n-2) - \Sigma i(i-1)$ or $n-2$ points, with freedom at least $\frac{1}{2}(n-2)(n+1) - \frac{1}{2}(n-1)(n-2)$, or $n-2$; and, therefore, with exactly this freedom (so that the multiple points furnish independent conditions for the adjoint curve). This is the series g_{n-2}^{n-2} in question.

For a rational curve of order n, in space $[r]$, we may assign rationally a space $[r-3]$; there are then the equivalent of $\frac{1}{2}(n-1)(n-2)$ chords of the curve which meet this space, because the curve can be projected from this space $[r-3]$ into a plane curve of order n with the equivalent of $\frac{1}{2}(n-1)(n-2)$ double points. We may then consider the most general cone of order $(n-2)$, having this space $[r-3]$ as vertex, described to pass through these $\frac{1}{2}(n-1)(n-2)$ chords. Thereby we obtain a g_{n-2}^{n-2} upon the curve. This series is evidently simple and without fixed points, because its freedom is equal to its grade.

Application of the lemma. Provided $n-2 \geqslant 2$, the series can be used (as in Vol. v, Chap. i, p. 7) to make a $(1, 1)$ birational transformation of the curve to one of order $n-2$; and this process can be continued until the given curve is transformed either into a rational cubic curve, or into a conic. Thus we have the inference: The expression of a rational curve in terms of a variable parameter involves no irrationality in the coefficients of the equations which determine the curve if we assume the coordinates of a single point upon the curve to be given, when necessary (as in the case of a conic).

Surfaces with a rational pencil of rational curves. That such surfaces are rational was proved by Noether (*Math. Ann.* iii, 1870, p. 161; wherein the preceding lemma for plane curves is given). His proof consists in establishing that there is upon the surface a rational curve, determinable rationally, which does not belong to the pencil of curves, but meets every curve of the pencil in one point. The points of the surface can then be expressed rationally (and reversibly) by two parameters, one determining the position of the point upon a curve of the pencil, the other determining the position of the intersection of this curve with the rational director curve, upon this director curve.

To illustrate the argument consider the case of a surface of order n, in ordinary space, which has an $(n-2)$-fold line; this surface contains the linear pencil of conics obtained by planes through the axis. Its equation may be supposed of the form

$$z^2 a + 2zth + t^2 b + zA + tB + C = 0,$$

wherein a, h, b, A, B, C are homogeneous polynomials in x, y, the first three of order $n-2$, while A, B are of order $n-1$, and C is of order n. A provisional proof of the existence of a suitable rational director curve upon this surface, which can be confirmed by more detailed examination, is easily given. Let u, v be homogeneous polynomials in x, y of order $n-2$, and w a similar polynomial of order $n-3$. The equations $z = u/w$, $t = v/w$ then represent a rational curve of order $n-2$, having the axis $x = 0 = y$ as an $(n-3)$-fold chord. That this curve lies on the surface requires the identical vanishing of a homogeneous polynomial, in x, y, of order $3n-6$; and there are in u, v, w just $3n-5$ (non-homogeneous) disposable coefficients with which to effect this, by the solution of linear equations. With u, v, w so determined, put $y = 1$ for simplicity of statement, denote u/w by ζ and v/w by τ; and substitute, in the equation of the surface, $t = \tau + \theta(z - \zeta)$. Then, by use of

$$C = -[a\zeta^2 + 2h\zeta\tau + b\tau^2 + A\zeta + B\tau],$$

after removing the factor $z - \zeta$, it is possible to express z as a rational function of x and θ; hence t is also such a function, and the surface is proved to be rational.

Surfaces with an irrational pencil of rational curves. In accordance with a paper of Enriques (*Math. Ann.* LII, 1899) there exists also a rational director curve, cutting each curve of the pencil in one point, when this pencil of curves is not rational. The point of the surface is then expressible once more in terms of two parameters, one expressing the position upon a rational curve of the pencil, the other expressing (irrationally) the curve of the pencil on which the point is taken. This is the general character of a ruled surface. Reference may be made (as Enriques remarks) to Painlevé's Stockholm lectures on Differential Equations (1895, lithographed, Paris, 1897, pp. 255–341), where a surface is under consideration which is capable of birational transformation into itself by a finite continuous group of transformations. Upon this surface there are rational curves, the locus of a point under a one-parameter subgroup; and upon each such curve are two united points. The locus of one of these gives the unisecant curve required for Enriques' demonstration, whose existence then, in Painlevé's case, does not require the proof which is necessary in general.

CHAPTER IV

PRELIMINARY PROPERTIES OF SURFACES IN THREE AND FOUR DIMENSIONS

WE consider an algebraic surface of order n, in ordinary space, in which the coordinates are x, y, z, t; and certain properties of the tangent cone drawn to this surface from an arbitrary point. We suppose, unless the contrary be stated, that the multiple points of the surface consist at most of the points of a double curve, whereon are triple points of the surface, through which pass three sheets of the surface; so that such points are also triple points of the double curve, with tangents not generally lying in a plane. Reasons will be given later for the conclusion that any surface can be birationally transformed to a surface of this character, so that the supposition involves no essential limitation of generality in the surfaces dealt with. At the same time, it is often convenient to suppose that, beside the double curve, the surface considered has isolated multiple points; and this possibility will also be taken into account in some of the results given.

Some familiar facts may be recalled first, particularly for the case when there is no such double curve, the surface being expressed by a polynomial equation of complete generality in the coordinates. At a general point of the surface there is a plane, called the *tangent plane*, which is the locus of lines through the point which have two intersections with the surface coinciding at this point; such a line has then $n-2$ further intersections with the surface. Thus the tangent plane meets the surface in a curve having a double point at the point of contact, whereof the tangents, each meeting the surface in three points coinciding at the point, are in general distinct. These two lines are the generating lines, at this point, of the quadric surface—called the *polar quadric* of the point of contact (x_0, y_0, z_0, t_0)—whose equation is $(x \, \partial/\partial x_0)^2 f(x_0, y_0, z_0, t_0) = 0$, where $f(x, y, z, t) = 0$ is the equation of the surface, and $x \, \partial/\partial x_0$ stands for $x \, \partial/\partial x_0 + y \, \partial/\partial y_0 + z \, \partial/\partial z_0 + t \, \partial/\partial t_0$; this quadric surface touches the tangent plane of the given surface at the point (x_0, y_0, z_0, t_0). The two particular lines are called the *asymptotic* or the *inflexional* lines of the surface at this point. The statement is easily proved by considering the equation in λ, $f(\lambda x_0 + x, \ldots, \lambda t_0 + t) = 0$, which defines the intersections of the surface with the line joining the points (x_0, y_0, z_0, t_0), (x, y, z, t). Further, if (x', y', z', t') be another point of the surface, the line in which the tangent plane at (x_0, y_0, z_0, t_0) is met by the tangent plane at (x', y', z', t'), has for limit, as the

latter point approaches (x_0, y_0, z_0, t_0), a line through this, whose
direction is the harmonic conjugate, in regard to the inflexional
lines thereat, of the direction in which (x', y', z', t') approaches
(x_0, y_0, z_0, t_0). Two such lines, in the tangent plane at (x_0, y_0, z_0, t_0),
which are harmonic conjugates in regard to the inflexional lines,
are often called simply *conjugate* lines at the point. At a less
general point of the surface it may happen that the inflexional lines
coincide, the section by the tangent plane, at this point, having here
a cusp; then the polar quadric of this point, having two intersecting
generators coincident, must be a cone (and conversely). If the polar
quadric be written $a_{11}x^2 + \ldots + 2a_{12}xy + \ldots = 0$, the condition for this
is the vanishing of the determinant $|a_{ij}|$, of which each element is
a partial differential coefficient, of the second order, such as
$\partial^2 f/\partial x_0^2$ or $\partial^2 f/\partial x_0 \partial y_0$, if the point in question be (x_0, y_0, z_0, t_0).
Thus this point lies on a surface of order $4(n-2)$; this surface is
called the *Hessian* of the original; its intersection with the original
surface is called the *parabolic curve* on the surface; at a point of this
curve the tangent plane is said to have *stationary contact* with the
surface. We may speak of the line, in which the two inflexional
lines at a parabolic point coincide, as the *parabolic* line, to be dis-
tinguished from the tangent line of the parabolic curve at this point.
At a parabolic point one of two conjugate directions, defined as
above by a harmonic pencil, must be along the parabolic line; so
that the tangent planes, at *all* points (x', y', z', t') of the surface,
ultimately pass through this parabolic line as (x', y', z', t') approaches
the parabolic point.

Ex. Of the inflexional lines of the general surface $f = 0$, at points of
the curve in which this surface is met by another surface of order n',
there are $nn'(3n-4)$ which meet an arbitrary line, and $nn'(3n+2n'-8)$
which touch the common curve, in general.

We enquire now: (i), what is the number of the inflexional lines
of the surface which pass through an arbitrary point of space, remarking that, as there are ∞^2 inflexional lines and two conditions
are required for a line to contain a given point, we may expect a
definite number of such lines; and (ii), what is the number of tangent planes of the surface, having stationary contact with this,
which pass through an arbitrary point of space, remarking that
there are ∞^1 such tangent planes and one condition is required for
a plane to contain a given point?

As to the enquiry (i), in order that an inflexional line, at a point
(x, y, z, t) of the surface, may pass through a point (ξ, η, ζ, τ), this
last must lie both on the tangent plane and on the polar quadric of
(x, y, z, t). This point must then satisfy the two equations denoted
by $(\xi \partial/\partial x)f = 0$ and $(\xi \partial/\partial x)^2 f = 0$; these equations, when (x, y, z, t)
are regarded as current coordinates, which represent what are called

respectively the *first* and *second polars* of (ξ, η, ζ, τ) in regard to the surface, are of orders $n-1$ and $n-2$; the common curve of these polars thus meets $f=0$ in $n(n-1)(n-2)$ points. Conversely, when the surface has no multiple points, the conditions ensure that an inflexional line, at the point (x) of the surface, passes through the point (ξ), and the number of such lines is $n(n-1)(n-2)$. But, if the surface has a double curve, there is a difference; in this case, the equation $(\xi \partial/\partial x)f=0$ is satisfied, wherever (ξ) may be, for all points (x) of the double curve; and the enveloping cone, of tangent planes of the surface which pass through (ξ), breaks up into the cone joining (ξ) to the double curve, and another *proper* enveloping cone; the first polar of (ξ) contains the double curve, and its remaining intersection with the surface constitutes the *proper curve of contact* of the enveloping cone from (ξ); thus if the order of this proper curve of contact be denoted by μ_1, and the order of the double curve by ϵ_0, we have $n(n-1)=\mu_1+2\epsilon_0$. There is likewise change as to the intersection of the surface with the second polar of (ξ): Evidently, if the second polar of (ξ) pass through a point (x), then the polar quadric of (x) passes through (ξ). Now, at any point of the double curve, the polar quadric of this point breaks up into the two tangent planes of the surface at this point, as may be seen by taking particular coordinates with origin at this point, it being known that the polar surfaces are unaffected by linear transformation of the coordinates; and, of the ∞^1 tangent planes of the surface at points of the double curve, there will be a number passing through an arbitrary point (ξ). Thus, the second polar of (ξ) passes through as many points of the double curve. If the number of such points be denoted by ρ, it appears on examination that the number of intersections of the second polar of (ξ), with the proper curve of contact, at ordinary points of this, is lessened precisely by ρ; and hence that the number of points of the surface at which an inflexional tangent passes through (ξ) is $\mu_1(n-2)-\rho$. In general we denote the number of points of the surface at which an inflexional tangent line passes through an arbitrary point, (ξ), by κ. Thus, when there is a double curve, we have $\kappa=\mu_1(n-2)-\rho$. This will be proved below. For the case when there is no such curve we have shewn that $\kappa=n(n-1)(n-2)$.

Now, (ii), consider the points of the surface at which the tangent plane, from the arbitrary point (ξ), has stationary contact with the surface. At such a point consider the two lines, both in the tangent plane, one joining the point to (ξ), the other the tangent line of the curve of contact of tangent planes from (ξ) to the surface. By the harmonic property spoken of, one of these must coincide with the parabolic line at the point. It can be shewn (see the analysis below) that, in general, it is the tangent of the curve of contact which co-

incides with the parabolic line, the other possibility involving that
the curve of contact has a multiple point at the point considered.
We shall denote by i the number of tangent planes of the surface,
drawn from an arbitrary point (ξ), which have stationary contact
with the surface.

The two kinds of points of the surface we have considered, of
numbers respectively κ and i, correspond to characters of the enveloping cone drawn from the point (ξ). Take an arbitrary plane
section of this cone; we desire to make it clear that, in general, on
a generator of this enveloping cone which is an inflexional line of
the surface, there is a cusp of this plane section; and, on a generator containing a point of stationary contact, there is an inflexion
of this plane section; which thus has κ cusps and i points of inflexion. In fact, at a point (x) of the surface for which the joining
line (x, ξ) is an inflexional line, the curve of contact of the enveloping cone with the surface touches the line (x, ξ) at (x), and we
have the same generating line of the cone for two coinciding tangent
planes; while, at a point (x) of stationary contact, we have the
same tangent plane for two coinciding points of the curve of
contact.

Consider the two possibilities analytically. For the case where
an inflexional line passes through the point (ξ), take non-homogeneous coordinates x, y, z, with origin at the point of the surface,
and $z = 0$ for the tangent plane. The equation of the surface may be
represented by

$$z + xy + (xz, yz, z^2) + (x^3, x^2y, \ldots, z^3) + \ldots = 0,$$

the axes x, y being along the inflexional lines. The first polar of a
point $x = a, y = 0, z = 0$, on the inflexional line $z = 0, y = 0$, has an
equation of the form

$$a\{y + (z, 0, 0) + (x^2, xy, \ldots) + \ldots\} + (n-1)\,z + (n-2)\,xy + \ldots = 0;$$

this meets the tangent plane $z = 0$ in a curve given by

$$a\{y + (x^2, xy, \ldots) + \ldots\} + (n-2)\,xy + \ldots = 0,$$

of which the first approximation $(a \neq 0)$ is of the form $y - kx^2 = 0$.
This verifies what was said above, that, when an inflexional line,
at a point (x) of the surface, passes through an external point (ξ),
the curve of contact, of the tangent planes from (ξ) to the surface,
touches the line (x, ξ) at the point (x). Next, for the other case,
when a tangent plane to the surface from (ξ) has stationary contact
with the surface at a point (x), the equation of the surface, expressed
by non-homogeneous coordinates, with origin at the point of
contact, is of the form

$$z + \tfrac{1}{2}y^2 + (xz, yz, z^2) + (x^3, x^2y, \ldots) + \ldots = 0;$$

thus the first polar of a point $x = a$, $y = b$, $z = 0$ is given by an equation

$$a\{\dots (z, 0, 0) + (x^2, xy, \dots) + \dots\} + b\{y + (0, z, 0) + (x^2, \dots) + \dots\}$$
$$+ (n-1)\, z + \tfrac{1}{2}(n-2)\, y^2 + \dots = 0,$$

and the intersection of this with $z = 0$ is given by

$$a\{\dots(x^2, xy, \dots) + \dots\} + b\{y + (x^2, \dots) + \dots\} + \tfrac{1}{2}(n-2)\, y^2 + \dots = 0;$$

provided $b \neq 0$, that is, in the original description, provided the point (ξ) is not on the parabolic line at the point (x), this is also, to the first approximation, of the form $y - kx^2 = 0$; so that the curve of contact has the parabolic line for tangent. When $b = 0$, the curve of contact has a double (or higher) point, at the point of stationary contact.

We may consider the matter from another point of view. The sections of the surface by all possible planes through the external point (ξ), form a linear system of curves on the surface. This system is such that one curve goes through any two arbitrary points of the surface; it is given by the intersection of the surface with another system of surfaces, whose equation is of the form $\lambda \vartheta + \mu \phi + \nu \psi = 0$, in which $\vartheta = 0$, $\phi = 0$, $\psi = 0$ represent definite surfaces (in our case planes through (ξ)), and λ, μ, ν are variable parameters. We may consider the more general case when $\vartheta = 0$, $\phi = 0$, $\psi = 0$ are any surfaces of the same order; a system of curves on the surface so determined is called a *net of curves*. Of such a net of curves we may consider the *Jacobian curve*; this is defined as the locus of points of the surface at which a curve of the net has a double point. For an ordinary point of the surface $f = 0$, the condition for this is that the surface should have the same tangent plane as the surface $\lambda \vartheta + \mu \phi + \nu \psi = 0$; the Jacobian is then part (or the whole) of the intersection of $f = 0$ with the surface expressed by the vanishing of the four-rowed determinant $\partial(\vartheta, \phi, \psi, f)/\partial(x, y, z, t)$. The curves of the net which pass through a point of the Jacobian curve, this being supposed to be an ordinary point of $f = 0$, form a pencil of curves, obtainable by the intersection of $f = 0$ with surfaces whose equation is of the form $pU + qV = 0$, in which p, q are variable, but U and V are definite; we may suppose that $U = 0$ gives the curve of the net which has a double point at the point considered. Thus the point is such that all curves of the net which pass through it have the same tangent line thereat, that of the curve given by $V = 0$. For the sake of comparison with the simple case considered above, we may describe this common tangent, at the point (x) of the Jacobian curve, as the (x, ξ) line; in that particular case, the locus of double points of plane sections of $f = 0$ made by planes through the external point (ξ), is the curve of contact of tangent planes of the surface drawn from (ξ). As in that case, we may prove in general that the

(x, ξ) line is the harmonic conjugate of the tangent line of the Jacobian curve, C_J, with respect to the tangent lines of the curve, C_0, of the net which has a double point at (x). We return to this general point of view later.

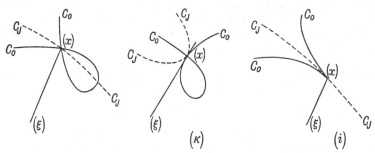

(κ) (i)

Consider now in further detail the case when the net of curves is formed by the plane sections of the surface whose planes contain an external point (ξ); and suppose the surface has no double curve, or other multiple points. The enveloping cone has then an order, N, equal to $n(n-1)$, the same as the order of the curve of contact; this curve is now the complete intersection of the surface with the first polar of (ξ), which is a surface of order $n-1$. The *class* of the enveloping cone, defined by the number of its tangent planes which contain an arbitrary line through its vertex, and thus equal to the number of tangent planes of the *surface* through this line, is also equal to the number of intersections, on the surface, of the two curves of contact of tangent planes to the surface drawn from any two points of this line. It is thus given by $N' = n(n-1)^2$. These numbers N, N' are respectively the order and class of any plane section of the enveloping cone. This plane section has also, we have proved, a number K of cusps given by $K = n(n-1)(n-2)$. Thus we have $N' - N = K$, a relation not generally true for a plane curve. If, further, this plane section have D double points, T double tangents, and I inflexions, and no other singularities, we can employ the formulae of Plücker

$$N' = N^2 - N - 2D - 3K, \quad I - K = 3(N' - N),$$
$$2(T - D) = (N' - N)(N' + N - 9),$$

and thence deduce

$$I = 4K = 4n(n-1)(n-2), \quad D = \tfrac{1}{2}n(n-1)(n-2)(n-3),$$

together with

$$T = \tfrac{1}{2}n(n-1)(n-2)(n^3 - n^2 + n - 12).$$

Of these, the number I, by what we have said, will be the number of tangent planes with stationary contact which pass through an arbitrary external point (ξ); but an independent proof of this arises from the form of I itself, which is the number of intersections of three surfaces, namely the given surface, of order n; the Hessian of this, of order $4(n-2)$; and the first polar of (ξ), of order $n-1$. The number D will be the number of tangent lines from (ξ) which touch the surface in two distinct points. The number T will be the number of tangent planes from (ξ) which have two distinct points of contact with the surface; for instance, in the case of a cubic surface, $T=27$, a plane from (ξ) to any one of the 27 lines of the surface having, we know, two distinct points of contact with the surface, lying on this line. The numbers D, K are such that

$$D+K=\tfrac{1}{2}n(n-1)^2(n-2);$$

we also obtain this equation directly, in an independent way, thus: If we denote $(x\,\partial/\partial\xi)^r f(\xi)/r!$ by f_r, this being of order r in (x, y, z, t), the result of the substitution of $(\xi+\lambda x, \eta+\lambda y, \ldots)$ for $(x, y, \ldots)$, in $f(x, y, z, t)=0$, is $f_0+\lambda f_1+\lambda^2 f_2+\ldots+\lambda^n f_n=0$; this determines the intersections with $f=0$ of the line joining the points (ξ), (x). In order that this line may touch the surface, the two first derivatives of this equation in λ must have a common root; for the line to have two distinct contacts with the surface these derivatives must have two different common roots, and for the line to have three coincident intersections with the surface, these derivatives must have two coincident common roots. To express this, we form (as in Vol. v, Chap. viii) a determinant of $2(n-1)$ rows and columns, of which the elements of the first $n-1$ rows are, beside zeros, the coefficients, $nf_0, (n-1)f_1, (n-2)f_2, \ldots, f_{n-1}$, of one derivative of the λ equation, beginning respectively in the first, second, $\ldots$, $(n-1)$th place; and the elements of the last $n-1$ rows are, beside zeros, the coefficients, $f_1, 2f_2, 3f_3, \ldots, nf_n$, of the other derivative, beginning likewise in the first, second, $\ldots$, $(n-1)$th place of these rows, respectively. The vanishing of this determinant expresses that the λ equation has a repeated root; the terms in the expansion of this determinant are however isobaric in the suffixes, and thus of equal order in (x, y, z, t); this order is then equal to that of the single term $(nf_0.nf_n)^{n-1}$ in this expansion, and so is $n(n-1)$. This gives the equation of the enveloping cone to the surface drawn from (ξ). To form the condition that the first derivatives should have two common roots, we form from this determinant another, of $2(n-2)$ rows and columns, which can be obtained from this one by leaving out the $(n-1)$th and the $(2n-2)$th rows of it, as well as the last two columns. The new determinant has, in its expansion, the term $(nf_0.(n-1)f_{n-1})^{n-2}$, and is of order $(n-1)(n-2)$ in x, y, z, t. The vanishing of this gives the cone containing both the double and the cuspidal edges of the previous enveloping cone drawn from (ξ).

Hence we have $2(D+K)=n(n-1).(n-1)(n-2)$, as was deduced. We may likewise obtain a direct proof of the value of D, by supposing (ξ) to be on the surface, and considering the lines drawn tangent to the surface at this point. Then we consider the λ equation, obtained from the former by replacing f_0 and f_1 by zero, $f_2+\lambda f_3+\ldots+\lambda^{n-2}f_n=0$. For the tangent line (x, ξ) to have further contact with the surface, the two first derivatives of this must have a common root; the determinant whose vanishing expresses this fact contains the term $(f_2.f_n)^{n-3}$, and is of order $(n+2)(n-3)$ in (x), and of order $(n-2)(n-3)$ in (ξ). By this latter fact, the point of contact (ξ), when (x) is a given point on the double tangent of the surface, lies on a surface of order $(n-2)(n-3)$, beside lying on $f_0=0, f_1=0$, which are of respective orders $n, n-1$ in (ξ); as the double tangent contains two such points, the number of such tangents is $\tfrac{1}{2}n(n-1)(n-2)(n-3)$; this proves independently the value of D deduced above. That the determinant is of order $(n+2)(n-3)$ in (x) expresses that this is the number of tangents which can be drawn to a general plane curve of order n from a double point of the curve.

We may also consider the genus, π, of the plane section of the enveloping cone. By the Plücker formulae this is given by $2\pi-2=n(n-1)(2n-5)$; and, by the correspondence between the generators of the cone, and the points of the curve of contact, π is also the genus of this curve. We may find this genus, also, from the general formula (Vol. v, Chap. viii) for the genus of the complete curve of intersection of two surfaces, namely, here, the original surface and the first polar of the external point (ξ). Incidentally, we thus have an independent determination of $D+K$, and thence of D when K is found. By the same theory, any canonical set of points on the curve of contact lies on a surface of order $2n-5$; and a canonical set of points is obtained by taking a canonical set of generators of the enveloping cone, which can be found by the intersection of this cone with a cone of order $n(n-1)-3$ drawn through its double and cuspidal generators.

Ex. 1. A very simple example is the case of an ordinary cubic surface, to which no double tangent lines can be drawn from an arbitrary external point (ξ). The curve of contact is of order 6 and genus 4, and has 6 tangent lines which pass through the vertex (ξ) of the enveloping cone, these being cuspidal generators of the cone. The points of contact of these tangents lie on a plane, the second polar of the vertex (ξ), and on a conic. A cubic cone, put through these 6 tangents, meets the curve of contact in 6 other points, and these lie in a plane; in particular there is a cubic cone meeting the enveloping cone only along its 6 cuspidal generators.

Ex. 2. By the formulae given above (Vol. v, Chap. viii), for the curve which is the complete intersection of a general surface of order n, and the first polar of an arbitrary external point, of order $n-1$, the rank, r; the number, h, of chords of the curve from an arbitrary point; the class,

n'; and the number, ν', of planes containing two tangent lines of the curve which pass through an arbitrary point, are given, respectively, by

$$r = n(n-1)(2n-3), \quad h = \tfrac{1}{2}n(n-1)^2(n-2), \quad n' = 6n(n-1)(n-2),$$

$$\nu' = \tfrac{1}{2}n(n-1)(2n-3)\{n(n-1)(2n-3)-10\} + 4n(n-1).$$

If a curve in space be projected from a point at which χ tangents of the curve intersect, this curve having no double points, or cusps, or inflexions, or double tangents, prove that, *for the resulting plane curve,*

$$N' = r - \chi, \quad D = h - \chi, \quad K = \chi, \quad T = \nu' - (r-4)\chi + \tfrac{1}{2}\chi(\chi-1), \quad I = n' - 2\chi.$$

Shew that, when $\chi = n(n-1)(n-2)$, the values for N', D, K, T, I which are found in the text for the curve of contact, result from these two sets of formulae.

Ex. 3. For a cubic surface, $f_x{}^3 = 0$, with the notation $u_0 = f_\xi{}^3$, $u_1 = f_\xi{}^2 f_x$, $u_2 = f_\xi f_x{}^2$, $u_3 = f_x{}^3$, prove that the enveloping cone from the point (ξ) has the equation $(u_0{}^2 u_3 - 3u_0 u_1 u_2 + 2u_1{}^3)^2 + 4(u_0 u_2 - u_1{}^2)^3 = 0$.

Ex. 4. If u_r be a homogeneous polynomial in x, y, z, of order r, the discriminant in regard to θ of the form $u_0 \theta^n + u_1 \theta^{n-1} + \ldots + u_n = 0$, represents, when equated to zero, with x, y, z as coordinates, a plane curve of order $n(n-1)$, with $\tfrac{1}{2}n(n-1)(n-2)(n-3)$ double points, and $n(n-1)(n-2)$ cusps lying on a curve of order $(n-1)(n-2)$.

Ex. 5. If a plane sextic curve have 6 nodes which lie on a conic, then the sextic is touched by another conic in 6 points which lie on a cubic curve through the nodes (Rohn, *Math. Ann.* xxv, 1885, p. 598). Any plane sextic curve with 6 nodes can be birationally transformed to a sextic curve with 6 nodes which lie on a conic.

Ex. 6. In general, the Hessian of a surface contains the points of the surface for which the polar quadric is a cone, and the points at which the inflexional lines coincide. A ruled surface (see below, Ex. 7, p. 165) has generators, called *torsal* generators, each with the property that the tangent plane of the surface is the same at every point of the generator; and the ruled surface has a double curve. Salmon has proved that the Hessian of a ruled surface meets the surface only in its torsal generators and its double curve, having the former as double lines, and the latter as a four-fold curve. For example, the Hessian of the ruled cubic surface $x^2 z + y^2 t = 0$ is $x^2 t^2 = 0$.

Ex. 7. A related theorem, for a surface of order n, containing a line, is that all the intersections of the Hessian with this line are contacts, so that their number is $2(n-2)$. If the equation of the surface be $x\phi + y\psi = 0$, the tangent plane of the surface at any point $(0, 0, z_1, t_1)$ of the line is given by $xu_1 + yv_1 = 0$, where u_1, v_1 are the polynomials, homogeneous of order $n-1$ in z_1, t_1, obtained by substituting $(0, 0, z_1, t_1)$ in ϕ, ψ; thus any plane containing the line $x = 0 = y$ touches the surface in $n-1$ points, where the section of the surface by this plane meets the line. And the points of the line whereat the second inflexional line of the surface coincides with this (inflexional) line, are the points where this plane section, for a suitably chosen plane, touches the line. These points arise for values of z_1/t_1 satisfying the two equations $x\,\partial u_1/\partial z_1 + y\,\partial v_1/\partial z_1 = 0$, $x\,\partial u_1/\partial t_1 + y\,\partial v_1/\partial t_1 = 0$, and are $2(n-2)$ in number. Only in these points can the Hessian meet the line; so that all its intersections are contacts. The argument assumes that the tangent plane of the surface is not the same at all points of the line. When this is so, the equation of the surface may be taken to be $x\phi + y^2 \vartheta = 0$; and it can be verified that the line lies wholly on the Hessian, which has, further, the tangent plane of the surface for its tangent plane at every point of the line (Salmon).

The surface with a double curve. In what now follows we deal in some detail with surfaces having a double curve, but with no other multiple points beside triple points, which are triple points for this double curve, at which three sheets of the surface meet. This apparent limitation is suggested by the following considerations: It has been proved (Vol. v, Chap. II) that any curve may be changed, by birational correspondence, to one with no singular points, lying possibly in space of higher dimension than that of the original curve. It is believed, similarly (see the references given below) that any surface may be obtained, by birational transformation, from one without multiple points; and that such a representative non-singular surface may be supposed to be in space of five (or less) dimensions. If such a surface, in space [5], be projected into space [4], the resulting surface will also have no multiple points (for general projection) other than double points of a particular kind, at which two sheets of the surface cross with only point-intersection, like the intersection of two planes in this space. There are in fact ∞^4 chords of a surface, and, for a line in space [5] to pass through a given point, 4 conditions must be satisfied. Thus, in the general projection of a non-singular surface in [5], there will be a finite number of chords of the surface passing through the centre of projection, and these will give rise to the double points spoken of, on the surface in [4]. Such double points, of a surface in [4], with, in general, only point-intersection, at which there are two distinct tangent planes of the surface, will be spoken of as *accidental* double points. They have also been called *improper* double points*, for a reason which will appear. If, now, a surface in [4], with no other singularity than such accidental double points, be projected, from a general point of its space, into space [3], we shall obtain therein a surface with a double curve. For, now, though the surface still has ∞^4 chords, only 3 conditions are necessary for a line to pass through a given point. Thus, through an arbitrary point of the space [4], there pass ∞^1 chords of the surface, forming a conical sheet, whose intersection with the space [3] will give the double curve of the projected surface in that space. As two conditions are necessary for a plane, in space [4], to pass through a given point, a certain number of the ∞^2 tangent planes, of the surface in [4], will pass through an arbitrary point; the line from this point to the point of contact of such a plane is the limit of chords of the surface;

---

* Severi, "Intorno ai punti doppi impropri di una superficie generale dello spazio a quattro dimensioni...", *Rend. Palermo*, xv, 1901. When the surface in space [5] is the Veronese surface, of order 4 (whose general point has co-ordinates ξ^2, η^2, ζ^2, $\eta\zeta$, $\zeta\xi$, $\xi\eta$, where ξ, η, ζ are parameters), no chord of the surface can be drawn from an *arbitrary* point, and the projected surface in [4] has no double point. But this surface, and cones, are the only surfaces.in space [5] for which this exception arises.

158	*Chapter IV*

thus, on the double curve on the surface in [3], there will be a certain number of points at which the two tangent planes of the surface are not different; these are called (after Cayley) the *pinch points* of the double curve (Cayley, *Papers*, VI, p. 123, of date 1868). There will also in general be a finite number of triple chords of the surface in [4], passing through an arbitrary point; these give rise to triple points of the surface in [3], triple both for the surface and the double curve. The accidental double points, of the surface in [4], will project into points of the double curve, of the surface in [3], which will be ordinary points of this curve. Thus plane sections of the surface in [3] will all have the same genus; and prime sections of the surface in [4], made by linear spaces of dimension 3 (primes), will be curves of the same genus; in particular the genus of a prime section through an accidental double point is the same as of a general prime section. This was the reason for the name *improper*, applied to these double points*.

The formulae which we proceed to give for surfaces in ordinary space [3], with only a double curve and triple points thereon, were originally given by Salmon (cf. Cayley, *Papers*, VI, p. 329), in connexion with investigations in regard to reciprocal surfaces, and on the hypothesis of the existence of a cuspidal curve as well as a double curve. The proof given by Salmon, in part tentative, will be found in his *Solid Geometry* (1882, p. 581), and is indicated below. The formulae were revised, and extended in great detail, by Zeuthen, *Math. Ann.* x, 1876, pp. 446–546. The order of the surface we denote most often by μ_0 (instead of n). The enveloping cone to the surface from an arbitrary point will break up into the cone standing on the double curve, and a *proper cone of contact*, as has already been said; the order of the proper cone of contact, which is also the order of the proper curve of contact, and the class of a general plane section of the surface, we denote by μ_1. The class of the proper cone of contact, which is also the class of an arbitrary plane section of this cone, and is also the number of tangent planes of the surface passing through an arbitrary line (the *class* of the surface), we denote by μ_2. As in the earlier part of this chapter, we denote by κ the number of inflexional lines of the surface which pass through an arbitrary point, this being in general the number of cusps of an arbitrary plane section of the proper enveloping cone. And we denote by i the number of tangent planes of the surface

* For the reduction of a surface to one without multiple points, the following references may be of use: (1), Segre, *Ann. d. Mat.* xxv, 1897; also, *Rend. Palermo*, xxx, 1910, and *Atti...Torino*, xxxvi, 1901, p. 635; (2), Beppo Levi, *Atti...Torino*, xxxiii, 1897, p. 66; also, *Ann. d. Mat.* xxvi, 1897, and *Torino Mem.* xlviii, 1898, and *Rend. Lincei*, vii, 1898; (3), O. Chisini, *Bologna Mem.* viii, 1921; (4), Albanese, *Rend. Palermo*, xlviii, 1924; also *Rend. Lincei*, xxxiii, 1924, p. 13, and Severi, *Atti...Veneto*, lxxix, 1920, p. 929.

having stationary contact, which pass through an arbitrary point; this is in general, as we have seen, the number of inflexions of an arbitrary plane section of the proper enveloping cone. The genus of a general plane section of the surface will often be denoted by p, and the genus of the proper curve of contact of the enveloping cone, from an arbitrary point, by π. The order of the double curve we denote by ϵ_0, the rank of this curve by ϵ_1, the number of its triple points by t, and the number of its pinch points by ν_2. As an auxiliary number, we use ρ for the class of the developable surface formed by the tangent planes of the surface at the points of the double curve, namely the number of such tangent planes passing through an arbitrary point. The surface is supposed to be irreducible; but evidently the corresponding characters can be formed for a composite surface by addition of the characters for the component surfaces, if a proper correction be introduced for the intersections of the components.

With these notations we have the following five equations:

(I), $\mu_0(\mu_0-1)=\mu_1+2\epsilon_0$;　　(II), $\mu_1(\mu_0-2)=\kappa+\rho$;

(III), $\epsilon_0(\mu_0-2)=\rho+3t$;　　(IV), $2\rho-2\epsilon_1=\nu_2$;

(V), $\mu_2+2\nu_2=\mu_1+\kappa$,

of which we proceed to give proofs. The equation (I) is obvious by considering the curve of intersection of the surface with the first polar of an arbitrary point; this is of order μ_0-1, and contains the double curve. It is obtainable also by the formula for the class of a plane section of the surface, namely $\mu_1=2\mu_0+2p-2$, compared with the definition of p, given by $2p-2=\mu_0(\mu_0-3)-2\epsilon_0$. The equation (II) expresses that the number of intersections of the proper curve of contact, of tangent planes of the surface from an arbitrary point O, with the second polar of this point O, which is of order μ_0-2, is equal to the sum of: (a), the number, κ, of points of the surface at which an inflexional tangent line passes through O —and this is justified just as in the case discussed earlier when the surface has no double curve—and, (b), the number of points of the double curve for which one of the two tangent planes of the surface passes through O. In regard to this, the point in question is a point of contact of a tangent plane at O and thus lies on the proper curve of contact; but it is also on the second polar of O, because O is on the polar quadric of the point, which breaks up into the two tangent planes of the surface at this point. The matter is made clearer by considering the surface as arising by projection from a surface in space [4]. Suppose that this surface is projected from a point L, and that l is a line through L meeting the space [3], on to which the projection is made, in a point O: of the ∞^2 tangent planes of the surface in [4], there will be ∞^1 which meet the line l, and their points of contact will describe a curve, say σ, on this surface. The projection of the curve σ is the proper curve of contact, of tangent planes of the projected surface in [3], which pass through O. If a point P, of the curve σ, be one of the two intersections, with the surface in [4], of a chord of this surface from L, and the tangent plane of this surface at P meet the line l in a point other than L, we obtain, on projection, a point of the double curve of the surface in [3], at which one of the two tangent planes passes through O, this point being on the proper curve of contact of tangent planes from O; but, if the tangent plane of the surface in [4], at the point P, passes through L, then P projects into a pinch point of the double curve in space [3], and the tangent plane of the surface in [3] does not, as a rule, pass through O. Thus the pinch points of the surface in [3] lie on the proper curve of contact of tangent planes from every point O (a fact we shall need below), but are in general different from the points of the double curve contributing to ρ. The formula (II) states that the only ordinary points of the surface in [3], for which the tangent plane and the polar quadric both pass through O, are the (κ) points; and the only multiple points are the (ρ) points; we assume that this is so. More detailed consideration of the behaviour of a first polar at a pinch point is found in Note I, at the end of this chapter. Coming now to equation (III), this expresses that the intersections of the second polar, of the point O, with the double curve, consist of the (ρ) points just considered, and the triple points, each counted triply. That these latter points are on the second polar follows because the polar quadric, of a triple point of the surface, vanishes identically. It appears clear that in general there are no other than these two kinds of intersection. Further justification of the equation arises below, Chap. VI, p. 254. The equation (IV) is in fact only the dual of a theorem which has been obtained previously (Chap. I, p. 25), that an algebraic curve of order ν, lying on a ruled surface of order n (in the present application a developable surface), which meets every generator in 2 points, touches $2\nu - 2n$ generators. The dual statement is that, for a developable surface of class ν (i.e. having ν tangent planes through an arbitrary point), which is such that two of its tangent planes pass through every tangent line of a certain curve of rank n, it arises $2\nu - 2n$ times that the two tangent planes of the developable surface, through a tangent line of the curve, coincide with one another. In the application to be made, the developable surface is that formed by the tangent planes of the given surface at the points of its double curve (so that ν is to be replaced by ρ), and the curve is the double curve (so that n is to be replaced by ϵ_1). Thus $2\rho - 2\epsilon_1$ is the number, ν_2, of the pinch points of the given surface. Equation (V) may be obtained by an easy generalisation of a result previously found

Surfaces, preliminary properties   161

(Vol. v, p. 218); this result was that, if two surfaces of orders m, m', both passing through a curve C, of genus π, have for their remaining intersection two curves C_1, C_2, simple on both surfaces, then a surface of order $m+m'-4$ drawn through C_1 and C_2, and the double points of C, if such exist, cuts on C, other than at the inter- sections of C with C_1 and C_2 and the double points of C, a canonical set of $2\pi-2$ points. The generalisation we require is that this re- mains true when the curves C_1 and C_2 coincide in a double curve of the surface of order m, the surface of order $m+m'-4$ having then this curve as a double curve. This is in fact a particular case of a theorem, due to Noether, given below. In the application we make, the curve C is the proper curve of contact of the tangent planes to the fundamental surface under consideration, drawn from a point O; the two surfaces passing through this are the funda- mental surface of order $(m=)\mu_0$, and the first polar of O, of order $(m'=)\mu_0-1$; and the curves C_1, C_2 coincide in the double curve of the fundamental surface. The proper curve of contact is of order μ_1 and of genus π; we have seen that this curve passes through the pinch points of the double curve, and also through the (ρ) points of this, for which one tangent plane of the surface passes through O. These points together will then absorb $2\nu_2+2\rho$ of the intersections of the surface of order $m+m'-4$ (here equal to $2\mu_0-5$) with the proper curve of contact; hence we have

$$\mu_1(2\mu_0-5)=2\nu_2+2\rho+2\pi-2.$$

The genus π of the proper curve of contact is obtainable by re- marking that the rank of this curve, the number of its tangents which meet an arbitrary line, which we may take to pass through O, is $\kappa+\mu_2$; for, an inflexional line of the surface which passes through O is a tangent line of the curve of contact; and a tangent line of the curve of contact which meets a line through O lies in a tangent plane of the surface passing through this line; therefore, by the ordinary formula for the rank of a curve, we have

$$\kappa+\mu_2=2\mu_1+2\pi-2.$$

Using this, and the equation (II), namely $\mu_1(\mu_0-2)=\kappa+\rho$, which enables us to eliminate ρ, we find the result stated, or

$$\mu_2+2\nu_2=\mu_1+\kappa.$$

This is a remarkable result, expressing the number of pinch points of the double curve in terms of the order, the class and the number of cuspidal edges of the proper enveloping cone. Other investi- gations of equations (I)–(V) arise below.

The general theorem of Noether ("Theorie des eindeutigen Ent- sprechens, u.s.w.," *Math. Ann.* VIII, 1875, § 7), of which we have used a particular case in proving equation (V), has already been referred to

(Vol. v, p. 218); it is as follows: If through a curve C there pass two surfaces F_1, F_2, of respective orders n_1, n_2, whose further intersection consists of curves, such that a j_1-fold curve of F_1 is a j_2-fold curve of F_2, while a k_1-fold node of F_1 is a k_2-fold node of F_2, then the complete canonical series on C is obtained by surfaces of order $n_1 + n_2 - 4$, which have every such common (j_1, j_2) curve as a $(j_1 + j_2 - 1)$-fold curve, and every such common (k_1, k_2) node as a node of such character that the surfaces are there met in $(k_1 + k_2 - 2)$ points by every branch of C passing through such node; which moreover pass through every point of contact of F_1 and F_2 occurring at simple points of these. Beside the simple application just made, for equation (V), an application is made in the next chapter (p. 227) to the case when the curve C is defined, on a surface F of order n, as the residual intersection of this surface with a surface, ϕ, of order $n-4$, which passes $(j-1)$ times through every j-fold curve of F, and $(k-2)$ times through every k-fold node of F. The canonical series on C is then determined by surfaces of order $n + n - 4 - 4$, or $2n - 8$, passing $j + j - 1 - 1$, or $2(j-1)$ times, through every j-fold curve of F, and $k + k - 2 - 2$, or $2(k-2)$ times, through every isolated k-fold node of F. A particular surface of this description is then given by the aggregate of two of the surfaces ϕ, of order $n-4$, which we have described. Wherefore, if π be the genus of the curve C, or (F, ϕ), and ν be the number of intersections of two such curves, (F, ϕ) and (F, ϕ'), we have $2\pi - 2 = 2\nu$. This will be found to be an important result. (Cf. p. 223.)

The equations (I)–(V) contain nine characters, μ_0, μ_1, μ_2, ν_2, ϵ_0, ϵ_1, t, κ, ρ, any five of which can be expressed thereby in terms of the other four. For instance, in terms of the properties of the double curve, and the order of the surface, that is, in terms of ϵ_0, ϵ_1, t, μ_0, we have $\mu_1 = \mu_0(\mu_0 - 1) - 2\epsilon_0$, and

$$\mu_2 = \mu_0(\mu_0 - 1)^2 - (7\mu_0 - 12)\,\epsilon_0 + 4\epsilon_1 + 15t.$$

And we may add the expression for the number of inflexions of an arbitrary plane section of the proper enveloping cone, in terms of the double curve. For, by Plücker's formula, we have $i - \kappa = 3(\mu_2 - \mu_1)$, which is equivalent here to either of $i = 4\kappa - 6\nu_2$, $i = 4(\mu_2 - \mu_1) + 2\nu_2$; thus we obtain

$$i = 4\mu_0(\mu_0 - 1)(\mu_0 - 2) - 24(\mu_0 - 2)\,\epsilon_0 + 12\epsilon_1 + 48t;$$

this is the number of tangent planes of the surface from an arbitrary point with stationary contact. We also have

$$\nu_2 = 2(\mu_0 - 2)\,\epsilon_0 - 2\epsilon_1 - 6t, \quad \kappa = \mu_0(\mu_0 - 1)(\mu_0 - 2) - 3(\mu_0 - 2)\epsilon_0 + 3t.$$

The equation (V) will be modified if the surface have isolated multiple points. Thus, if there be δ isolated ordinary double points, though the formulae for π and ρ are as before, the equation we have employed is changed to

$$\mu_1(2\mu_0 - 5) - 2\delta = 2\nu_2 + 2\rho + 2\pi - 2,$$

and equation (V) becomes $\mu_2 + 2\nu_2 = \mu_1 + \kappa - 2\delta$. As the Plücker formula $i - \kappa = 3(\mu_2 - \mu_1)$ remains true, we hence deduce

$$i = 4(\mu_2 - \mu_1) + 2\nu_2 + 2\delta.$$

More generally, if there be isolated ordinary multiple points, of which the typical order is denoted by k, we have

$$\kappa + \mu_2 = 2\mu_1 + 2\pi - 2, \quad \mu_1(\mu_0 - 2) = \kappa + \rho + \Sigma k(k-1)(k-2),$$

of which the latter is a modified form of equation (II); and we also have $\mu_1(2\mu_0 - 5) = 2\nu_2 + 2\rho + 2\pi - 2 + \Sigma k(k-1)(2k-3)$, so that equation (V) becomes $\mu_2 + 2\nu_2 = \mu_1 + \kappa - \Sigma k(k-1)$. Hence, again with $i - \kappa = 3(\mu_2 - \mu_1)$, we obtain $i = 4(\mu_2 - \mu_1) + 2\nu_2 + \Sigma k(k-1)$. From this, taking, for a plane section of the proper enveloping cone, the dual of a well-known result, namely $i + \mu_1 = 2\mu_2 + 2\pi - 2$, we deduce $2\pi - 2 = 2\mu_2 + 2\nu_2 - 3\mu_1 + \Sigma k(k-1)$.

For completeness and comparison it seems desirable to quote here still more general formulae given by Noether ("Sulle curve multiple di superficie algebriche," *Ann. d. Mat.* v, 1871–3; quoted in his paper in *Math. Ann.* viii, 1875, already referred to). For a surface with multiple curves which have no mutual intersections and no multiple points—typically of multiplicity j on the surface, of order ϵ_0 and rank ϵ_1—the surface having also isolated general multiple points (typically of multiplicity k), the formulae of Noether lead, as a particular case, to

$$\mu_1 = \mu_0(\mu_0 - 1) - \Sigma j(j-1)\epsilon_0, \qquad \nu_2 = \Sigma(j-1)\{2(\mu_0 - j)\epsilon_0 - j\epsilon_1\},$$

$$\mu_2 = \mu_0(\mu_0 - 1)^2 - \Sigma k(k-1)^2 - \Sigma(j-1)\{(3j+1)\mu_0 - 2j(j+1)\}\epsilon_0$$
$$+ \Sigma j^2(j-1)\epsilon_1,$$

$$i = 4\mu_0(\mu_0 - 1)(\mu_0 - 2) - 4\Sigma j(j-1)(3\mu_0 - 2j - 2)\epsilon_0 + 2\Sigma j(j-1)(2j-1)\epsilon_1.$$

For the general formulae see the Note (II, p. 180) at the end of this chapter.

For the case when the surface has no multiple points beside a double curve with triple points, if we use Plücker's formula for the double points of a plane section of the proper enveloping cone, namely $2D = \mu_1^2 - \mu_1 - \mu_2 - 3\kappa$, we find that the number of tangent lines to the surface, from an arbitrary point, which have two points of contact, is given by

$$D = \tfrac{1}{2}\mu_0(\mu_0 - 1)(\mu_0 - 2)(\mu_0 - 3) + 2\epsilon_0^2 - (\mu_0^2 - 5\mu_0 + 7)\epsilon_0 - 2\epsilon_1 - 12t;$$

and we can compute the number, T, of twice touching tangent planes, from an arbitrary point, from the well-known formula $2(T - D) = (\mu_2 - \mu_1)(\mu_2 + \mu_1 - 9)$. Again, the genus, π, of the proper curve of contact, whose rank is $\mu_2 + \kappa$, as we have remarked, computed with help of $2\pi - 2 = \mu_2 - 2\mu_1 + \kappa$, $= 2\mu_2 + 2\nu_2 - 3\mu_1$, is given, for this case of a surface with a double curve, by

$$2\pi - 2 = \mu_0(\mu_0 - 1)(2\mu_0 - 5) - 2(5\mu_0 - 11)\epsilon_0 + 4\epsilon_1 + 18t.$$

Reversely, in terms of $\mu_0, \mu_1, \mu_2, \nu_2$, the characters of the double curve are given by

$$2\epsilon_0 = \mu_0(\mu_0 - 1) - \mu_1, \quad 2\epsilon_1 = 2\mu_0\mu_1 - 2\mu_1 - 2\mu_2 - 5\nu_2,$$

$$6t = \mu_0(\mu_0 - 1)(\mu_0 - 2) - 3\mu_0\mu_1 + 4\mu_1 + 2\mu_2 + 4\nu_2,$$

and therefore, the genus, P, of the double curve, computed from $\epsilon_1 = 2\epsilon_0 + 2P - 2$, is given by

$$2(2P-2) = 2\mu_0\mu_1 - 2\mu_0(\mu_0 - 1) - 2\mu_2 - 5\nu_2.$$

Ex. 1. For the surface with only a double curve, and triple points thereon, triple for the surface, the formulae shew that the numbers μ_1, ν_2, ϵ_1, $\mu_2 + \frac{1}{2}\nu_2$ are all even; that the two numbers κ, $\mu_2 - \mu_1 - \nu_2$ divide by 3; and the number i divides by 24 (cf. Roth, *Camb. Phil. Soc. Proc.* xxv, 1929, p. 395).

Ex. 2. It has been shewn (p. 155) that, if a line from an arbitrary point meet the surface in $\mu_0 - 4$ simple points, and also in two points each a coincidence of two intersections, then these two points lie on a certain surface of order $(\mu_0 - 2)(\mu_0 - 3)$. Let then D, E denote, respectively for the proper curve of contact of tangent planes from an arbitrary point, and for the double curve of the surface, the number of chords of these curves passing through this point; so that

$$\mu_2 = \mu_1{}^2 - \mu_1 - 2D - 3\kappa, \quad \epsilon_1 = \epsilon_0{}^2 - \epsilon_0 - 2E - 6t;$$

also, let $[\mu_1\epsilon_0]$ denote the number of common generators of the two cones with vertex at this arbitrary point, (1), the proper cone of contact, (2), the cone through the double curve, excluding common generators through actual intersections of the proper curve of contact and the double curve; then prove that

$$[\mu_1\epsilon_0] = \mu_1\epsilon_0 - 2\rho - \nu_2, \quad \mu_1(\mu_0 - 2)(\mu_0 - 3) = 2D + 2[\mu_1\epsilon_0],$$

$$\epsilon_0(\mu_0 - 2)(\mu_0 - 3) = 4E + [\mu_1\epsilon_0];$$

subtracting $\mu_1\mu_0(\mu_0 - 1)$ and $\epsilon_0\mu_0(\mu_0 - 1)$, respectively, from the last two equations, using the equation $\mu_0(\mu_0 - 1) = \mu_1 + 2\epsilon_0$, and the values above given for D, E, and also the two equations

$$\mu_1(\mu_0 - 2) = \kappa + \rho, \quad \epsilon_0(\mu_0 - 2) = \rho + 3t,$$

infer the results $\mu_2 + 2\nu_2 = \mu_1 + \kappa$, $2\rho - 2\epsilon_1 = \nu_2$. This, essentially, is Salmon's deduction of equations (IV), (V) from (I), (II) and (III).

Ex. 3. For a ruled cubic surface with a double line, whose equation may be supposed to be $x^2z - y^2t = 0$, prove that $\mu_0 = 3$, $\mu_1 = 4$, $\mu_2 = 3$, $\nu_2 = 2$, $\epsilon_0 = 1$, $\epsilon_1 = 0$, $\kappa = 3$, $\rho = 1$, $i = 0$, obtaining the positions of the pinch points, and of the (κ) and (ρ) points in this case, for a given external point. Shew that the proper curve of contact of tangent planes, from a point (a, b, c, d), is the rational quartic curve given, in terms of a parameter θ, by $x = \theta(a\theta - b)$, $y = \theta^2(a\theta - b)$, $z = \frac{1}{2}\theta^2(d\theta^2 - c)$, $t = \frac{1}{2}(d\theta^2 - c)$, and find the intersections of this curve with the double line $x = 0 = y$. Prove that the (κ) points lie on the quadric surface expressed by

$$adyz - 3bdxz + 3acyt - bcxt = 0,$$

which meets the curve also on $x = 0 = y$ and on $z = 0 = t$; and that these (κ) points lie also on the plane given by $2acx - 2bdy + a^2z - b^2t = 0$, which also contains the (ρ) point.

Ex. 4. For a quartic surface with a double line, so that $\mu_0 = 4$, $\epsilon_0 = 1$, $\epsilon_1 = 0$, $t = 0$, prove that $\mu_1 = 10$, $\mu_2 = 20$, $\nu_2 = 4$, $\kappa = 18$, $\rho = 2$, $i = 48$.

Ex. 5. For a quartic surface with a double conic (a cyclide), from $\mu_0 = 4$, $\epsilon_0 = 2$, $\epsilon_1 = 2$, $t = 0$, deduce $\mu_1 = 8$, $\mu_2 = 12$, $\nu_2 = 4$, $\kappa = 12$, $\rho = 4$, $i = 24$.

Ex. 6. For the Steiner quartic surface, with three concurrent double lines, from $\mu_0 = 4$, $\epsilon_0 = 3$, $\epsilon_1 = 0$, $t = 1$, deduce $\mu_1 = 6$, $\mu_2 = 3$, $\nu_2 = 6$, $\kappa = 9$, $\rho = 3$, $i = 0$. These can be verified by considering the dual surface, a four-nodal cubic surface. The intersections of the proper curve of contact with the double curve may be computed directly, taking for the equation of the surface $y^2z^2 + z^2x^2 + x^2y^2 = 2xyzt$.

Ex. 7. For a general ruled surface, the characters can be expressed in terms of the order of the surface, μ_0, and the genus of the general plane section, p. A fundamental fact is that $\mu_2 = \mu_0$; for, every tangent plane is a plane through a generator, and every plane through a generator is a tangent plane; so that tangent planes through an arbitrary line, each containing a generator, must be those containing the generators which meet the line, and, conversely, all the planes through the arbitrary line and generators which meet it, are tangent planes. From this we have μ_1, the class of a plane section of the surface, equal to $2\mu_0 + 2p - 2$, or to $2\mu_2 + 2p - 2$. The order, ϵ_0, of the double curve is given by

$$\epsilon_0 = \tfrac{1}{2}\mu_0(\mu_0 - 1) - \tfrac{1}{2}\mu_1, \text{ or } \tfrac{1}{2}(\mu_0 - 1)(\mu_0 - 2) - p.$$

We consider now the direct deduction of the number of pinch points of the double curve, ν_2. On any generator, there is associated with any point a certain plane, the tangent plane of the surface at this point, which is the ultimate position of the plane joining this point to a consecutive generator; and there is associated with every plane through the given generator a certain point, the point of contact of this plane, which is the ultimate position of the intersection of this plane with a consecutive generator. But there are generators for which the associated plane is the same for every point of the generator, and the associated point is the same for every plane through the generator. Such may be described as *torsal* generators, or, more particularly, if not very clearly, as generators each of which intersects its consecutive. In general, the points of intersection of generators describe a curve on the surface, the double curve, but the pair of generators intersecting at any point of this are not consecutive; the two tangent planes of the surface at such a point of the double curve are the planes joining the tangent line of the curve to the two generators which intersect at this point. If the two tangent planes coincide (and the point is a pinch point of the double curve), it must be because the two generators are consecutive and ultimately coincident; and, conversely, a torsal generator meets the double curve in a pinch point. The number ν_2 is thus the number of torsal generators. Of the value of ν_2, which is $2\mu_0 + 4p - 4$, various proofs may be given; the simplest is, perhaps, as follows: Take an arbitrary line, and, through this, two arbitrary planes; establish a correspondence between points P, P' of the line, by drawing a tangent line PT to the section of the surface by the first plane, touching this in T; let the generator of the surface which contains T meet the section of the surface by the second plane in T'; and let the tangent line of the second section at T' meet the given line in P'. The correspondence on the line so established is evidently symmetrical, and the indices of this correspondence are both equal to the number, $2\mu_0 + 2p - 2$, of tangents PT which can be drawn from P to the first section. There are, therefore, $2(2\mu_0 + 2p - 2)$ cases of coincidence of P and P'. Among these are the coincidences which arise at the points where the given line meets the surface; through each of these μ_0 points both the first and second sections of the surface pass; we count each of these as two coincidences, so obtaining $2\mu_0$ in all. The other coincidences of P and P' arise when the tangent plane of the surface at T, defined by PT and the generator at T, coincides with the tangent plane of the surface at T', which is on the generator through T; in other words, TT' must be a torsal generator. Thus, if the number of such be ν_2, we have $\nu_2 + 2\mu_2 = 2(2\mu_0 + 2p - 2)$; which proves the formula cited. Another proof is obtained by representing the generators of the surface by the points of a curve lying on a quadric Ω in space of five dimensions (cf. Vol. IV, pp. 40 ff.). Then two non-consecutive intersecting generators determine a flat pencil of lines; this is represented in the space [5] by a line, lying on the quadric Ω, which is a chord of the curve thereon representing the generators of the ruled surface. Thus two consecutive intersecting generators give rise to a tangent line of this curve, itself lying wholly on Ω. To find the number of torsal generators of the ruled surface is thus to find the number of tangent lines of this curve on Ω which lie wholly thereon. The order of this curve, it is easily seen, is the number of generators of the ruled surface which meet an arbitrary line, namely is μ_0. Now, let P be any point of the representative curve on Ω; the lines through P which lie on Ω, lie, we know, on the tangent prime of Ω at P, which meets the curve in two coincident points at P, and hence in $\mu_0 - 2$ other points. There are thus $\mu_0 - 2$ chords of the curve, say PP', lying on Ω, through any point P of the curve. The ends P, P', of such a chord, are then in correspondence on the curve; and this correspondence, evidently symmetrical with both indices equal to $\mu_0 - 2$, has a valency 2, because the points P' which correspond to P are on a prime, with double intersection at P (see Chap. I, p. 3, preceding). The number of coincidences of P and P', namely the number of tangents of the curve which lie wholly on Ω, is thus $2(\mu_0 - 2) + 2p.2$; for the genus of the curve on Ω, whose points are in $(1, 1)$ correspondence with the generators of the ruled surface, and hence with the points of a plane section of this, is p. Thus $\nu_2 = 2\mu_0 + 4p - 4$, as before. Two other proofs of this result, also by the theory of correspondence, are given by Schubert; one of these will be found in Chap. II preceding (p. 102).

Having found ν_2, we deduce, for the rank of the double curve of the ruled surface,

$$\epsilon_1 = \mu_0\mu_1 - \mu_1 - \mu_2 - \tfrac{5}{2}\nu_2, \ = 2(\mu_0 - 2)(\mu_0 - 3) + 2(\mu_0 - 6)p.$$

The genus P, of this curve, is thence given by

$$2P - 2, \ = \mu_0\mu_1 - \mu_0(\mu_0 - 1) - \mu_2 - \tfrac{3}{2}\nu_2, \ = (\mu_0 - 5)(\mu_0 + 2p - 2), \ = \tfrac{1}{2}(\mu_0 - 5)\nu_2,$$

and the number of triple points, given by

$$t = \tfrac{1}{6}\mu_0(\mu_0 - 1)(\mu_0 - 2) - \tfrac{1}{2}\mu_0\mu_1 + \tfrac{2}{3}\mu_1 + \tfrac{1}{3}\mu_2 + \tfrac{2}{3}\nu_2,$$

is

$$t = (\mu_0 - 4)\{\tfrac{1}{6}(\mu_0 - 2)(\mu_0 - 3) - p\}.$$

These values lead to

$$i, \ = 4(\mu_2 - \mu_1) + 2\nu_2, \ = 0; \quad \kappa, \ = \mu_2 + 2\nu_2 - \mu_1, \ = 3(\mu_0 + 2p - 2), \ = \tfrac{3}{2}\nu_2,$$

$$\rho, \ = \epsilon_1 + \tfrac{1}{2}\nu_2, \ = (\mu_0 - 2)(2\mu_0 - 5) + 2p(\mu_0 - 5),$$

while the genus of the proper curve of contact of tangent planes from an arbitrary point, being given by $2\pi - 2 = \mu_2 + \kappa - 2\mu_1, \ = 2p - 2$, is that of the plane sections of the surface, as should evidently be the case. If we assume that for a ruled surface there are no tangent planes through an arbitrary point which have stationary contact, or $i = 0$ (cf. Salmon's result above, p. 156, Ex. 6), we have, from $\nu_2 = 2(\mu_1 - \mu_2)$, a further determination of ν_2. These general formulae are under the hypothesis that the double curve is irreducible; for cases of reducible double curve the reader should consult Edge, *Ruled Surfaces* (Cambridge, 1931).

Ex. 8. A torsal generator of a ruled surface is such that the tangent plane, at any point P of this generator, is the same; and the point of contact of any plane ϖ, through this generator, is the same. If the generator be given by the equations $x = t\xi + lz$, $y = t\eta + mz$, where x, y, z, t

are the homogeneous coordinates, ξ, η are connected by a certain algebraic equation, and l, m are algebraic functions of ξ and η, prove that the differential condition for the generator to be torsal is $d\eta . dl - d\xi . dm = 0$; further, that the unique tangent plane is then given by $(x - t\xi - lz)\,d\eta - (y - t\eta - mz)\,d\xi = 0$; and the unique point of contact has the homogeneous coordinates $\xi\,dl + l\,d\xi$, $\eta\,dl + m\,d\xi$, $d\xi$, dl.

Ex. 9. It will be of importance later to consider the intersection of a general surface with surfaces passing through the double curve. We may consider here the case of a ruled surface, though this is, in some respects, exceptional. Denote the order of the surface temporarily by n (instead of μ_0), and recall that the order, genus, and number of triple points of the double curve, respectively, are given by $\epsilon_0 = \frac{1}{2}(n-1)(n-2) - p$, $P - 1 = \frac{1}{2}(n-5)(n+2p-2)$ and $t = (n-2, 3) - (n-4)p$, where $(n-2, 3)$ is the binomial coefficient. Then, first, as an arbitrary plane through a generator meets the surface further in a curve of order $n-1$, of which one intersection with the generator is the point where the plane touches the surface, we infer that a generator of the surface meets the double curve in $n-2$ points. Hence there can be no surface of order $n-3$ passing through the double curve, since such a surface would contain every generator. But, second, we can prove that there are surfaces of order $n-2$ passing through the double curve, $n+p-1$ of which are linearly independent; and that the remaining intersection of one of these with the ruled surface consists of $n+2p-2$ generators; the aggregate of these surfaces meets an arbitrary plane in the complete system of curves of order $n-2$ adjoint to the section of the surface by this plane. For surfaces of order $n-2$ through the triple points of the double curve have $\epsilon_0(n-2) - 3t$ remaining intersections with the double curve; and this number, being $2P - 2 + n(n-2)$, is greater than $2P - 2$; such surfaces therefore cut on the double curve a series of freedom $\epsilon_0(n-2) - 3t - P - \xi$, where ξ is zero if this series is complete. Surfaces of order $n-2$ through the double curve thus have a freedom

$$\tfrac{1}{6}(n-1)n(n+1) - 1 - t - \{\epsilon_0(n-2) - 3t - P - \xi + 1\},$$

which is $n + p - 2 + \xi$. Such surfaces, however, determine, on an arbitrary plane, adjoint curves of order $n-2$, of this plane section; and there can be no surfaces of this kind containing the plane, since the residual surfaces would be surfaces of order $n-3$ passing through the double curve. The surfaces of order $n-2$ thus determine on the plane a system of curves of freedom $n + p - 2 + \xi$, which cannot, however, be greater than the freedom $n + 2p - 2 - p$, of the adjoint system of this order. Thus $\xi = 0$, the surfaces of order $n-2$ determine the complete adjoint system of curves of order $n-2$, and the system of surfaces is of freedom $n + p - 2$. Moreover, a surface of order $n-2$, through the double curve, will meet the generator passing through any general point, of a plane section, which lies on this surface of order $n-2$, in $n-1$ points in all; and will thus contain this generator. Such a surface thus contains the $n + 2p - 2$ generators passing through the points in which this surface meets a plane section of the ruled surface; and, since $n(n-2) - 2\epsilon_0 = n + 2p - 2$, these generators together form the complete residual intersection with the ruled surface beyond the double curve. If we now consider a surface of order $n-1$ passing through the double curve, this will meet the surface in a residual curve of order $2n + 2p - 2$, having one intersection with every generator, other than on the double curve. This residual curve can then be expressed, in the symbolical way previously explained (Vol. v, p. 216), in the form $\varpi + (n + 2p - 2)g$, where ϖ denotes a plane section and g denotes a generator; two such residual curves then meet in $(\varpi, \varpi) + 2(n + 2p - 2)(\varpi, g)$ points, that is $n + 2(n + 2p - 2)$ points. In the notation we have used above, this is $\mu_2 + \nu_2$. A particular surface of order $n - 1$ through the double curve is the first polar of an arbitrary external point, in regard to the ruled surface; the residual intersection of this is the proper curve of contact for this point; this agrees then with the number $\mu_2 + \nu_2$, since the proper curve of contact contains the pinch points. We may compute the freedom of surfaces of order $n - 3 + k$ passing through the double curve $(k \geqslant 1)$, as in the case above when $k = 1$; and find $\frac{1}{2}nk(k + 1) + k(p - 1) + \frac{1}{6}k(k - 1)(k - 2) - 1$. This number reduces to -1 for $k = 0$; and to $-p - 1$ for $k = -1$, of which an interpretation will occur in the next chapter. It can be shewn that these surfaces of order $n - 3 + k$ intersect a curve on the ruled surface which is given by the intersection of this with a general surface of order $k + 1$, in the canonical series of this curve (Enriques, *Mem. Soc. Ital. d. Scienze*, x, 1896, p. 56). The pinch points are further considered in Note I, at the end of this chapter (p. 176).

Ex. 10. The following three results (quoted from Severi, "Intersezioni...", *Mem. Torino*, LII, 1903, No. 19) are of importance in the sequel. A surface, of order μ_0, has a double curve of order ϵ_0 and rank ϵ_1, with ν_2 pinch points, and certain triple points triple for the surface. A surface of order μ_0' is drawn through the double curve, having as residual intersection with the given surface, a curve, of order ϵ_0' and rank ϵ_1', meeting the double curve in i points. Then

$$\epsilon_0' = \mu_0'\mu_0 - 2\epsilon_0, \quad i = 2(\mu_0' - \mu_0 + 1)\epsilon_0 + \epsilon_1 + \tfrac{3}{2}\nu_2,$$
$$\epsilon_1' = \mu_0'\mu_0(\mu_0' + \mu_0 - 2) - 2(3\mu_0' - \mu_0)\epsilon_0 - 2\epsilon_1 - 3\nu_2.$$

For example, for a quartic surface with a double line, a cubic surface through the line has, as residual intersection, a curve of order 10 and genus 10, meeting the line in 6 points.

Shew, in particular, that with $\mu_0' = \mu - 4$, $\nu_2 = 2(\mu_0 - 2)\epsilon_0 - 2\epsilon_1 - 6t$ (cf. p. 162, above) we obtain $i = 3(\mu_0 - 4)\epsilon_0 - 2\epsilon_1 - 9t$, and

$$\epsilon_1' = 2\mu_0(\mu_0 - 3)(\mu_0 - 4) - 2(5\mu_0 - 18)\epsilon_0 + 4\epsilon_1 + 18t.$$

Further, let two surfaces of orders m and n be drawn through the double curve; let their residual curves of intersection with the given surface, supposed irreducible, have $p_1^{(1)}$, $p_2^{(1)}$ for genera, respectively; let the residual curve of inter ection of the two surfaces of orders m, n, with one another, assumed to have in all $2\epsilon_0(m + n - 2) - 2\epsilon_1 - 9t$ intersections with the double curve, have $p_{12}^{(2)}$ further intersections with the given surface. Then prove that

$$p_1^{(1)} - 1 + p_2^{(1)} - 1 - 2p_{12}^{(2)} = \tfrac{3}{4}\mu_0(m - n)^2 + (2\mu_0 - 8 - m - n)\{\tfrac{1}{4}\mu_0(m + n) - \epsilon_0\}.$$

Thus, when $m = n$, $p^{(1)} - 1 - p^{(2)} = (\mu_0 - 4 - m)(\mu_0 m - 2\epsilon_0)$, which is a formula of subsequent interest (cf. p. 223).

Ex. 11. By means of the formulae given earlier in this chapter for a general surface with a double curve, and triple points thereon, we can prove that $\epsilon_0(\mu_0 - 4) - 3t = 2P - 2 + \tfrac{1}{2}\nu_2$. When $\nu_2 > 0$, this shews that surfaces of order $\mu_0 - 4$, or of higher order, passing through the triple points of the double curve, cut thereon a series which is not special. If, now, we define a number p_n by means of $12p_n = 2\mu_2 - 8\mu_1 + 12\mu_0 - 12 + \nu_2$, it can be proved, as in Ex. 9, that, if surfaces of order $\mu_0 - 3 + k$, through the triple points of the double curve, cut thereon, for $k = -1, 0, 1, \ldots$, a non-special series, of incompleteness ξ_k, then the tale ($=$ freedom $+ 1$) of such surfaces through the double curve is

$$\tfrac{1}{2}\mu_0 k(k + 1) + k(p - 1) + \tfrac{1}{6}k(k - 1)(k - 2) + p + p_n + \xi_k.$$

Of such surfaces, a certain number will wholly contain an arbitrary plane, the number, namely, of surfaces of order $n-3+k-1$ which exist passing through the double curve. Let this number be denoted by ϖ_{k-1}; then the tale of the system of curves, of order $n-3+k$, adjoint to the plane section of the surface, determined on this plane by surfaces of order $n-3+k$ passing through the double curve, is obtained by subtracting ϖ_{k-1} from the number just set down. In particular, for $k=0$, if the number, ϖ_{-1}, of surfaces of order $n-4$ which exist passing through the double curve, be denoted by p_g, then the surfaces of order $n-3$, through the double curve, determine, on an arbitrary plane, a system of curves adjoint to the section of the surface by this plane, of order $n-3$, with freedom $p-1-(p_g-p_n)+\xi_0$. Picard (Picard-Simart, *Fonctions algébriques*, II, 1906, p. 438) has proved, by transcendental methods, using a formula of Castelnuovo, a theorem which is equivalent to saying that $\xi_0 = 0$. Picard's theorem is thus equivalent to the result that *the non-special series, on the double curve of the surface, which is determined by surfaces of order $n-3$ put through the triple points of the curve, is complete.* It would be of great importance for the general theory of surfaces to have an elementary direct proof of this. For a ruled surface there are no surfaces of order $n-4$, or $n-3$, through the double curve, as we have shewn; but the results found for ruled surfaces, as to the completeness of series for surfaces of order $n-2$, or more, hold for surfaces in general, as Picard's theory shews. (Cf. Ex. 8, p. 269, below.)

Surfaces in space of four dimensions. We now consider a surface in space [4] which is the projection of a non-singular surface in space [5], as has been explained. As there are ∞^4 chords of a surface, and for a line to pass through a given point in space [5] requires 4 conditions, we expect that a finite number of chords of a surface in [5] will pass through a general point of this space. Further enquiry shews in fact that this is so save in two cases only, which are, when the surface is a cone, and when it is the surface, called the Veronese surface; this last is the surface of which the coordinates $(x_0, \ldots, x_5)$ of a general point may be taken, in terms of three arbitrary parameters, as $x_0 = \xi^2$, $x_1 = \eta\zeta$, $x_2 = \eta^2$, $x_3 = \zeta\xi$, $x_4 = \zeta^2$, $x_5 = \xi\eta$; the surface lies on the six quadrics whose equations are given by the vanishing of the quadratic minors in the symmetrical determinant whose rows consist in turn of the elements x_0, x_5, x_3; x_5, x_2, x_1; x_3, x_1, x_4; the surface itself, and all its chords, lie on the cubic primal expressed by the vanishing of this determinant. Thus no chords of the surface can be drawn from any point of the space [5] which does not lie on this primal; the surface contains (∞^2) conics, whose planes lie on this primal, and a chord of the surface is a chord of such a conic. We consider then a surface in [5] of which a finite number of chords passes through an arbitrary point of this space. If this surface be projected, from such an external point, into space [4], the surface obtained, which we shall denote by ψ, will have a finite number of isolated double points, each lying on one of the chords from the point of projection; at such a point of the surface ψ there will be two tangent planes, the projections of

the tangent planes of the surface in [5] at the points where this is
met by the chord. These two tangent planes will in general meet
only at this point, and every line through the point may be regarded
as a limit of a chord of the surface. Such double points we shall call
accidental. In particular it may happen that the two tangent
planes at the double point meet in a line, if the point in [5], from
which the projection is made, be suitably taken. Again, the chords
of the surface ψ, in [4], are ∞^4 in aggregate; thus, there pass ∞^1
such chords through an arbitrary point L of this space; these chords
form a conical sheet, a surface, with vertex at L. For the moment,
let the order of this surface, the number of points in which it meets
a general plane, be called S. Every generator of the conical sheet
meets the surface ψ in two points, forming a curve on this surface;
denote this curve by γ, and, for a moment, denote the order of this
curve, the number of its intersections with a general prime space
[3], by N. The accidental double points of the surface ψ, where two
sheets of the surface have point intersection, are double points of
the curve γ, where it crosses itself. The surface ψ is met by a prime
in a curve; let h be the number of chords of this curve which can
be drawn through an arbitrary point of this prime. In particular,
when this prime is taken through L, these chords are generators of
the conical sheet, of chords of ψ from L, and each meets the curve
γ in two points; thus $2h$ is the number of intersections of the curve
γ with the prime through L, this prime meeting the conical sheet
only in chords of ψ. Wherefore, $N = 2h$. Again, S, the order of this
conical sheet, being the number of its intersections with a plane, is
the number of chords of the surface ψ which lie in the prime joining
L to this plane; and, conversely, in a prime through L, a plane
meets every chord of ψ which lies in this prime. Wherefore, $S = h$,
and $N = 2S$. Now take an arbitrary general plane ϖ, and a pencil
of primes through this plane. Any one such prime meets the curve
γ in N points; and each of these points, say P, by projection from
L, gives another point, P', of the curve γ; while this point P'
determines another prime of the pencil through ϖ. We may thus
regard any prime of this pencil as corresponding to N other primes
of this pencil, namely there is a, clearly symmetrical, correspondence, of indices N, N, between the primes passing through the
plane ϖ. There are then $2N$ cases of coincidence of corresponding
primes. Such coincidences, we assume, are of only two kinds:
namely, such a coincidence arises when the points P, P', on a pair
of corresponding primes, and on a chord of the surface ψ passing
through L, coincide with one another; and, a coincidence arises
when the chord PP' meets the base plane ϖ. A coincidence of P
with P' arises for every tangent line of the surface ψ which passes
through L; let ν_2 denote the number of such tangent lines, which is

the same as the number of tangent planes of ψ which pass through L; a coincidence of P with P' also arises at a double point of the surface ψ; let d denote the number of such double points. At a coincidence of P and P' where the tangent line of ψ passes through L, the chord from L touches the curve γ; at a coincidence at a double point, the curve γ crosses itself, and not only does P coincide with P', but also P' coincides with P. Thus we say that the number of coincidences of corresponding primes which arise by coincidence of P with P' is $\nu_2 + 2d$. Coincidences of corresponding primes which arise because the chord PP' meets the plane ϖ, occur for each of the S points in which the plane ϖ meets the conical sheet of chords from L to the surface ψ; these, however, likewise involve, not only that P' coincides with P, but also that P coincides with P', and count for $2S$. Thus we have $\nu_2 + 2d + 2S = 2N$, or as $N = 2S$, we have $\nu_2 + 2d = N$. In what follows we constantly denote the order of the curve γ by ζ_0, so that $N = \zeta_0$; and we denote the order of the double curve, on the surface in [3], into which ψ projects from L, by ϵ_0; so that $\epsilon_0 = S = \frac{1}{2}\zeta_0$. Hence, the number of the accidental double points of the surface ψ is $d = \frac{1}{2}(\zeta_0 - \nu_2)$. Further, as the condition for a plane, in space [4], to meet a line, is one-fold, there will be a curve of points on ψ, at each of which the tangent plane of the surface meets a given arbitrary line; the order of such curve we denote by μ_1. Thus, μ_1 is also the number of tangent lines, of the curve section of ψ by a prime drawn through a given arbitrary line, which meet this line, namely the rank of such prime section. Hence, if the prime section have no cusps, by the properties of a curve in space [3], we have $\mu_1 = \mu_0(\mu_0 - 1) - 2h$, where h, as above, is the number of so-called apparent double points of such prime section, and we have seen that $h = S = \epsilon_0$. Thus we have $\mu_0(\mu_0 - 1) = \mu_1 + 2\epsilon_0$. In fact, if a line l, from L, meet the space [3] on which we project, in the point O, and μ_1 be defined by the tangent planes of the surface ψ which meet the line l, it is clear that μ_1 is the order of the proper enveloping cone from O to the surface in [3] into which ψ projects; thus the equation just obtained is the equation (I), which was proved above in space [3] by use of the first polar of the point O. With this equation the number of accidental double points of the surface ψ, which was found as $d = \epsilon_0 - \frac{1}{2}\nu_2$, can be expressed also by $d = \frac{1}{2}\mu_0(\mu_0 - 1) - \frac{1}{2}\mu_1 - \frac{1}{2}\nu_2$. The proof of the value of d we have given is due to Severi ("Intorno ai punti doppi...," already referred to, *Rend. Palermo*, xv, 1901). Another proof arises below (Chap. vi, p. 259, footnote), in connexion with the formulae for the intersection of manifolds in space [4].

We have now defined the characters μ_0, μ_1, μ_2 for a surface in space [4]; and these give rise, by projection to a surface in space [3], to the characters already denoted by the same symbols. We may also define μ_2, for the surface ψ in [4], so that, on projection, it gives the class of the surface in space [3]: That a plane, of the ∞^2 tangent planes of the surface ψ, should meet a given arbitrary plane in the space [4], not in a point, but in a line, requires two conditions; there is thus a finite number of the tangent planes of ψ which satisfy this condition. This number we denote by μ_2. When we take the arbitrary plane through the point L, from which we project to [3], the tangent planes in question become the tangent planes, of the resulting surface in [3], which pass through a line, namely the line in which the arbitrary plane meets the space [3]; or μ_2 is the class of the projected surface. Consider now in more detail the origin of the triple points on the double curve of the surface in [3], obtained by projection. Among the ∞^1 chords of the surface ψ which pass through the point L, there will be a finite number satisfying the one-fold condition of meeting ψ in a further point; thus, there is a finite number, say t, of trisecant chords of ψ through L. On projection from L to space [3], these give rise to triple points of the double curve of the surface obtained, which are evidently also triple points of this surface. Denote now by A, B, C the points of the surface ψ lying on a trisecant line from L; for the moment, denote the surface obtained from ψ, by projection from L, by ϕ. We may regard the single point of ϕ which arises by projection of both B and C as a point of the double curve of ϕ, the point, namely, where this curve is met by the sheet of ϕ which arises by projection of the neighbourhood of the point A on ψ; and a similar remark may be made in regard to each of the other two pairs C, A and A, B. The tangents at the triple point of the double curve on ϕ thus arise as the intersections of the three pairs of tangent planes of the sheets of ϕ which meet at this point. The conical surface formed by the chords of ψ from L will, correspondingly, have three sheets meeting in the trisecant line $LABC$; the chord LBC will be one of a continuum of chords, lying in one sheet of the cone, meeting ψ in two curves, respectively through B and C; say $B2$ and $C3$; likewise the chord LCA will be one of a continuum of chords lying in another sheet of the cone, meeting ψ in two curves, respectively through C and A; say $C3'$ and $A1'$; and the chord LAB will be one of a

continuum of chords, meeting ψ in two curves; say $A1''$ and $B2''$. The conical sheet of chords to ψ from L thus meets ψ in a curve having a double point at each of A, B, C, the two branches of the curve which cross at A being $A1', A1''$; those which cross at B

being $B2''$, $B2$; and those which cross at C being $C3$, $C3'$. The two
sheets $LA1'$, $LA1''$ have curve intersections with ψ at A, but the
sheet $LBC23$ has only point intersection with ψ at A. And similarly
at B and C.

As a concrete example of these relations which can be examined in
detail, we may take the surface which is the locus of intersection of pairs
of osculating planes of a rational quartic curve in [4]; this is in fact the
same as the projection from space [5] of the Veronese surface referred to
above. In terms of three independent parameters ξ, η, ζ, the coordinates
of a point of this surface may be represented by ξ^2, $\xi\eta$, $\frac{1}{3}(2\eta^2+\xi\zeta)$, $\eta\zeta$, ζ^2.
It can be shewn without difficulty that a single trisecant of this surface
passes through an arbitrary general point, that through this trisecant
there pass three planes each meeting the surface in a conic, and that each
conic contains two of the three points of the surface which lie on the
trisecant.

Without much difficulty we can state five equations, for the surface ψ in space [4], which contain the equations (I)–(V) which were
found for the surface ϕ in [3] into which this projects from the point
L. The order of the curve γ, in which the surface ψ is met by the
conical sheet of chords from L, we denote by ζ_0; it is given by
$\zeta_0 = 2\epsilon_0$, where ϵ_0 is the order of the double curve on the surface ϕ.
By the *rank* of a curve, in space [4], we mean the number of its
tangent lines which meet an arbitrary plane; for the curve γ we
denote the rank by ζ_1. Taking a plane, which we denote by λ for the
moment, which passes through L, a tangent line of the curve γ,
which meets the plane λ, will project from L, into a tangent line,
of the double curve of the surface ϕ, which meets the line of intersection of λ with the space [3] in which ϕ exists—unless the tangent
line in question passes through L. The tangent lines of the surface
ψ which pass through L, we have remarked, are tangent lines of the
curve γ, and we have used ν_2 for their number. Using ϵ_1, as before,
for the number of tangent lines of the double curve of the surface
ϕ which meet an arbitrary line of its space, we thus have the relation $\zeta_1 = 2\epsilon_1 + \nu_2$, every pertinent tangent line of the double curve
arising from two tangent lines of the curve γ. Next, from the tangent planes of the surface ψ at points of the curve γ, which are ∞^1
in aggregate, denote the number which meet an arbitrary general
line of the space [4], by R. In particular, take this line, say l, to
pass through L, and let it meet the space [3], on which we project,
in the point O. There will be ν_2 such tangent planes of ψ meeting l
in L; any other will project into a tangent plane of the surface ϕ, at
a point of the double curve of this, which has the property of passing
through the point O. We have denoted the number of such tangent
planes of ϕ by ρ. Thus we have the equation $R = \rho + \nu_2$. Evidently
this equation is in accord with the fact we have previously remarked that, for the surface ϕ, in space [3], the proper curve of

contact of tangent planes, from an arbitrary point, passes through the (ρ) points, and through the pinch points, of the double curve of ϕ. We may also define, for the surface ψ, the number called κ for the surface ϕ, which was the number of stationary generators of the proper enveloping cone, or the number of inflexional lines of ϕ passing through an arbitrary point of the space [3]. In the space [4], the prime joining L to the tangent plane at a point P of ψ, meets ψ in a curve having a double point at P; the number κ is the number of points P for which one of the tangent lines of this curve, at P, meets an arbitrary line l drawn through L. With the slight changes thus described, the equations found (p. 159) for a surface in [3] lead to the following equations for the surface ψ:

(I), $\mu_0(\mu_0-1)=\mu_1+\zeta_0$;   (II), $\mu_1(\mu_0-2)=\kappa+R-\nu_2$;

(III), $\epsilon_0(\mu_0-2)=R-\nu_2+3t$;   (IV), $2\rho=\zeta_1$, or $2R=\zeta_1+2\nu_2$;

(V), $\mu_2+2\nu_2=\mu_1+\kappa$.

We can also obtain an interesting result in regard to the genus of the curve γ; this genus we denote by Q. As in the case of curves in space [3], we have, of course, $\zeta_1=2\zeta_0+2Q-2$, which is obtained by considering the coincidences of pairs of points of the sets of ζ_0 points, obtained on γ by primes through an arbitrary plane. Another form of this equation is $2Q-2=2\epsilon_1+\nu_2-4\epsilon_0$, and hence

$$2Q-2=2\mu_0(\mu_1-\mu_0+1)-2\mu_2-4\nu_2.$$

This result is obtainable also by considering the (2, 1) correspondence between the curve γ, and the double curve of the surface ϕ; Zeuthen's formula (p. 19 above) then gives $2Q-2-\nu_2=2(2P-2)$, which, with the value of P given above (p. 164) leads to the same form for Q. Another formula for Q, in terms of κ, is

$$Q-1+\kappa=(\mu_1-\mu_0)(\mu_0-1).$$

The genus Q satisfies the equation $2Q-2+6t=\zeta_0(\mu_0-4)$, as the formulae shew. Geometrically, this expresses that a primal of order μ_0-4, drawn through the double points of the curve γ, which lie on the trisecants of ψ drawn from L, meets γ, beside, in a set of $2Q-2$ points. It will appear (see p. 243, below) that such a set is a canonical set on γ. This is an interesting result.

Ex. 1. For the surface in [4], referred to above (p. 173), which is the projection of a Veronese surface in [5], obtain the values $\mu_0=4$, $\mu_1=6$, $\mu_2=3$, $\nu_2=6$, $\zeta_0=6$, $\zeta_1=6$, $t=1$, $d=0$, $Q=-2$, $\kappa=9$, $\rho=3$, and obtain these geometrically. This surface projects from a point into the Steiner quartic surface in space [3].

Ex. 2. For a ruled surface in [4], the number of accidental double points, in general $\frac{1}{2}(\zeta_0-\nu_2)$, is $\frac{1}{2}(\mu_0-2)(\mu_0-3)-3p$. Also $2Q-2=(\mu_0-4)\nu_2$, and $3t=(\mu_0-4)d$, while ζ_1, or $2\epsilon_1+\nu_2$, is $2(\mu_0-2)(2\mu_0-5)+4(\mu_0-5)p$. Segre (*Enzyk. d. Math. Wiss.* III, C 7, p. 913) states that the lines having 3 coincident intersections with the ruled surface form a ruled surface of order N, and their points of contact form a curve of order E, for which $N = 6\{\mu_0 + 3(p-1)\}$, and $E = 3\mu_0 + 8(p-1)$. Formulae giving these numbers for any surface in [4] are found by Roth, *Proc. Camb. Phil. Soc.* xxv, p. 395, with references to Severi.

Ex. 3. We may compute directly the characters of a surface in [4] which is the complete intersection of two general primals of orders m, m'. For this surface, $\mu_0 = mm'$; and μ_1, which is the rank of the curve in which the surface is met by an arbitrary prime, is the rank of the complete curve of intersection of two surfaces of orders m, m' in space [3]; so that $\mu_1 = mm'(m + m' - 2)$. It is convenient to use symbols for $m-1$, and $m'-1$, say k and k'; thus $\mu_1 = mm'(k+k')$. To find the number, μ_2, of tangent planes of the surface which meet a given arbitrary plane in a line, let $f = 0, f' = 0$ be the equations of the primals which define the surface, and $a_0 x_0 + \ldots + a_4 x_4 = 0, b_0 x_0 + \ldots + b_4 x_4 = 0$, be two primes containing the given plane; then, if the tangent plane of the surface, at the point (x), meet the plane in a line of which (ξ) is any point, the four equations
$$\xi_0 \partial f / \partial x_0 + \ldots + \xi_4 \partial f / \partial x_4 = 0, \quad \xi_0 \partial f' / \partial x_0 + \ldots = 0, \quad \xi_0 a_0 + \ldots = 0, \quad \xi_0 b_0 + \ldots = 0,$$
are satisfied, and, as equations for the determination of (ξ), are equivalent only to three equations. Thus, in the matrix of four rows and five columns formed with the coefficients of $\xi_0, \ldots, \xi_4$ in these equations, all the determinants of four rows and columns must vanish. Conversely, this is a sufficient condition for the geometrical relation. Thus (x) lies on a locus, a surface, whose order can be found by a rule, previously explained (Chap. II, pp. 75, 109), which we have associated with a symbol $[(c)/(-r)]_t^{q+1-p}$. In this case, the orders of the elements in any row of the matrix are the same, these orders being respectively k, k', 0, 0; thus the order of the surface is equal to the coefficient of t^2 in the ascending expansion of $(1 - kt)^{-1}(1 - k't)^{-1}$, namely is $k^2 + kk' + k'^2$. This surface intersects the given surface in $mm'(k^2 + kk' + k'^2)$ points, and this is the value of μ_2. Similarly, to find ν_2, the tangent plane at any point of the surface is the intersection of the tangent primes of $f = 0, f' = 0$, and the points of $f = 0$ whereat the tangent prime passes through an arbitrary point are on the first polar of this point in regard to $f = 0$. Thus, the number ν_2, of tangent planes of the surface (f, f') which pass through an arbitrary point is ν_2, $= mm'kk'$. This result also furnishes an example of the rule quoted; for, if the arbitrary point be given as the intersection of the four primes $(a_0, \ldots, a_4), \ldots, (d_0, \ldots, d_4)$, the matrix, of five rows and six columns, of which the general row consists of the elements $\partial f / \partial x_i, \partial f' / \partial x_i$, a_i, b_i, c_i, d_i, for $i = 0, \ldots, 4$, has its columns respectively of orders k, k', 0, 0, 0; and the coefficient of t^2 in the expansion of $(1 + kt)(1 + k't)$ is kk'. From the values of $\mu_0, \mu_1, \mu_2, \nu_2$, by the preceding formulae, we find
$$\zeta_0 = mm'kk', \quad d, = \tfrac{1}{2}(\zeta_0 - \nu_2), = 0, \quad \zeta_1 = 2mm'kk'(k + k' - 1),$$
$$2Q - 2 = 2mm'kk'(k + k' - 2), \qquad t = \tfrac{1}{6}mm'kk'(k-1)(k'-1),$$
as also $\quad \kappa - \nu_2 = mm'(k+k')(k + k' - 1), \quad \rho + \nu_2 = mm'kk'(k + k'),$
$$i + 2\nu_2 = 4mm'(k + k')(k + k' - 1).$$
Of these, for instance, t is also easily obtained directly by the known determinantal method (Vol. v, Chap. VIII) of finding the condition that the two equations in λ, $f_0 + \lambda f_1 + \lambda^2 f_2 + \ldots + \lambda^m f_m = 0, f_0' + \lambda f_1' + \ldots = 0$, should have three common roots—where $f_i = (x \partial / \partial \xi)^i f(\xi)/i!$, etc.

It may also be remarked (see p. 244 below) that the number of chords, from an arbitrary point lying on the surface (f, f'), which have three other intersections with the surface, is $6[m_3 m_3' - \tfrac{1}{3} m_2 m_2' + \tfrac{1}{3} m_1 m_1' - 1]$, where $m_i = m!/i!(m-i)!$.

Note I; in regard to pinch points. It has appeared in this
chapter (p. 162) that the presence, on a surface of order n, in space
of three dimensions, of a double curve of order ϵ_0, and class ϵ_1, with
t triple points, involves a diminution of the class $n(n-1)^2$ of the
surface of amount $(7n-12)\,\epsilon_0 - 4\epsilon_1 - 15t$. We now remark a relation
of this formula with that for the pinch points of the double curve,
whose number is $\nu_2,\ = 2(n-2)\,\epsilon_0 - 2\epsilon_1 - 6t$, by means of which, in
terms of the double curve, either μ_2 or ν_2 can be found when the
other is known. Take an arbitrary general line, m, and consider the
curve of intersection of the first polars, in regard to the surface, of
two points of the line. For both polars the double curve of the
surface is a simple curve; the curve, θ, which we study, is the remaining curve of intersection, of order $(n-1)^2 - \epsilon_0$. We assume, the
proof being considered below, that this curve θ passes, (i), simply
through each point of the double curve at which the tangent line
of this curve meets the given line m; (ii), simply through each triple
point of the double curve; (iii), through each pinch point in such
a way as to have three coincident intersections with the surface at
this point. At the points under (i) and (ii), the curve θ will have,
respectively, 2, and 3, intersections with the surface. The number of
intersections of θ with the surface, at points not on the double curve,
is thus $n\{(n-1)^2 - \epsilon_0\} - 2\epsilon_1 - 3t - 3\nu_2$. Each of such intersections is,
however, a point of contact with the surface of one of the μ_2
tangent planes which can be drawn through the line m. Thus we
have $\mu_2 + 3\nu_2 = n\{(n-1)^2 - \epsilon_0\} - 2\epsilon_1 - 3t$; this is the relation between μ_2 and ν_2 which we had in view. The present direct proof of
this may replace the proof of one of the equations (I)–(V) previously
considered (p. 159). With the double curve itself, the curve θ has a
number of intersections given by $\epsilon_1 + 3t + \nu_2$, which, by the value of
ν_2 found above, is the same as $2(n-2)\,\epsilon_0 - \epsilon_1 - 3t$. Consider now the
assumptions we have made. It can be shewn that the first polar of
a point (ξ,η,ζ,τ) has, for tangent plane, at a point (x,y,z,t) of
the double curve, the plane, through the tangent line of the double
curve at this point, which is the harmonic conjugate of the point
(ξ) in regard to the tangent planes of the surface at (x); this is then
the same for any two points (ξ) which lie on a plane through the
tangent line of the double curve; wherefore, the first polars of two
points (ξ), (ξ_1), whose join meets this tangent line, meet in a curve
having a double point at (x); one branch of this curve is the double
curve; thus, the residual curve of intersection, θ, of the two polars
passes through (x). This proves the assumption (i). The assumption
(ii) is also very easy to justify. For, both first polars have a conical
double point at the triple point of the double curve; and, of the four
common generators of the asymptotic cones of these surfaces at
this point, three are the tangent lines of the double curve. The

residual intersection, θ, of the two polars, thus passes simply through the triple point. Assumption (iii) is less obvious: at a pinch point, the common tangent plane of the two sheets of the surface, which we may call here the *pinch plane*, meets the surface in a curve having a triple point; of the three tangent lines of this curve, at this point, two coincide, namely along the tangent line of the double curve; to the third tangent we apply the name used by Cayley (*Papers*, VI, p. 335), and call it the *cotangent*. Every general line, in the pinch plane, through the pinch point, has three coincident intersections with the surface at this point; but the tangent line of the double curve, and the cotangent, have each four intersections. The *proper curve of contact*, of tangent planes to the surface from an arbitrary point, O, *passes through the pinch point, and touches the cotangent*; the first polar of O has the pinch plane for tangent plane (as follows from the harmonic relation remarked above for an ordinary point of the double curve). One of the inflexional lines of the first polar surface, at the pinch point, is the tangent line of the double curve; in general the cotangent is *not* the other inflexional line. The curve θ, the residual curve of intersection of two first polars, may have, in general, any direction in the pinch plane, and has then just three coincident intersections with the surface at the pinch point. We first prove these statements directly in space of three dimensions, and then consider the proof which arises from regarding the surface as obtained by projection from [4]. For the first, we assume that it is sufficient to regard the double curve as a line. Let this line be taken for $x=0$, $y=0$, the pinch plane being $x=0$. In non-homogeneous coordinates the equation of the surface, $f=0$, can then be supposed to be of the form

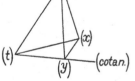

$$x^2(1+ax+cz+u)+2xy(Ax+Cz+v)$$
$$+y^2(Px+Rz+Qy+w)=0,$$

where u, v, w are of the second and higher orders in x, y, z. The pinch plane $x=0$ then meets the surface doubly in $y=0$, and also in the line $x=0$, $Rz+Qy=0$; by modifying the coordinate z we can suppose the second line to be $x=0$, $z=0$, so that $Q=0$; this line is then the cotangent. The point (x), or $y=z=t=0$, of the coordinate tetrahedron, may be supposed to be an arbitrary point of space; it will then be general to consider the first polar of this point, $(1, 0, 0, 0)$. This is given by $f_1=0$ where, with u_1 for $\partial u/\partial x$, etc.,

$$f_1=2x(1+ax+cz+u)+2y(Ax+Cz+v)+x^2(a+u_1)$$
$$+2xy(A+v_1)+y^2(P+w_1);$$

this surface has an ordinary point at the origin (t), the tangent plane being the pinch plane $x = 0$. Remarking the identity

$$2f - xf_1 = -ax^2 + xy(2Cz + Py) + 2Ry^2z$$
$$+ \text{terms of fourth order at least,}$$

we see that the proper curve of contact, of tangent planes of the surface drawn from the point (x), touches the cotangent $x = 0$, $z = 0$. Further, not only is the point (x) arbitrary, but the points (z), (y) are also arbitrary upon the tangent line of the double curve, and upon the cotangent, respectively. We can therefore suppose a second arbitrary point of space to be $(1, \eta, \zeta, 0)$. To terms of the second order in x, y, z, the first polar of this point is given by

$$f_1 + \eta\{2x(Ax + Cz) + 2y(Px + Rz)\} + \zeta\{cx^2 + 2Cxy + Ry^2\} = 0,$$

and has also $x = 0$ for tangent plane at the origin. But the curve of intersection of this with $f_1 = 0$ has tangent lines, lying in $x = 0$, given by $Ry(2\eta z + \zeta y) = 0$, so that the curve θ, the residual of the double line in the intersection of the two first polars we have considered, has the line $x = 0$, $2\eta z + \zeta y = 0$ for tangent at the origin, and has three coincident intersections with the surface at the pinch point. Only when the line, which joins the two points of which the first polars have been taken, meets the cotangent (or $\zeta = 0$), does the curve $\theta = 0$ touch the cotangent. With this analysis the reader should also consult Zeuthen, *Math. Ann.* x, 1876, where *close* points are considered as well as pinch points. It is instructive further, to consider the pinch point of the surface ϕ, in [3], as arising when we project a surface ψ, in space [4], from a point L, as above (p. 170). Any chord LPP', of the surface ψ, through L, gives rise to a single point on the double curve of ϕ, at which, for a general position of P, the tangent planes are the projections of the tangent planes of the surface ψ at P and P'; the tangent line of the double curve of ϕ, at the corresponding point, is the intersection of the space [3] with the plane through L which contains the tangent lines, at P and P', of the curve γ which projects into the double curve. If T be a point of the curve γ at which P and P' coincide, the tangent line of γ at T passing through L, then, as P and P' approach to T, the tangent planes of ψ at P and P' tend to coincidence with the tangent plane of ψ at T; this plane, however, passes through L, and does not project into the pinch plane, at the point S, of the double curve of ϕ, into which T projects. This pinch plane is the limit of the intersection, with the space [3], of the prime which may be defined as containing the tangent plane of ψ at T and a chord LPP' which is approaching the line LT; the tangent line, of the double curve of ϕ at S, is the limit of the intersection of the space [3] with the plane containing LT and LPP'; the cotangent at the pinch point S is the intersection of this space [3] with the tangent plane of ψ at T. If l be an arbitrary line through L, meeting the space [3] in O, and σ be the curve on ψ, defined by the property that the tangent plane of ψ at a point of this curve meets the line l, then σ projects into the proper curve of contact, of tangent planes to ϕ from O. The curve σ passes through T, because the tangent planes of ψ, at points near to T, ultimately meet the line l in L; moreover, the curve σ touches the tangent plane of ψ at T, so that, on ϕ, *the proper curve of contact touches the cotangent*. It does not follow, however, that two curves, σ and σ', on ψ, arising from two lines, l and l', through L, touch one another at T. For, consider a (1, 1) correspondence, between the points of ψ in the neighbourhood of T, and points of the plane (l, l') in the neighbourhood of L, determined by the point intersection of the plane (l, l') with a tangent plane of ψ at a point near T; then, to two lines l, l' through L, in the plane (l, l'), will correspond different directions from T on ψ, in general. This remark will be of importance below.

Ex. 1. We have already considered (Ex. 3, p. 164) the cubic ruled surface $x^2z - y^2t = 0$, of which the points are expressible by two parameters by $x = \phi$, $y = \theta\phi$, $z = \theta^2$, $t = 1$. For this surface the point (t) is a pinch point, with $y = 0$ as pinch plane, and the line (tx) as cotangent. Similarly (z) is a pinch point, with $x = 0$ as pinch plane and the line (zy) as cotangent. The proper curve of contact, of tangent planes from the point (a, b, c, d), whose points are there expressed in terms of θ, meets the line (xy) in two points, touches the line (tx) at (t), touches the line (zy) at (z), and meets the double line (tz) in a further point. This surface is the projection, from the point (u), in the four-fold space in which the coordinates are x, y, z, t, u, of the surface whose points are given, in terms of two parameters, by $x = \phi$, $y = \theta\phi$, $z = \theta^2$, $t = 1$, $u = \theta$. The tangent plane of this surface in [4], at the point (θ, ϕ), is given by $2\theta(\theta x - y) + \phi(z - \theta^2 t) = 0$, $z - 2u\theta + t\theta^2 = 0$; in particular, at the point (t), for which $\theta = \phi = 0$, the tangent plane is the plane (tux). The various statements made above can be easily tested in this case.

Ex. 2. Shew that if X, Y, Z respectively denote $yt - z^2$, $yz - xt$, $zx - y^2$, the elimination of θ from the two equations
$$X + hY + \theta Z = 0, \quad aX + bY + (\theta^2 + c)Z = 0,$$
leads to the equation of a rational ruled quartic surface; and, if these equations lead to $X/\lambda = Y/\mu = Z/\nu$, so that λ, μ, ν are definite polynomials in θ, shew that the generator (θ) of the quartic surface is the intersection of the planes $\lambda x + \mu y + \nu z = 0$, $\lambda y + \mu z + \nu t = 0$. Shew further that, on the double curve of this surface, which is given by $X = Y = Z = 0$, the point $(0, 0, 0, 1)$ is a pinch point, at which the (torsal) generator is $y - hx = 0$, $z - hy = 0$. Also, that the proper curve of contact, for tangent planes to this surface from the point (ξ, η, ζ, τ), is the (sextic) locus of the point where the generator (θ) is met by the plane
$$(2\theta - a)A + (2h\theta - b)B + (\theta^2 - c)C = 0,$$
wherein, for brevity, A, B, C denote, respectively $y\tau - 2z\zeta + t\eta$, $-x\tau + y\zeta + z\eta - t\xi$, $x\zeta - 2y\eta + z\xi$; and shew that this curve touches the generator at $(0, 0, 0, 1)$, and that this generator is the cotangent at this pinch point (cf. Cayley, *Papers*, VI, p. 127).

Ex. 3. And so in general, on a ruled surface, to use a crude description, let O be a pinch point on the double curve, and Q, Q' be the points, on the proper curve of contact, which lie on two generators PQ, PQ' through a point P of the double curve; then, as P approaches O, the generators PQ, PQ' "open out" until, when P is at O, they are in the same line—the (torsal) generator at O, and the points Q, Q' both approach to P, and coincide, at O, in the point of contact of the proper curve of contact with the torsal generator.

Ex. 4. Verify the results, for the neighbourhood of a pinch point, given above in this Note, by forming the equation of the enveloping cone drawn from $(1, 0, 0, 0)$, namely, by forming the discriminant, in regard to θ, of the equation. It is useful to notice that the θ-discriminant of an equation of the form $a + b\theta + c\theta^2 + \ldots = 0$ has a form $aU + b^2V$, where U, V are polynomials in a, b, c, ... This is easily seen from the Sylvester determinantal form of this discriminant.

Note II, to p. 163 preceding. Noether's formulae, quoted by himself ("Zur Theorie des eindeutigen Entsprechens algebraischer Gebilde, Zw. Auf.," *Math. Ann.* VIII, 1875), from his own paper, "Sulle curve multiple di superficie algebriche," *Ann. d. Mat.* V, 1871–3, are as follows:

The surface, of order n, has j-fold curves C_j, of order m_j, and rank r_j; a curve C_j has s_{jj} actual double points, and meets a curve $C_{j'}$ in $s_{jj'}$ further points $(j \leqslant j')$. The surface also has k-fold nodal points, P_k, through which a curve C_j passes with h_{jk} branches, at which the osculating cone (with no further multiple generators) has class ν_k. Then the class of the surface, say μ_2, and the class of the envelope of its stationary tangent planes, say i, are given by

$$\mu_2 = n(n-1)^2 - \sum_j (j-1)\{(3j+1)n - 2j(j+1)\}\, m_j + \sum_j j^2(j-1)r_j$$

$$- \sum_k (k-1)\nu_k + \sum_{jj'} (j-1)\{(3j+1)j' - j(j+1)\}s_{jj'}$$

$$+ \sum_{jk} (j-1)(2j+1)(k-j)h_{jk},$$

$$i = 4n(n-1)(n-2) - 2\sum_j j(j-1)\{2(3n-2j-2)m_j - (2j-1)r_j\}$$

$$- \sum_k (4k-5)\nu_k + 4\sum_{jj'} j(j-1)(3j'-j-1)s_{jj'}$$

$$+ 8\sum_{jk} j(j-1)(k-j)h_{jk},$$

where the $\sum_{jk}$ extend to all the branches of the curves C_j which pass through P_k, and the $\sum_{jj'}$ extend to all the intersections of two curves C_j outside the P_k, and to all the actual double points of a curve C_j.

For convenience of reference, we add here, also from Noether's paper, the following formulae, of which the meaning appears in the following chapter:

$$p_n = \tfrac{1}{6}(n-1)(n-2)(n-3)$$
$$-\tfrac{1}{6}\sum_j j(j-1)\{(3n-2j-5)m_j - \tfrac{1}{2}(2j-1)r_j\} - \tfrac{1}{6}\sum_k (k-2)\nu_k$$
$$+\tfrac{1}{6}\sum_{jj'} j(j-1)(3j'-j-1)s_{jj'} + \tfrac{1}{3}\sum_{jk} j(j-1)(k-j)h_{jk},$$

$$p^{(1)} = n(n-4)^2 + 1 - \sum_j (j-1)\{(3j-1)n - 2j(j+3)\}m_j + \sum_j j(j-1)^2 r_j$$
$$-\sum_k \{k + (k-3)\nu_k\} + \sum_{jj'} (j-1)\{(3j-1)j' - j(j+1)\}s_{jj'}$$
$$+\sum_{jk} (j-1)(2j-1)(k-j)h_{jk}.$$

The reader may also consult Zeuthen, *Lehrbuch der abzählenden Methoden der Geometrie*, Leipzig, 1914, pp. 160–70.

CHAPTER V

INTRODUCTION TO THE THEORY OF THE IN-VARIANTS OF BIRATIONAL TRANSFORMATION OF A SURFACE, PARTICULARLY IN SPACE OF THREE DIMENSIONS

Preliminary; a general survey. The genus of a plane curve, it is well known, may be defined by drawing lines, in the plane of the curve, through an arbitrary fixed point, O, to meet the curve, say, in m variable points, and considering the coincidences of two (consecutive) points, belonging to the intersections of one of these lines, with the curve, which may arise as the line varies. If w be the whole number of such coincidences, for all the lines through O for which such coincidences occur, with suitable allowance for multiple coincidences, the genus of the curve is $\frac{1}{2}w - m + 1$. In regard to this, it is proved, (i), that this number is the same for all positions of O in the plane, whether O be on the curve or not, and however intricate may be the multiplicities existing on the curve, if w be properly defined in each case; and, (ii), that this number is the same for any other plane curve which is in (1, 1) birational correspondence with the original curve. This definition of the genus can, it is known, be generalised, by considering, instead of the point series obtained on the curve by variable lines through a fixed point, any linear series of freedom 1 on the curve. The aggregate of the points of coincidence of two (consecutive) points of a set of the series, for all the sets in which coincidences occur, forms a set of points which is often called the Jacobian set of the series; if m be the number of points in any set of the series, and w the number of points in the Jacobian set, with suitable allowance for points at which more than two points of a set of the original series coincide, the genus of the curve is still $\frac{1}{2}w - m + 1$. And more generally still, if $g_r{}^m$ denote any linear series of sets of m points on the curve, with freedom r, we may choose from this a linear series of freedom 1, say $g_1{}^m$, and consider the Jacobian set of this series; all the Jacobian sets so obtainable, for all possible series $g_1{}^m$ contained in the series $g_r{}^m$, are then known to belong to the same linear series; this we may speak of as the Jacobian series of $g_r{}^m$. If w be the number of points in any set of the Jacobian series, $\frac{1}{2}w - m + 1$ is the genus of the curve.

When we seek to generalise these ideas to the case of surfaces, the primary definition of the genus of a curve, as here considered, which is effectively by the number of tangent lines which can be drawn to the curve from an arbitrary point, suggests the study of the enveloping cone to the surface, drawn from an arbitrary point, with a view to finding, if possible, numbers defined by the cone, or by the curve sections of the surface lying on planes through its vertex, which, (i), shall be independent of the position of the vertex of the cone, and, (ii), shall be equal to the numbers similarly defined for enveloping cones to another surface which is in (1, 1) birational correspondence with the original surface. This procedure, with the enveloping cone, was followed by Zeuthen, *Math. Ann.* IV, 1871, p. 21; and, with acknowledgments to Zeuthen and Clebsch, by Noether, *Math. Ann.* VIII, 1875, p. 495. A carefully revised introductory account is given by Zeuthen, *Lehrbuch der abzählenden Methoden der Geometrie*, 1914, pp. 160–7. But the sections of a surface, by planes through a fixed point, are a particular linear ∞^2 system, say a *net*, of curves on the surface; and it may be hoped to obtain invariants of the surface by consideration of any linear ∞^2 system, or net, of curves on the surface. For one invariant, I, it appears indeed that it is sufficient to consider a linear ∞^1 system, or *pencil*, of curves on the surface; this was shewn by Segre, *Atti... Torino*, XXXI, 1895–6, p. 485. For another invariant, ω, a definition was obtained with a net of curves by Severi, with special acknowledgments to Segre, in the *Atti...Torino*, XXXVII, 1901–2, p. 625. In many respects, the arguments followed by Segre and Severi are a direct generalisation of those of Zeuthen and Noether, and the results are the same. But they have a more geometrical form, and an elementary introduction will be given here from this point of view. The analytical work of Noether, however, provides a proof that the only possible invariants are those which are obtained here; and provides material for the consideration of intricate possibilities in the singularities of the surface considered; the account in Zeuthen's *Lehrbuch*, referred to above, provides the geometrical aspect of Noether's formulae, for the simpler cases. For greater complexity, Zeuthen's paper, *Math. Ann.* x, 1876, p. 446, and Noether's paper, *Ann. d. Mat.* v, 1871–3, p. 162, should be consulted. For the treatment of the canonical curves of a surface contained in this chapter, special acknowledgment is due to Enriques' paper, "Intorno ai fondamenti della Geometria sopra le superficie algebriche," *Atti...Torino*, XXXVII, 1901–2; as Enriques points out, the use of the Jacobian determinant is essentially involved in the detailed formulae of transformation of Clebsch's everywhere finite double integral, as developed by Noether. For the treatment of the invariant ω similar acknowledgment is due to Severi's paper, in the same volume of the *Atti...Torino*, referred to above, "Il genere aritmetico ed il genere lineare in relazione alle reti di curve tracciate sopra una superficie algebrica." The approach we make to the theory here is designedly of a tentative character, by a method of successive approximation, beginning with the simplest cases. Since this account was prepared (and given in lectures, substantially as here), there has appeared the volume of Enriques-Campedelli, *Lezioni sulla teoria delle superficie algebriche*, Parte prima, 1932, which contains a brilliant summary of the results obtained by Enriques, since his first " Ricerche " in the *Mem....Torino* of 1893, as well as by Castelnuovo and Severi; but, in the writer's view, an elementary introduction still remains desirable.

It may seem that it should be possible to evade the consideration of complex singularities on a surface, in a manner analogous to that followed for a plane curve, by the consideration of algebraic integrals associated with the surface. It was in fact the suggestion, by Clebsch (*Compt. rend.* LXVII, Dec. 1868, p. 1238), that we may consider everywhere finite *double* integrals for a surface, which gave the first assured hope of invariants of the birational transformation of surfaces. But more detailed consideration has shewn that the invariant so suggested (the number, p_g, of everywhere finite double integrals) does not always coincide with, and cannot be used to replace, another invariant p_n, which arises by such considerations as we have sketched; curiously enough, it is the difference of these two invariants, $p_g - p_n$, which finally emerges as the most important invariant of birational transformation. The existence of two invariantive genera, p_g and p_n, for a surface is one important difference of the theory from that of a curve. Another vital difference arises from the fact that, when one surface is transformed into another, a point of the original surface may fail to give a definite corresponding point of the other; the various methods of approach to the point on the first surface may all give rise separately to different points on the new surface, whereon there thus arises a *curve* to correspond to the *point* of the original. This possibility is familiar in the transformation of one plane into another plane. Let e' be the number of curves of the new surface which thus arise from points of the original surface; and e the number of curves of the original surface which similarly transform into points of the new surface. Curves thus arising by transformation from points are called *exceptional* curves; the name is often applied only to the case when the points in question are *simple* points of the surface on which they lie. We shall define invariants I, ω for the original surface, and, correspondingly, invariants I', ω' for the new surface, such that, subject to appropriate modification when either surface possesses multiple points, we have $I - e = I' - e'$ and $\omega + e = \omega' + e'$. Because of the corrective terms e, e' here, which depend on the particular transformation, I and ω are spoken of as relative invariants; from these an absolute invariant p_n can be defined by means of the equation $I + \omega = 12p_n + 9$. A form of this equation which is geometrically significant is $(I+4)+(\omega-1)=12(p_n+1)$. In regard to I and ω, we consider, apart, the two questions: first, of shewing that either can be defined in terms of systems of curves on a particular surface, and is the same whatever system be used, so that it is an invariantive property of this surface; and second, of finding the modification which arises by a birational transformation to another surface. The surfaces considered first of all are supposed to lie in ordinary space of three dimensions, and to have no multiple points other than a double curve with triple points, as explained in the preceding chapter; but the aim is to present the theory so that the definitions may be capable of generalisation to any surfaces whatever.

Definition and illustration of the invariant I. Consider, upon the surface considered, any linear ∞^1 system, or pencil, of curves; denote by p the genus of the general curve of the system, which is supposed to be irreducible; such a pencil of curves is given by the intersection of the surface with a system of surfaces having an equation of the form $U + \lambda V = 0$, where λ is variable, but U, V are definite (of the same order). The curves of the system may have assigned base points, common to all; these are supposed, unless the contrary is said, to be at simple points of the surface, and the base point has an assigned multiplicity for the curves of the pencil, all the tangents of the curve thereat being free. The number of points of the surface at which such base points occur is denoted by σ. Among the curves of the pencil there will generally be a finite number possessing a double point, other than at the base points of the pencil, and other than on multiple curves of the surface; in such case let the whole number of double points so arising be denoted by δ. There is difficulty in case there are curves in the pencil which are multiple, or contain a multiple part; and the question arises whether in such case a consistent rule can be given for estimating the corresponding contribution to δ. In the simplest cases, however, it can be proved that the number $\delta - \sigma - 4p$ is the same for any two linear pencils of curves on the surface; it is this number which is the invariant I spoken of. According to the plan we adopt, we consider various simple cases before passing to the proof of the invariance of I in general. The most elementary example is that of a pencil of plane sections of the surface, made by planes through a fixed line of general position. In this case, as the base points of the curves of the system are the intersections of the line with the surface, σ is the order (μ_0) of the surface; likewise δ is the number of planes through the line which touch the surface, the class (μ_2) of the surface; if p is the genus of a general plane section of the surface, the rank (μ_1) of the surface, equal to the class of an arbitrary plane section of the surface, is such that $\mu_1 = 2\mu_0 + 2p - 2$, there being no cusps on an arbitrary plane section when the surface has no cuspidal curve. Hence we have $I_1 = \mu_2 - \mu_0 - 4p$, $= \mu_2 - 2\mu_1 + 3\mu_0 - 4$. In particular, for a general surface of order n, with no multiple curve (and no multiple points), so that $\mu_2 = n(n-1)^2$, and $p = \frac{1}{2}(n-1)(n-2)$, this gives $I = (n-2)(n^2 - 2n + 2)$. Thus, for a general cubic surface $I = 5$; for a quadric surface $I = 0$, and, for a plane $I = -1$.

It will be interesting to verify at once, as an example of the general theorem of the invariance of I for a given surface, that the value $I = -1$, just obtained, essentially by a pencil of lines in the plane, is obtained also, from the original definition of I, by considering a general pencil of curves in the plane; we take the case in which the assigned base points are the intersections of any two curves of the pencil, supposed, first, all to be simple intersections. The curves being then given by an equation $\phi + \lambda\psi = 0$, where ϕ and ψ are of order n, the double points arise for values of the parameter λ for which the three curves $\phi_1 + \lambda\psi_1 = 0$, $\phi_2 + \lambda\psi_2 = 0$, $\phi_3 + \lambda\psi_3 = 0$ meet in a point, where $\phi_1 = \partial\phi/\partial x$, etc. The order in λ, of the eliminant* of these three equations, is $3(n-1)^2$, so that there are $3(n-1)^2$ curves of the pencil which have double points; taking then $\sigma = n^2$, $p = \frac{1}{2}(n-1)(n-2)$, we obtain

$$I = 3(n-1)^2 - n^2 - 2(n^2 - 3n + 2),$$

which is -1 whatever n may be. If, however, second, we suppose that all the curves of the pencil have a base point which is multiple, of order s, the remaining base points being simple, then $\sigma = n^2 - s^2 + 1$ and $p = \frac{1}{2}(n-1)(n-2) - \frac{1}{2}s(s-1)$; the value $I = -1$ is then only obtained if we take $\delta = 3(n-1)^2 - (s-1)(3s+1)$. Thus, when we have proved that the value of I is independent of the pencil of curves with which it is defined, it will follow that the number of curves having double points, in a pencil which has a single s-fold common point, is less than the general number $3(n-1)^2$ by $(s-1)(3s+1)$; or, as we easily see, in the more general case of several multiple points, by the sum of the separate terms $(s-1)(3s+1)$ arising for these singly. For example, in the simple case of a pencil

---

* This is obtainable as an example of a rule which is often useful in general cases: Let $f_i(x, a)$ be a rational polynomial homogeneously of order n_i in the coordinates $x_0, \ldots, x_r$, and homogeneously of degree e_i in the parameters $a_0, \ldots, a_s$. Then the eliminant, in regard to $x_0, \ldots, x_r$, of the $r+1$ equations $f_i(x, a) = 0$, for $i = 0, \ldots, r$, is homogeneously of degree $n_0 n_1 \ldots n_r \Sigma e_i/n_i$ in the parameters $a_0, \ldots, a_r$. The eliminant may in fact be formed, for example, by taking the product of the $n_1 \ldots n_r$ factors $f_0(x^{(k)}, a)$, in which $(x^{(k)})$ denotes one of the $n_1 \ldots n_r$ points common to the r loci $f_1(x, a) = 0 = \ldots = f_r(x, a)$; this procedure shews the degree in which the parameters $a_1, \ldots, a_r$ enter from $f_0(x, a)$; the other similar methods of forming the eliminant shew the influence of the other functions $f_1(x, a), \ldots, f_r(x, a)$, in the aggregate entry of the parameters in this eliminant.

Invariants of surfaces 187

of plane cubic curves with one given double point, and five simple points common, a cubic with a double point is obtained, as a composite curve, by the conic through the double point and four of the simple points, taken with the line joining the simple point to the double point; and all the $12 - 7$, or 5, cubics of the pencil which have double points, not at the base points, are accounted for in this way.

Ex. If $u = 0$, $v = 0$, $w = 0$ be three general quadric surfaces, a linear pencil of elliptic curves, with 8 simple base points, is determined on the quadric surface $w = 0$, by means of the system of quadric surfaces $u + \lambda v = 0$. Find the number of curves of this pencil which have double points.

As another illustration, consider the case of a surface, of order n, which has a j-fold line, but is in other respects general. We can find directly the number of tangent planes to the surface through a general line, or the number through the j-fold line, and compute the value of I in both ways. But the assumption of the identity in the values of I in the two cases would enable us to infer the number of tangent planes to the surface from either line from the number from the other line. We shall see that the value of I is less than the value found above, in the general case, namely $(n-2)(n^2 - 2n + 2)$, by $(j-1)\{(3j+1)n - 2j(j+2)\}$. Take first the case of the tangent planes which can be drawn to the surface from the j-fold line. This line being taken to be $x = 0 = y$, the equation of the surface is $f = 0$ where f is of the form $x^j u + x^{j-1} yv + \ldots + y^j w$, where $u, v, \ldots, w$ are of order $n - j$ in the coordinates x, y, z, t. The tangent plane at a point (x) of the surface, having an equation of the form $(x' \partial/\partial x)f = 0$, in which (x', y', z', t') are current coordinates, will contain the multiple line when the point of contact (x) is such that $\partial f/\partial z = 0$ and $\partial f/\partial t = 0$. The points of contact of tangent planes, from the multiple line, thus lie on two surfaces of order $n - 1$, with equations of the forms $x^j u_z + x^{j-1} yv_z + \ldots = 0$, $x^j u_t + x^{j-1} yv_t + \ldots = 0$, where $u_z = \partial u/\partial z$, etc., these representing the first polars of the two points $(0, 0, 1, 0)$, $(0, 0, 0, 1)$ of the multiple line. These surfaces meet, outside the multiple line, in a curve of order $(n-1)^2 - j^2$. At a point where this curve of intersection meets the multiple line, one of the j tangent planes of the former of the two polars coincides with one of the j tangent planes of the latter. The values of z/t at these points of meeting are therefore obtainable by putting $x = 0$ and $y = 0$ in the coefficients $u_z, \ldots, u_t, \ldots$, and then forming the eliminant, in regard to x/y, of the reduced equations. As $u_z, \ldots, u_t, \ldots$ are of order $n - j - 1$, we thus see that the curve of intersection of the polars meets the j-fold line in $2j(n - j - 1)$ points. But, further, at such an intersection, these two polars have a common tangent plane with $f = 0$; for, when $x = 0$, $y = 0$, we have identities such as

$$zu_z + tu_t = (n-j)u, \quad zv_z + tv_t = (n-j)v, \quad \text{etc.};$$

the common curve of the two polars thus touches one of the j sheets of $f=0$ at the intersection, and meets the other $j-1$ sheets, thus having $j+1$ intersections with $f=0$. The number of intersections of this curve with $f=0$, other than on the multiple line, is thus $A-B$, where $A=n[(n-1)^2-j^2]$ and $B=2(j+1)j(n-j-1)$; it is easy to see that $A-B$ is the same as $(n-j-1)^2(n+2j)$, or as $C-D$, where $C=n(n-1)^2$ and $D=j[(3j+2)n-2(j+1)^2]$; this gives the number of tangent planes to the surface $f=0$ from the multiple line. Assuming now that the general plane section of $f=0$, through the multiple line, has no fixed intersections therewith, putting therefore $\sigma=0$ in the defining formula for I, we obtain

$$I=(n-j-1)^2(n+2j)-2(n-j-1)(n-j-2);$$

this is the same as that stated above. For instance, for a quartic surface with a double line, $I=8$. The number of tangent planes to the surface from an arbitrary line may be found in a similar way (given below) and is $C-E$, where $C=n(n-1)^2$ and

$$E=(j-1)\{(3j+1)n-2j(j+1)\}.$$

Calculating then I from this as the value of δ, with $\sigma=n$, and $2p=(n-1)(n-2)-j(j-1)$, we easily find the same value of I as before.

The proof of this value of δ may be given: The first polars of two points of an arbitrary line have the multiple line of the surface as $(j-1)$-fold line, and meet, beside, in a curve of order $(n-1)^2-(j-1)^2$; an arbitrary plane section of these polars, by a plane through the multiple line, gives curves of order $n-j$, with $(n-j)^2$ intersections; thus the curve of intersection of the two polars meets the multiple line in a number of points given by $(n-1)^2-(j-1)^2-(n-j)^2$, which is $2(n-j)(j-1)$. There will be no loss of generality in assuming the two first polars to be those of $(1, 0, 0, 0)$ and $(0, 1, 0, 0)$; they will then have equations of which one is $U+U_x=0$, in which U denotes $jx^{j-1}u+(j-1)x^{j-2}yv+\dots=0$, and U_x denotes $x^j u_x+x^{j-1}yv_x+\dots=0$; thus their curve of intersection meets the multiple line ultimately on planes, through this line, satisfying both the equations $(U)=0$, $(V)=0$, where $(U)=jx^{j-1}(u)+(j-1)x^{j-2}(v)+\dots$, $(V)=x^{j-1}(v)+\dots+jy^{j-1}(w)$, where (u), (v), ..., (w) are u, v, ..., w, with x, y both replaced by zero; namely, this curve meets the multiple line at points where two of the j tangent planes of $f=0$ coincide; the number of such points is $2(j-1)(n-j)$. And, as before, the two polars have a common tangent plane at this point, which coincides with the pair of coincident tangent planes of $f=0$; so that the curve of intersection of the two polars meets $f=0$ in 3 coincident points on this plane, and in $j-2$ other points, or, in all, in $(j+1)$ points. The total number of intersections, of the common curve of the two polars, with $f=0$, on the multiple line, is thus $2(j+1)(j-1)(n-j)$; whence, the number of remaining intersections of this curve with $f=0$, each corresponding to a tangent plane to the surface from the arbitrary line, is $n\{(n-1)^2-(j-1)^2\}-2(j^2-1)(n-j)$; this is the same as the number $C-E$ stated above. For the number of tangent planes to the surface through the multiple line, Berzolari (in Pascal, *Repertorium*, ii, 2, 1922, p. 665) refers to Fouret, *Rend. Palermo*, viii, 1894, p. 202, and to Godeaux, *Nouv. Ann.* ix, 1909, p. 162.

We may illustrate the definition of I, also, by considering a pencil of curves on a perfectly general surface, given by an equation $\phi + \lambda\psi = 0$, where λ is variable, in which the surfaces $\phi = 0$, $\psi = 0$ have no common base curve or common point on $f = 0$. The double points of the curves of the pencil, being such that

$$(\phi_1 + \lambda\psi_1)/f_1 = \ldots = (\phi_4 + \lambda\psi_4)/f_4,$$

where $\phi_1 = \partial\phi/\partial x$, etc., are the intersections of $f = 0$ with the curve common to the surfaces each represented by the vanishing of a determinant, of three rows and columns, in the matrix of three rows and four columns of which the general column consists of the elements ϕ_i, ψ_i, f_i. If m be the order of the surfaces $\phi = 0$, $\psi = 0$, this curve, in accordance with a known rule (p. 109), which we associate with a formula $[(c)/(-r)]_{t^{q+1-p}}$, has an order equal to the coefficient of t^2 in the ascending expansion of $(1 - \mu t)^{-2}(1 - \nu t)^{-1}$, where $\mu = m - 1$ and $\nu = n - 1$; this order is thus $3\mu^2 + 2\mu\nu + \nu^2$; the number, δ, of double points of curves of the pencil is n times this. The value of I is given by $\delta - m^2 n - 4p$, where

$$p = 1 + \tfrac{1}{2}mn(m + n - 4);$$

this is easily verified to be independent of m, and the same as the value first given, $(n-2)(n^2 - 2n + 2)$. It would be proper to verify the value of I in this way when the surfaces $\phi = 0$, $\psi = 0$ have a common base curve on $f = 0$; or conversely, the assumption of the invariance of I would enable us to compute the number of double points of curves of such a pencil. For instance, if a pencil of surfaces of order m be drawn through a fixed base curve, of order N and genus P, on a cubic surface, the resulting pencil of curves on the cubic surface has a number of double points given by

$$\tfrac{3}{2}m(5m - 3) - N(5m - 4) + 5P + 3,$$

when m is large enough in regard to N.

Note. Definition of I in terms of an irrational pencil of curves. In order to obtain a formula which it is interesting to compare with the results of this chapter, we anticipate here ideas which are more fully explained later. The fundamental surface, which we suppose to be without multiple points, and possibly lying in space of higher than three dimensions, may be such that there exists upon it a system of curves which, though not given by the intersection of it with a linear pencil of primals $\phi + \lambda\psi = 0$, still has the property that a single curve of the system passes through an arbitrary point of the surface. The generators of a ruled surface, when this is not a rational surface, form an obvious example of this possibility. More generally, if ξ, η be variables connected by a definite irreducible polynomial equation $\chi(\xi, \eta) = 0$, which may then be regarded as representing a plane curve, there may be rational functions of the coordinates, say u and v, such that the primals given by $u=\xi$, $v=\eta$, have a common curve of intersection with the fundamental surface when ξ, η are connected by $\chi(\xi, \eta)=0$; this requires that the equation $\chi(u, v)=0$ should be identically satisfied in virtue of the equations of the fundamental surface. When this is so, an arbitrary point of the surface, by determining the values of ξ and η, determines a definite curve of the system passing through this point. Such a system of curves is called an *irrational pencil of curves*, under the hypothesis that the curve $\chi(\xi, \eta)=0$ is not a rational curve; and the genus of $\chi(\xi, \eta)=0$ is called the *genus of the pencil of curves*. It can be shewn that when $\chi(\xi, \eta)=0$ is a rational curve, the curves of the pencil are curves of a linear pencil as defined before. We may suppose that there is a definite $(1, 1)$ correspondence between the curves of the irrational pencil, and the points of the representative curve $\chi(\xi, \eta)=0$. Denote the genus of this curve by ρ', and the genus of each curve of the pencil by ρ. Under the hypothesis $\rho' > 0$, we can suppose that the curves of the irrational pencil have no common base point, since they would else be in $(1, 1)$ correspondence with the rational pencil of lines formed by the tangents of the fundamental surface at this (necessarily simple) point. We prove that the invariant I of the surface can be defined by $I = \delta + (2\rho-2)(2\rho'-2) - 4$, where δ is the number of double points of curves of the irrational pencil. If ρ' were zero, this would be $\delta - 4\rho$, but would then be replaced by $\delta - \sigma - 4\rho$, where σ is the number of base points of curves of the pencil. This general formula is to be deduced from the case when I is defined by a linear pencil of curves, by remarking that a linear series g_1^m, of sets of m points, on the representative curve $\chi(\xi, \eta)=0$, of freedom 1, corresponds to a series of *sets of m curves* of the irrational pencil on the fundamental surface; and the composite curve, formed of these m curves of the pencil, is one of a linear system of freedom 1, that is of a linear pencil of composite curves, on the surface. From the known result, for a linear series g_1^m on $\chi(\xi, \eta)=0$, it follows that, among the sets of m curves of the pencil, there are $2m + 2\rho' - 2$ cases in which the two components of a set coincide (or, more than two curves may coincide, a possibility here disregarded). In the definition of I by a linear pencil of curves, we have supposed the curves of the pencil to be irreducible, and have definitely excluded the case in which particular curves of the pencil contain multiple parts; for our present purpose, we must assume a value for the number of double points to which the occurrence of a repeated part, in a curve of the linear system, is to be counted equivalent. Suppose that the preceding definition of I by a linear system of curves remains valid when, as here, the curves of the system are formed by sets of m curves of an irrational pencil, each of genus ρ; suppose further that when two curves of the set coincide, this is counted as equivalent to the occurrence of h double points in the curve of the linear system. Then, as there are $2m + 2\rho' - 2$ such coincidences, we have thus a contribution of $h(2m + 2\rho' - 2)$ to the number of double points; if δ be the number of actual double points of curves of the irrational pencil, the total number of double points in the linear system of reducible curves is therefore $\delta + h(2m + 2\rho' - 2)$. The genus of the composite curve of the linear pencil is found from a principle, arising below (p. 223), that the genus of a curve which degenerates into two curves of genera p_1, p_2, intersecting in i points, is $p_1 + p_2 + i - 1$; as two curves of the irrational pencil do not intersect, the composite curve, formed of m of these, has a genus $m\rho - (m - 1)$. The value of I is therefore given by $\delta + h(2m + 2\rho' - 2) - 4m(\rho - 1) - 4$, or $\delta + h(2\rho' - 2) - 4 + 2m(h - 2\rho + 2)$. In this argument, m is arbitrarily great $(> \rho')$; and it is natural to suppose that the number, h, of double points, to which two coincident curves of the irrational pencil are to be counted equivalent, is independent of m; if, therefore, I have a definite value, from whatever linear series it be computed, it must be independent of m, and thus the value to be taken for h is $2\rho - 2$. This gives the value of I stated above. The direct justification of this value of h belongs below, where we consider the proof that I is an invariant. We compare here a result previously found (Chap. I, p. 19), that if, on a curve of genus ρ, there be an irrational series, of genus π, of sets of k points, of freedom 1, the number of coincidences, of pairs of points of a set of this series, is $2k + 2\rho - 2 - 2k\pi$. Two curves of the irrational pencil here considered have no intersection; but, if two curves had k intersections, then, upon any one curve of the pencil, the other curves would determine an irrational series, of sets of k points, of freedom 1 and genus ρ', equal to the genus of the irrational pencil; there would then be $2k + 2\rho - 2 - 2k\rho'$ points, on this one curve, at which two of the intersections of another curve of the pencil coincided; this number is $2\rho - 2$ when, as is the case, $k = 0$. The extension of the definition of I to the case here considered is given by Castelnuovo and Enriques, *Ann. d. Mat.* VI, 1901, pp. 21–6 (see below, p. 209). If we apply it to the case of a ruled surface of genus p, the irrational pencil consisting of its generators, taking $\delta = 0$, $\rho = 0$, $\rho' = p$, we thus find $I = -4p$. This is true also for $p = 0$, and can be obtained independently, by considering a linear pencil of plane sections through an arbitrary line, from $\delta - \sigma - 4p$; since then $\delta = \sigma$, the class of the surface being equal to its order.

Introductory definition of the invariant ω. Consider, on the fundamental surface, $F = 0$, a *net* of curves, that is, a linear system of curves of freedom 2. Such a net will be determined by the intersection of $F = 0$ with a linear system of surfaces having an equation of the form $\lambda f + \mu \phi + \nu \psi = 0$, in which λ, μ, ν are variable, and $f = 0$, $\phi = 0$, $\psi = 0$ are definite surfaces, of the same order, which may have common points and common curves, on $F = 0$. Then consider the locus of the double points of curves of the net, other than at the base points common to all of $f = 0$, $\phi = 0$, $\psi = 0$, and other than those which may arise on multiple curves, or at multiple points, of $F = 0$, if such exist. We have already, in the preceding chapter, been concerned with such a net of curves. As has been said, through such a double point of a curve of the net, there will pass a pencil of curves of the net, of which one curve may be taken to be the curve which has a double point thereat; all curves of the net through this point will thus have the same tangent line at this point (the particularity that all curves of the net have a double point thereat being excluded). Conversely, if, at a simple point of $F = 0$, all curves, of the pencil of curves of the net which pass through this point, have the same tangent line, there is a curve of the net which has a double point at this point. The locus of such double points is a curve which we call the *Jacobian* of the net. More generally, if we have any linear system of curves on $F = 0$, given by a system of surfaces of the same order expressed by an equation $\lambda_0 \vartheta_0 + \lambda_1 \vartheta_1 + \ldots + \lambda_r \vartheta_r = 0$, of freedom $r(\geqslant 3)$, common points or curves of the surfaces $\vartheta_0 = 0, \ldots, \vartheta_r = 0$ lying on $F = 0$ not being considered, then we may form a net of curves $\lambda f + \mu \phi + \nu \psi = 0$ from this system, in which $f = 0$, $\phi = 0$, $\psi = 0$ determine particular curves of the system, given say by such equations as

$$f = \sum_{i=0}^{r} a_i \vartheta_i, \quad \phi = \sum_{i=0}^{r} b_i \vartheta_i, \quad \psi = \sum_{i=0}^{r} c_i \vartheta_i,$$

in which a_i, b_i, c_i are definite constants; and may then consider the Jacobian of this net. It can be proved, as below, that all Jacobian curves so obtainable, by taking different systems of values of a_i, b_i, c_i, belong to one linear system of curves on $F = 0$; this we call the *Jacobian system* of the original linear system of curves.

We can form an equation satisfied by the Jacobian curve of a net of curves on $F = 0$, given by $\lambda f + \mu \phi + \nu \psi = 0$, by expressing that the latter surface touches $F = 0$, and eliminating λ, μ, ν. The result is the vanishing of a determinant, say Δ, which we call the Jacobian determinant, of four rows and columns, whose rows consist respectively of the elements f_1, f_2, f_3, f_4; $\phi_1, \phi_2, \phi_3, \phi_4$; $\psi_1, \psi_2, \psi_3, \psi_4$; F_1, F_2, F_3, F_4, where $f_1 = \partial f / \partial x$, etc.; but from the complete intersection of $F = 0$, $\Delta = 0$, must be omitted points expressly excepted in the definition of the Jacobian curve. And it is easy to see that if, in Δ, we suppose f to mean $\Sigma a_i \vartheta_i$, with similar values for ϕ and ψ, as explained above, then Δ is a sum of Jacobian determinants, formed with triads from $\vartheta_0, \ldots, \vartheta_r$, with determinant coefficients depending on the constant coefficients a_i, b_i, c_i; this shews that the Jacobian curves of all nets from a general linear system of curves belong to a linear Jacobian system. It can be shewn directly that, at a simple point of $F = 0$, where the curves $\vartheta_i = 0$ have a common base point of multiplicity s, the Jacobian curves have a base point of multiplicity $3s - 1$. By the *complete* Jacobian system, of a given linear system of curves on $F = 0$, is to be understood the most general linear system of curves which includes the Jacobian curves formed from all nets of the given linear system, and has a $(3s - 1)$-ple point at every s-fold point, simple on $F = 0$, of this given system*. This complete system may well include curves not expressed by the vanishing of the determinant we have used. For example, when the given linear system consists of all plane sections of the surface, the equation $\Delta = 0$ is that of the first polar of a point in regard to the surface $F = 0$; if the surface have a double curve, this first polar passes through the double curve, and a Jacobian curve, according to the definition, consists of the residual inter-section; that is, of the curve which, in the last chapter, we have called a *proper* curve of contact. We have seen, however, that all proper curves of contact pass through the pinch points of the double curve; these, not being base points of the general plane section of the surface $F = 0$, are not to be regarded as conditions for the complete Jacobian system of the system of plane sections, unless this be explicitly stated. The system of all first polars is of freedom 3; the freedom of the complete Jacobian system may well be greater than this. Another fact, in regard to which clearness is desirable, presents itself when we consider the Jacobian curve of a net of curves given by an equation of the form $u(\lambda f_1 + \mu \phi_1) + \nu \psi = 0$; for such a net there exists a curve, given by $u = 0$, which meets the curves of the net only at base points, common to all the curves of the net; and the curves of the net which pass through an arbitrary point which lies on $u = 0$ are those for which the coefficient ν is zero; all these curves are composite, consisting of the curve given by $u = 0$, together with those of the pencil given by $\lambda f_1 + \mu \phi_1 = 0$; of this pencil, one curve passes through the point in question, so that the point is a double point of a curve of the net. Correspondingly, it may be seen directly (see below, p. 218) that u is a factor of the Jacobian determinant of the net. More generally, for a system of curves given by an equation of the form $u(\lambda_0 \vartheta_0 + \ldots + \lambda_{r-1} \vartheta_{r-1}) + \lambda_r \vartheta_r = 0$, the

* It is understood that the tangents at a base point of the net are all variable, and no two base points are consecutive, in Noether's sense. For example, if two base points of the first order be consecutive, that is, if all the curves of the net have a given tangent line, the Jacobian curve has in fact a triple point, of which one tangent coincides with that of the curves of the net. We may agree, in this case, that, for the complete Jacobian system, only this tangent line is prescribed.

complete Jacobian system, as defined above, is the linear system defined by particular Jacobian curves which have u as a part. Such a curve, given by $u = 0$, which meets the curves of the given linear system only at base points of the system, may be called a *fundamental* curve of the system. We postpone the general discussion of the matter; for the present, we make the assumption that the original linear system of curves is such that the general curve of its complete Jacobian system is irreducible.

We may give, however, some introductory consideration to the case of systems of *plane* curves for which a fundamental curve exists; in the Jacobian determinant above, F is then to be replaced by t, and the determinant reduces to one of three rows and columns, in the plane coordinates x, y, z. (i), For a system of conics in a plane which have two points in common, and are expressible by an equation $z(2gx + 2fy + cz) + x^2 + y^2 = 0$, the line $z = 0$, joining the base points, is evidently fundamental. Now, for a system of plane curves of order n, with s-fold base points, the Jacobian determinant would give rise to curves of order $3(n-1)$, with $(3s-1)$-fold base point at an s-fold base point of the system; when $n = 2$, and there are two simple base points, this leads to cubic curves with two double points; such a system consists of the line $z = 0$ joining the base points, together with all conics through the base points (if we consider the complete Jacobian system). (ii), Consider, next, the system of quartic curves in a plane which have two given double points. The general curve of this system is expressed by an equation $z(\lambda_0 \vartheta_0 + \dots + \lambda_7 \vartheta_7) + \lambda_8 \vartheta_8 = 0$, where $\vartheta_0 = 0, \dots, \vartheta_7 = 0$ are cubic curves through the two base points, which are supposed to lie on $z = 0$, and $\vartheta_8 = 0$ is a particular quartic curve with the two given double points. The Jacobian system will then consist of curves of order 9, with a 5-ple point at each base point. Such curves necessarily have the fundamental line $z = 0$ as part; and the system of all curves of order 9 with two 5-ple points consists of this line together with all the curves of order 8 having these two as 4-ple points. (iii), Consider now the system of ∞^3 cubic curves which have 6 common base points so situate as to lie on a conic; the conic is then a fundamental curve, the cubic curves being given by an equation $u(ax + by + cz) + dU = 0$, where $u = 0$ represents the conic, and $U = 0$ is a particular cubic curve of the system. The Jacobian system then consists of sextic curves with the 6 base points as double points, and the conic $u = 0$ is fundamental also for the system of these sextic curves; for sextic curves with 6 double base points are of tale $\frac{1}{2}7.8 - 18$, or 10, while quartic curves with 6 simple points are of tale $\frac{1}{2}5.6 - 6$, or 9; the complete Jacobian system is thus expressed by an equation $u(\lambda_0 \psi_0 + \dots + \lambda_8 \psi_8) + \lambda_9 \psi_9 = 0$, where $\psi_0 = 0, \dots, \psi_8 = 0$ are quartic curves, and $\psi_9 = 0$ is a particular sextic. Evidently, in this case, the six base points, though lying on the conic, are not independent conditions for the conic; and this is the reason why the fundamental curve of the original system is a fundamental curve of the Jacobian system, instead of being part of every Jacobian curve, as in the two former examples. Of this statement it is easy to give a preliminary enumerative justification. For this, consider a system of plane curves of order n, with base points typically of multiplicity s. Let $u = 0$ be a curve, of order k ($k \leqslant n$), having a base point of the original system multiple of order s_1 ($s_1 \leqslant s$), but having no intersection with the curves of the original system other than at these base points, so that $\Sigma s s_1 = nk$. Let λ be the redundance of the base points for the curves of the original system, so that the tale of this system is

$\frac{1}{2}(n+1)(n+2) - \frac{1}{2}\Sigma s(s+1) + \lambda$, say L. Similarly, let μ be the redundance of the base points, supposed now of typical multiplicity $s - s_1$, for curves of order $n - k$; so that the tale of such curves, with these multiple points, is

$$\tfrac{1}{2}(n-k+1)(n-k+2) - \tfrac{1}{2}\Sigma(s-s_1)(s-s_1+1) + \mu,$$

say M. It is then easy to compute that if the genus of the curve $u = 0$, defined by $\frac{1}{2}(k-1)(k-2) - \frac{1}{2}\Sigma s_1(s_1-1)$, be equal to $\lambda - \mu$, then $L = M + 1$. In this case, the system of curves of order n has an equation of the form $u\Theta + \lambda_\rho \vartheta_\rho = 0$, where Θ is of the form $\lambda_0 \vartheta_0 + \dots + \lambda_{\rho-1}\vartheta_{\rho-1}$, and $\Theta = 0$ represents the complete system of curves of order $n - k$ having the base points with multiplicity $s - s_1$. Thus $u = 0$ is a fundamental curve for the former system. The Jacobian system of this system consists of curves of order $3(n-1)$, with $(3s-1)$-ple base points; let N be the tale of the complete system of such curves. Similarly, let N_1 be the tale of the complete system of curves of order $3(n-1) - k$, with base points, typically of multiplicity $3s - 1 - s_1$ at the original base points. Then, when $N = N_1$, the Jacobian system of the former system breaks up into $u = 0$ taken with all curves of the latter system. But, if $N = N_1 + \xi$, with $\xi > 0$, this will not be so; the complete Jacobian system will be more ample than the system directly defined as the locus of double points, or by the Jacobian determinants. We have, in fact,

$$N = \tfrac{1}{2}(3n-3+1)(3n-3+2) - \tfrac{1}{2}\Sigma 3s(3s-1) + \lambda',$$
$$N_1 = \tfrac{1}{2}(3n-3-k+1)(3n-3-k+2) - \tfrac{1}{2}\Sigma(3s-s_1)(3s-s_1-1) + \mu',$$

where λ', μ' are the redundancies of the base points for the curves of order $3(n-1)$ and $3(n-k)$, respectively. Wherefore ξ, or $N - N_1$, is equal to $\frac{1}{2}\Sigma s_1(s_1+1) + \lambda' - \mu' - \frac{1}{2}k(k+3)$; thus, if both λ' and μ' be zero, ξ will be greater than zero when the conditions to be satisfied by the fundamental curve, $u = 0$, at the base points, are not independent conditions for this curve, but exceed the number $\frac{1}{2}k(k+3)$ of conditions which a curve of order k can satisfy. The case of a conic through 6 points, cited above, is an example. For the theory of fundamental curves, here indicated, the reader should consult Castelnuovo's paper, *Memorie Torino*, XLII, 1892, p. 3.

We pass now to the definition of the invariant ω. We consider, on the fundamental surface $F = 0$, a linear system of curves, of freedom at least 2, with base points only at simple points of the surface; and then the complete Jacobian system of this. We suppose this Jacobian system to consist of generally irreducible curves (in particular excluding the consideration of the case where the Jacobian curves have a common part arising from a fundamental curve of the original system). At an s-ple base point of the original system, the Jacobian curves are prescribed to have a $(3s-1)$-ple point. We denote by β the number of (simple) points of the surface at which base points of the original system of curves are found. We denote also by p the effective genus of a general curve of the original system, and by π the effective genus of a curve of the Jacobian system. Then we define ω in the first instance by

$$\omega - 1 = \pi - 1 - 9(p-1) + \beta;$$

it can be shewn that this gives a value independent of the system of curves used for its computation.

As a simple illustration, suppose $F=0$ to be without multiple
points, of order n, and the original system of curves to be given by
$\lambda f+\mu\phi+\nu\psi=0$, the surfaces $f=0$, $\phi=0$, $\psi=0$ being of order m, and
the curves to have multiple points typically s-fold. The Jacobian
determinant is of order $3m+n-4$, and thus (recalling the formula
for the genus of the complete intersection of two surfaces)

$$p-1=\tfrac{1}{2}nm(n+m-4)-\tfrac{1}{2}\Sigma s(s-1),$$

and $\pi-1=\tfrac{1}{2}n(3m+n-4)(3m+2n-8)-\tfrac{1}{2}\Sigma(3s-1)(3s-2);$

hence we find $\omega-1=n(n-4)^2$, which is thus independent of the
system of curves employed. It is in fact equal to the *grade* of the
system of curves in which $F=0$ is met by the general system of
surfaces of order $n-4$, namely the number of intersections of any
two such curves. And we notice that if, with the value of I found
above for a non-singular surface, $I=(n-2)(n^2-2n+2)$, we define
a number p_n by the equation $I+\omega-1=12p_n+8$, then

$$p_n=\tfrac{1}{6}(n-1)(n-2)(n-3),$$

which is the tale of all surfaces of order $n-4$. In particular the
values of I and $\omega-1$, respectively for a plane $(n=1)$, a quadric
surface, and a general cubic surface, are $(-1, 9)$, $(0, 8)$ and $(5, 3)$,
the value of p_n being zero in all these cases. It may indeed be proved
directly, for a net of plane curves of genus p, without fundamental
curves, having base points at β points of the plane, that the genus
of the Jacobian curve of the net is $9p-\beta+1$.

More generally, when the surface $F=0$ has a double curve, with
triple points triple for the surface, but no other multiple points, consider the system of plane sections of the surface. The Jacobian
curves are then, as we have seen, the proper curves of contact of
tangent planes to the surface drawn from points. With the notation
employed in the last chapter we then have (p. 161), for $p-1$ and
$\pi-1$ respectively, the values $\tfrac{1}{2}\mu_1-\mu_0$ and $\mu_2+\nu_2-\tfrac{2}{3}\mu_1$. Thus $\omega-1$
has the value $\mu_2+\nu_2-6\mu_1+9\mu_0$. In this chapter, also, we have
found for $I+4$ the value $\mu_2-2\mu_1+3\mu_0$. Thus, for the number p_n
defined by $\omega-1+I+4=12(p_n+1)$, we have $12(p_n+1)$ equal to
$2\mu_2+\nu_2-8\mu_1+12\mu_0$. If p denote the genus of an arbitrary plane
section of the surface, and i denote the number of stationary tangent
planes of the surface which pass through an arbitrary point, that is
the number of cusps in the net of curves formed by plane sections of
the surface which pass through the point, this (see p. 162) is the
same as $i=24(p_n+p)$. As a particular verification of this result it
may be proved directly that among the curves of a net of plane
curves (without fundamental curve), of genus p, there are $24p$
curves which have a cusp. (More general results are quoted below,
p. 241; references in Pascal, *Repertorium*, II, 1, 1910, p. 340.)

The formula $\omega - 1 = \mu_2 + \nu_2 - 6\mu_1 + 9\mu_0$, just obtained, has an interpretation which suggests a more general definition of $\omega - 1$ than that we have given. We explain this, first, when, as in the derivation of the formula, the net of curves used for the definition of ω consists of the sections of the surface by planes through an arbitrary point. Use the notation (P, Q) for the number of intersections of two curves P, Q on the surface, other than at the base points prescribed for these curves; in particular use (P, P) for the number of intersections of two curves of a linear system, other than at the prescribed base points, this being the *grade* of the linear system. Then, if C, C' denote plane sections of the surface, this being of order μ_0, the term $9\mu_0$ in the expression for $\omega - 1$ is the number $(3C, 3C')$; if J, J' denote the proper curves of contact of tangent planes to the surface from two different points, each of order μ_1, the term $6\mu_1$ is $(J, 3C') + (J', 3C)$. Further, μ_2, the number of tangent planes to the surface from an arbitrary line, is the number of intersections of two curves J, J' at ordinary points of the surface; we have seen, however, that all curves J, J' pass through the pinch points on the double curve of the surface, and touch the cotangent at this point (Chap. IV, p. 178); but we have also seen (*ibid.* p. 179) that of the two intersections of the curves J, J' thus arising, only one arises on the surface in space [4] by whose projection the double curve is generated. For this reason we may regard the complete value of (J, J') as being $\mu_2 + \nu_2$. The formula for $\omega - 1$ thus has the interpretation $(J, J') - (J, 3C') - (J', 3C) + (3C, 3C')$. This we may agree to write in the form $(J - 3C, J' - 3C')$, or the form $(3C - J, 3C' - J')$. As in the theory of linear series of sets of points on a curve, if $|A|$, $|B|$ denote two linear systems of curves on a surface, we denote by $|A + B|$ the complete linear system which contains all the composite curves, of which each is the aggregate of a curve of the system $|A|$ with a curve of the system $|B|$, the base points prescribed for $|A + B|$ being the aggregate of those prescribed for $|A|$ and $|B|$ singly. Similarly, if there exist a linear system $|C|$, such that $|C + B| = |A|$, then $|C|$ may be denoted by $|A - B|$. Now, let $|C_J|$ denote the complete linear system of curves containing the proper curves of contact J of tangent planes to the surface from all arbitrary points; the system $|C_J|$ will, if we maintain our definition, not have the pinch points of the double curve as base points, but the grade (C_J, C_J), of this system, will be the same as the number (J, J'), in which we have counted these points as simple intersections of curves of J and J'; the number (C_J, C) will also be μ_1. Thus it appears that, if one of the two complete systems $|C_J - 3C|$, $|3C - C_J|$ exist, then $\omega - 1$ is the grade of this system. This important result is proved when $|C|$ is the system of plane sections of the surface; we shall see below that it can be extended to the case when $|C|$

is any linear system of curves, of which $|C_J|$ is the Jacobian system.

In terms of the characters of the double curve of the surface (see p. 162 of the preceding chapter) the value of $\mu_2 + \nu_2 - 6\mu_1 + 9\mu_0$ can be expressed in the form

$$\omega - 1 = \mu_0(\mu_0 - 4)^2 - 5(\mu_0 - 4)\epsilon_0 + 2\epsilon_1 + 9t;$$

similarly the value of $\mu_2 - 2\mu_1 + 3\mu_0$ can be expressed by

$$I + 4 = \mu_0(\mu_0^2 - 4\mu_0 + 6) - (7\mu_0 - 16)\epsilon_0 + 4\epsilon_1 + 15t.$$

From these the value of p_n, given by $\omega - 1 + I + 4 = 12(p_n + 1)$, is such that $p_n + 1 = (\mu_0 - 1, 3) - (\mu_0 - 4)\epsilon_0 + 2t + P$, where $(\mu_0 - 1, 3)$ is the binomial coefficient, and P is the genus of the double curve. Thus p_n is the theoretical tale of surfaces of order $\mu_0 - 4$ which pass through the double curve, valid when μ_0 is sufficiently great in respect to ϵ_0 (Vol. v, p. 236). We have remarked (p. 168, Ex. 11) that surfaces of order $\mu_0 - 4$, put through the triple points of this curve, cut thereon a non-special series; when this series is incomplete, the number, p_g, of actual surfaces of order $\mu_0 - 4$ through the double curve, is greater than p_n. A surface of order $\mu_0 - 4$, through the double curve, has, as residual intersection with the given surface, a curve of order $\mu_0(\mu_0 - 4) - 2\epsilon_0$, which is $\mu_1 - 3\mu_0$. This is the order of a curve of the system $|C_J - 3C|$, when this system exists.

Effect of isolated nodal points of the fundamental surface upon the definition of I and ω. The values of the invariants I and ω have been given under the hypothesis that the only multiple points of the surface are a double curve with triple points. The singularities of a surface may be of great complexity, and, in general, as has been suggested, it must be supposed that the surface has been transformed into the form adopted above, before the theory can be applied. But, when the only singularities beside a double curve consist of isolated nodes of general kind, a modified definition of I and ω can be applied directly, which is interesting because of the frequent occurrence of such surfaces. We limit ourselves to supposing I defined by a pencil of sections by planes through an arbitrary line and ω defined by a net of sections by planes through an arbitrary point. The modified definitions must be such that they include the case when no such nodes occur; they must be invariant for birational transformation into other surfaces with isolated nodes, save for corrections indicative of these nodes; and we choose them such that the value p_n, arising by the equation $I + 4 + \omega - 1 = 12(p_n + 1)$, is the theoretical tale of surfaces of order $\mu_0 - 4$, passing through the double curve, which have a $(k - 2)$-ple node at each isolated k-ple node of the given surface. It has in fact been shewn by Noether and Zeuthen that the number p_n so defined is an absolute invariant for birational transformation of the surface; but we do not give the algebraical proof of this fact, since a more general proof emerges at a later stage of the theory of surfaces. As in what has been said, our present aim is introductory and provisional.

Consider first the invariant I. It may be shewn (for the literature, cf. Pascal, *Repertorium*, II, 2, 1922, pp. 663, 690; also Segre, *Annal. d. Mat.* XXV, 1897, p. 26) that, for a surface with isolated k-fold nodes, at which the osculating cone is of class k', the number of actual tangent planes which can be drawn from an arbitrary general line is less than $\mu_0(\mu_0-1)^2$ by the sum of the values of $(k-1)k'$ at these nodes; we consider only the general case, in which $k'=k(k-1)$. Then the class of the surface is given by $\mu_2=\mu_0(\mu_0-1)^2-\Sigma k(k-1)^2$. We define then a number I_0, to be subsequently modified to I, by the equation $I_0=\delta-\sigma-4p+\Sigma(k-1)$, where, as before, p is the genus of a general plane section of the surface, σ is the number of base points of a section of the surface by planes through an arbitrary line, δ is the number of double points, that is of actual contacts with the surface, on such plane sections, and the summation $\Sigma(k-1)$ extends to all the isolated nodes of the surface not lying on the arbitrary line. If this line be quite general we thus have

$$I_0=\mu_0(\mu_0-2)^2-\Sigma k(k-1)^2-\mu_0-4p+\Sigma(k-1),$$
$$=\mu_0^2(\mu_0-2)-4p-\Sigma(k-1)(k^2-k-1).$$

And we can prove that the same value is obtained if we take a line which contains a single k_0-fold node, say P_0. For the line joining P_0 to an arbitrary point, A, is $k_0(k_0-1)$-fold on the enveloping cone drawn from A to the surface, so that there are $2k_0(k_0-1)$ tangent planes of the osculating cone at P_0 which pass through the line AP_0. The number of proper tangent planes to the surface from this line is therefore $\mu_0(\mu_0-1)^2-\Sigma k(k-1)^2-2k_0(k_0-1)$; this is the value to be taken for δ when I_0 is computed by planes through the line AP_0; the values for σ and the genus are respectively μ_0-k_0+1 and $p-\frac12 k_0(k_0-1)$. Thus we have

$$I_0=\delta-(\mu_0-k_0+1)-4[p-\tfrac12 k_0(k_0-1)]+\{\Sigma(k-1)-(k_0-1)\},$$

where the $\Sigma(k-1)$ extends to *all* the isolated nodes; with the value of δ put down, this is the same as was obtained by sections through a general line. A similar argument shews that the same result is obtained if we take sections by planes through a line which contains two of the isolated nodes. In particular for a surface with isolated nodes but without a double curve,

$$I_0+4=\mu_0(\mu_0^2-4\mu_0+6)-\Sigma(k-1)(k^2-k-1).$$

In general, the value given by the definition is

$$I_0+4=\mu_2-2\mu_1+3\mu_0+\Sigma(k-1),$$

where μ_2 is the actual class of the surface, as estimated above.

Consider next the invariant ω. For a surface with a double curve (and triple points thereon), and k-ple isolated quite general nodes, we define a number $\omega_0 - 1$ given by

$$\omega_0 - 1 = \pi - 1 - 9(p-1) + \beta + \Sigma(k-1)(2k-1) - \tfrac{1}{2}\Sigma k(k-1),$$

where p is the genus of the curves of the net used for the definition (which we take to be given by planes through an arbitrary point), β is the number of (simple) points of the surface where curves of this net have prescribed base points of any orders (in the case taken, $\beta = 0$), and π is the genus of the Jacobian curve of this net (in this case the proper curve of contact of tangent planes from the arbitrary point). From the formulae given in the last chapter (p. 163), we have $2p - 2 = \mu_1 - 2\mu_0$, and $2\pi - 2 = 2(\mu_2 + \nu_2) - 3\mu_1 + \Sigma k(k-1)$; thus we deduce for $\omega_0 - 1$ the value $\mu_2 + \nu_2 - 6\mu_1 + 9\mu_0 + \Sigma(k-1)(2k-1)$. The correction for the isolated nodes is introduced so that the value of p_n deduced from $I_0 + 4 + \omega_0 - 1 = 12(p_n + 1)$, taking account of $\mu_2 = \mu_0(\mu_0 - 1)^2 - \Sigma k(k-1)^2$, should be

$$(\mu_0 - 1, 3) - (\mu_0 - 4)\epsilon_0 + 2t + P - 1 - \Sigma(k, 3),$$

where $(\mu_0 - 1, 3)$ and $(k, 3)$ denote binomial coefficients. This is the theoretic tale of surfaces of order $\mu_0 - 4$, when μ_0 is large enough, which pass through the double curve, and have a $(k-2)$-ple point at every isolated k-ple point of the surface; we have in fact $k(k-1)^2 - k(k-1) = 6(k, 3)$. The value of p_n is such that

$$12(p_n + 1) = 2\mu_2 + \nu_2 - 8\mu_1 + 12\mu_0 + 2\Sigma k(k-1).$$

Another point of view is to define ω by its value on another surface, in birational correspondence with the given one, in which the k-ple point of the original is replaced by a curve of order k. See below, p. 234.

Ex. 1. Prove directly, in accordance with Noether's formula quoted in Note II to the preceding chapter, that, for a general surface of order n, the presence of a j-ple line involves a diminution in the value of p_n given by $\tfrac{1}{6}j(j-1)(3n-2j-5)$. Cf. also Cayley, *Papers*, viii, p. 395, for a direct calculation of p_n which includes the case of a cone.

Ex. 2. For sufficiently large μ_0, that a surface of order μ_0 should have a given curve of order n, and genus p, with δ double points, as a j-ple line, involves a number of conditions given by

$$[\tfrac{1}{2}j(j+1)\mu_0 - \tfrac{2}{3}j(j^2-1)]n - \tfrac{1}{6}j(j+1)(2j+1)(p+\delta-1).$$

Cf. Noether, *Ann. d. Mat.* v, 1871–3, § 7.

Ex. 3. For sufficiently large μ_0, that a surface of order μ_0 should touch a given surface of order m, along a given curve of order n and genus p, lying thereon, involves a number of conditions given by

$$(2\mu_0 + m - 4)n - 4(p-1).$$

Cf. Hudson, *Cremona Transformations* (Cambridge, 1927), p. 253.

We now illustrate the definitions of I_0, ω_0 which we have given, by applying them to several simple examples. These definitions are $I_0 + 4 = \mu_2 - 2\mu_1 + 3\mu_0 + \Sigma(k-1)$, and

$$\omega_0 - 1 - \Sigma(k-1)(2k-1) = \pi - 1 - 9(p-1) - \tfrac{1}{2}\Sigma k(k-1)$$
$$= \mu_2 + \nu_2 - 6\mu_1 + 9\mu_0.$$

(a) For a general cubic surface we have seen that $I_0 = 5$, $\omega_0 - 1 = 3$ (p. 196). Consider a cubic surface with a single double point; then $k = 2$, $\mu_0 = 3$, $\mu_1 = 6$, $\nu_2 = 0$, $\mu_2 = 3 \cdot 4 - 2$, $= 10$, and $p = 1$, the value of the class, μ_2, being found from the general formula

$$n(n-1)^2 - \Sigma k(k-1)^2.$$

The value of π, the genus of an enveloping cone having one double generator, is 3. Thus we find $I_0 = 4$, $\omega_0 - 1 = 4$, and $p_n = 0$ as for a general cubic surface. Consider the correspondence of this nodal cubic surface with a plane: In general, a cubic surface has its plane sections representable by plane cubic curves passing through 6 points of the plane; but, if these 6 points lie on a conic, given, say by $u' = 0$, such cubic curves are given by an equation

$$u'(ax + by + cz) + U' = 0,$$

where a, b, c are arbitrary constants, and $U' = 0$ is a particular cubic curve through the 6 base points. The cubic surface, with x, y, z, t as coordinates, is then obtained by the transformation

$$x/x'u' = y/y'u' = z/z'u' = t/U',$$

and the conic $u' = 0$ corresponds to the neighbourhood of the double point on the cubic surface; the transformation is obtainable by projection from the double point. There is thus a *curve* on the plane corresponding to a *point* on the cubic surface, namely the conic $u' = 0$; this we may call an *exceptional* curve, and, if e' denote the number of exceptional curves in the plane, we have $e' = 1$. As in the case of the general cubic surface, there are 6 lines on the cubic surface corresponding each to one of the 6 base points in the plane; in this case they lie on the osculating quadric cone at the double point of the cubic surface; if e denote the number of exceptional curves on the cubic surface, we thus have $e = 6$. The values of I_0, $\omega_0 - 1$ for the plane are, we have seen, given by $I' = -1$, and $\omega' - 1 = 9$. Thus we have the equations $I_0 - e = I' - e'$, $\omega_0 + e = \omega' + e'$, for which $I_0 + \omega_0 = I' + \omega'$.

More generally, for any surface of order n, with a single node of multiplicity $n-1$, given by an equation of the form

we have
$$t(x, y, z)_{n-1} + (x, y, z)_n = 0,$$
$$\mu_0 = n, \quad \mu_1 = n(n-1), \quad \nu_2 = 0,$$
$$\mu_2 = n(n-1)^2 - (n-1)(n-2)^2 = (n-1)(3n-4), \quad k = n-1,$$

and hence $I_0 = n^2 - n - 2$, and $\omega_0 - 1 = -n(n-1) + 10$; while, for the transformation of this surface to a plane, which is obtainable by projection from the node, we have $e = n(n-1)$ and $e' = 1$, as the respective numbers of exceptional curves on the surface and on the plane. Thus, in this case also, $I_0 - e = I' - e'$ and $\omega_0 + e = \omega' + e'$. In view of a subsequent theorem, we may remark further that if two surfaces of order $n-4$ be taken, each having a node of multiplicity $n-3$ at the node of the given surface, then their curve of intersection meets the surface, other than at the node, in

$$n(n-4)^2 - (n-1)(n-3)^2$$

points, which is $\omega_0 - 2$. This is then the number of free intersections of any two curves of the system determined, on the surface, by all surfaces of order $n-4$ having a $(n-3)$-fold node at the given node.

(*b*) Next consider the cubic surface with four nodes (of multiplicity 2). For this surface we have $\mu_0 = 3$, $\mu_1 = 6$, $\mu_2 = 4$, $\nu_2 = 0$, $p = 1$, $\pi = 0$ (the enveloping cone from an arbitrary point being the dual of a plane quartic curve with three nodes, since the dual of the surface is the Steiner quartic surface, considered in the following example); also $\Sigma(k-1) = 4$, $\Sigma(k-1)(2k-1) = 12$, $\frac{1}{2}\Sigma k(k-1) = 4$. Hence we have $I_0 = 1$, $\omega_0 - 1 = 7$, $p_n = 0$. The plane representation of this surface is by cubic curves through the six intersections of four lines; if such lines be $x' = 0$, $y' = 0$, $z' = 0$, $u' = 0$, such a cubic curve has an equation of the form

$$u'(ay'z' + bz'x' + cx'y') + x'y'z' = 0,$$

where a, b, c are parameters, and the formulae of transformation are $x/u'y'z' = y/u'z'x' = z/u'x'y' = t/x'y'z'$. There are e, $= 6$, exceptional lines on the surface, joining the nodes in pairs, and e', $= 4$, exceptional lines in the plane, corresponding to the nodes of the surface. Thus, with $I' = -1$, $\omega' - 1 = 9$ for the plane, we have $I_0 - e = I' - e'$; $\omega_0 + e = \omega' + e'$.

(*c*) For the Steiner quartic surface, with three double lines meeting in a triple point of the surface, we have

$$\mu_0 = 4, \quad \mu_1 = \mu_0(\mu_0 - 1) - 2\epsilon_0 = 6, \quad \nu_2 = 6,$$

there being two pinch points on each double line; further $\mu_2 = 3$, this being the order of the dually corresponding surface considered in the preceding example. There are no isolated nodes, and the formulae give $I_0 = -1$, $\omega_0 - 1 = 9$, as for a plane. The formulae of transformation to a plane (x', y', z') may be taken to be

$$x'/yz = y'/zx = z'/xy$$

with $x/y'z' = y/z'x' = z/x'y' = t/(x'^2 + y'^2 + z'^2)$,

there being no exceptional lines on the surface, or on the plane.

(d) The Kummer quartic surface, with 16 nodes and 16 tropes (tangent planes each having contact along a conic), has $\mu_0 = 4$, $\mu_1 = 12$, $\nu_2 = 0$, $\mu_2 = 4$, $p = 3$, $\pi = 3$, the values of μ_2 and π being the same as the order and genus of the plane section of the dually corresponding surface, which is also a Kummer surface. Thus we find $I_0 = 4$, $\omega_0 - 1 = 16$, $p_n = 1$.

(e) For the Weddle surface, a quartic surface with six nodes, the locus of the vertices of quadric cones passing through six points, we have $\mu_0 = 4$, $\mu_1 = 12$, $\nu_2 = 0$, $\mu_2 = 24$, $p = 3$, $\pi = 13$; and hence $I_0 = 14$, $\omega_0 - 1 = 6$, $p_n = 1$. It can in fact be shewn (see Note III at the end of this chapter) that the Kummer surface can be birationally transformed to the Weddle surface so that ten of the sixteen nodes of the Kummer surface correspond to ten lines lying on the Weddle surface (which has also fifteen other lines, the joins of the nodes in pairs); the other six nodes of the Kummer surface correspond to the six nodes of the Weddle surface. Thus, in this transformation, the Kummer surface has no exceptional curves, or say $e = 0$; but the Weddle surface has 10 exceptional curves, say $e' = 10$. From $I_0 = 4$, $\omega_0 - 1 = 16$ for the Kummer surface, and $I_0' = 14$, $\omega_0' - 1 = 6$ for the Weddle surface, we thus have $I_0 - e = I_0' - e'$, $\omega_0 + e = \omega_0' + e'$.

(f) Next consider the general surface, ψ^{2k}, of order $2k$, which has a k-ple node at each of the 8 intersections of three general quadric surfaces $u = 0$, $v = 0$, $w = 0$. If f_k be a general homogeneous polynomial of order k in the three quantities u, v, w, the equation $f_k(u, v, w) = 0$ represents such a surface; we prove conversely that every surface ψ^{2k} is expressible by an equation of this form. Such a surface was considered for $k = 2$ by Cayley (*Papers*, VII, p. 147); and the general case by Castelnuovo (*Rend. Ist. Lombardo*, 1891, Nota I). Let $U = 0$ be a general quadric surface containing the 8 nodes; consider the curve determined thereon by the general surface ψ^{2k}; by definition this surface has a k-ple node at each of the 8 points, and it meets the quadric $U = 0$ in a curve of order $4k$. Take k arbitrary points, P_i, on the quadric $U = 0$, and let c_i denote the elliptic quartic curve on $U = 0$ through the 8 given base points and through P_i; this curve c_i is the intersection of $U = 0$ with another quadric surface through the 8 base points and P_i. This curve c_i meets any surface ψ^{2k} in $8k$ points at the base points, and is thus contained in any surface ψ^{2k} which meets it in a point other than the base points. The general surfaces ψ^{2k} have at least the tale of the particular surfaces $f_k(u, v, w) = 0$, namely $\frac{1}{2}(k+1)(k+2)$, which is certainly $> k$; such a surface can therefore be put through the k points P_i, and will then, by what we have said, contain the k curves c_i, whose aggregate order, $4k$, is equal to that of the curve in which a ψ^{2k} meets $U = 0$. The complete curve of intersection of a surface ψ^{2k} with $U = 0$ is thus determined by k points of $U = 0$. Among surfaces

ψ^{2k}, however, will be $(k \geqslant 2)$ surfaces consisting of surfaces ψ^{2k-2} taken with the quadric surface $U=0$; and the freedom of the curves cut by surfaces ψ^{2k} on $U=0$ will be less than that of all surfaces ψ^{2k} only in consequence of this possibility. Thus, if τ_{2k} denote the tale of all surfaces ψ^{2k} we have $\tau_{2k}-1-\tau_{2k-2}=k$. As $\tau_2=3$, we thus have τ_{2k} equal to $(k+1)+k+\ldots+3+3$, or $\frac{1}{2}(k+1)(k+2)$, which is the tale of surfaces $f_k(u,v,w)=0$. This last equation thus represents the general surface ψ^{2k}. From this also it follows that the redundance of the 8 k-ple nodes, for surfaces of order $2k$, is $\frac{1}{2}k(k+1)$; for, if we subtract, from the general tale of surfaces of order $2k$, eight times the conditions necessary for a k-ple node, whose number is $\frac{1}{6}k(k+1)(k+2)$, we find $k+1$, which is then the *virtual* or theoretical number of surfaces ψ^{2k}. If the actual number of surfaces ψ^{2k-4}, having then a $(k-2)$ point at each of the k nodes, namely $\frac{1}{2}k(k-1)$, be denoted by p_g, and the virtual number, $k-1$, of these surfaces, be denoted by p_n (cf. p. 198 above), we have $p_g-p_n=\frac{1}{2}(k-1)(k-2)$. This is equal to the genus of the plane curve $f_k(\xi,\eta,1)=0$. The surface ψ^{2k} possesses an irrational pencil, of this genus $\frac{1}{2}(k-1)(k-2)$, of elliptic curves, the quartic curves through the 8 base points, given by the equations $u-\xi w=0$, $v-\eta w=0$ (cf. p. 190, preceding). For this surface we have $\mu_0=2k$, $\mu_1=2k(2k-1)$,

$$\mu_2=2k(2k-1)^2-8k(k-1)^2=2k(4k-3);$$

thus the numbers I_0, ω_0-1, defined as above, are $I_0=12(k-1)$, $\omega_0-1=8$, leading to $p_n=k-1$, as we have found. It can be computed that the genus, π, of the proper curve of contact for ψ^{2k}, is $(k-1)(6k-1)$.

We may remark at once that this result for I_0 is in accord with the definition of I given above (p. 190) for the case of a surface, with an irrational pencil of curves, which was supposed to be without multiple points. For we shall see (p. 235), that, for a surface with κ isolated general nodal points, the value of I is, under certain conditions, to be taken to be $I=I_0+\kappa$. This would give in this case, $(\kappa=8)$, $I+4=12(k-1)+8+4=12k$; and this is the value of $\delta+(2\rho-2)(2\rho'-2)$, where ρ, $=1$, is the genus of each curve of the irrational pencil, and ρ', $=\frac{1}{2}(k-1)(k-2)$, is the genus of the pencil, provided $\delta=12k$. Now a curve on ψ^{2k}, given by $u-\xi w=0$, $v-\eta w=0$, has a double point if the quadric surfaces $u-\xi w=0$, $v-\eta w=0$ have contact. The condition of contact of two quadric surfaces is however of degree 12 in the coefficients of either. Thus the curves of the pencil which have a double point correspond to the intersections of $f_k(\xi,\eta,1)=0$ with a curve of order 12 in $(\xi,\eta,1)$ and are $12k$ in number. This result is also clear because the locus of vertices of quadric cones contained in the net of quadric surfaces $lu+mv+nw=0$ is a curve of order 6; this will have $12k$ intersections with ψ^{2k}. It is convenient also to make another remark at this place. The general surface ψ^{2k-4}, with a $(k-2)$-ple node at each of the 8 base points, meets ψ^{2k} in a curve of order $2k(2k-4)$ or $4k(k-2)$; this is evidently the aggregate of the $k(k-2)$ quartic curves $u-\xi_i w=0$, $v-\eta_i w=0$, where $(\xi_i,\eta_i,1)$ are the points common to the curve $f_k(\xi,\eta,1)=0$, which determines ψ^{2k}, whose equation is $f_k(u,v,w)=0$, and to the curve $f_{k-2}(\xi, \eta, 1) = 0$ which determines ψ^{2k-4}. Thus two surfaces ψ^{2k-4} meet ψ^{2k} in curves which have no common intersection save at the 8 nodes. And (recalling p. 197 above), this is in accord with $\omega - 1 = \omega_0 - 1 - \kappa = 0$; we shall in fact be led, correspondingly to $I = I_0 + \kappa$, to $\omega = \omega_0 - \kappa$. The case when, on a surface of order n, the surfaces of order $n - 4$, passing $(j - 1)$ times through every j-ple curve of the given surface, and $(k - 2)$ times through every k-ple isolated node, have as intersection with the given surface an aggregate of elliptic curves, is considered by Noether, *Math. Annal.* VIII, 1875, § 11. Further information in regard to such surfaces is found in Enriques, *Rend. Lincei*, Dec. 1906, Jan. 1912, Feb. 1914; also *Rend. Bologna*, Dec. 1906. The result $p_g - p_n = \frac{1}{2}(k-1)(k-2)$ is also an illustration of a general theorem; cf. Castelnuovo, "Alcuni risultati...," *Mem. Soc. Ital. d. Sci.* 1896, p. 102.

(*g*) Consider a septimic surface ψ^7, with a single triple point, and a triple conic. By the formulae given (on p. 163), in the preceding chapter, putting $\mu_0 = 7$, $j = 3$, $\epsilon_0 = 2$, $\epsilon_1 = 2$, $k = 3$, we find $\mu_1 = 30$, $\mu_2 = 92$, $\nu_2 = 20$; and hence $I_0 = 51$, $\omega_0 - 1 = 5$, $p_n = 4$. The equation of the surface can be supposed of the form

$$t^4(x, y, z)_3 + t^3(x, y, z)_4 + t^2[x, y, z]_3 u + t(x, y, z)_2 u^2$$
$$+ (lx + my + nz)u^3 = 0,$$

where the triple conic, in $t = 0$, is given by $u = 0$; we can suppose $u = yz + zx + xy$. If we now put $x'/xt = y'/yt = z'/zt = t'/u$, which, with $u' = y'z' + z'x' + x'y'$, lead to $x/x't = y/y't = z/z't = t/u'$, the given surface is transformed to the quintic surface expressed by

$$t'^4(lx' + my' + nz') + t'^3(x', y', z')_2 + t'^2[x', y', z']_3 + t'(x', y', z')_4$$
$$+ (x', y', z')_3 u' = 0,$$

having neither multiple point nor multiple curve. For this surface, $\mu_0' = 5$, $\mu_1' = 20$, $\mu_2' = 80$, $\nu_2' = 0$ and therefore $I_0' = 51$, $\omega_0' - 1 = 5$. There is, however, an exceptional line on the septimic surface, given by $t = 0$, $lx + my + nz = 0$, which transforms to the simple point $x' = y' = z' = 0$, $t' = 1$ on the quintic surface; and there is an exceptional curve on the quintic surface, namely the conic $u' = 0$, $t' = 0$, which arises from the cubic node of the septimic surface. Thus $e = e' = 1$, and $I_0 = I_0'$, $\omega_0 = \omega_0'$. The septimic surface is considered by Noether (see below, p. 231).

(*h*) For a sextic surface ψ^6 with a double elliptic quartic curve and a double line not meeting this, by the same formulae (p. 163 of preceding chapter), we find $\mu_0 = 6$, $\mu_1 = 20$, $\mu_2 = 32$, $\nu_2 = 24$, and hence $I_0 = 6$, and $\omega_0 - 1 = -10$, $p_n = -1$. The surface can in fact be birationally transformed into an elliptic ruled surface (cone). Cf. Noether, "Ueber eine Fläche 6-ter Ordnung," *Math. Ann.* XXI, 1883.

(*i*) For a quintic surface ψ^5 with a double conic, and two triple nodes (Cayley, *Papers*, VIII, p. 397), we find by the same formulae (p. 163 of preceding chapter) $\mu_0 = 5$, $\mu_1 = 16$, $\mu_2 = 18$, $\nu_2 = 8$, and hence $I_0 = 1$, and $\omega_0 - 1 = -5$, $p_n = -1$. This surface can in fact also be birationally transformed to an elliptic cone.

(j) We may add the values of I_0 and $\omega_0 - 1$ for (i), a ruled surface whose plane section is of genus p; (ii), a quartic surface with a double line; (iii), a quartic surface with a double conic. These are, respectively, for (i), $-4p$ and $-8(p-1)$; for (ii), 8 and 0; for (iii), 4 and 4.

Sketch of the proof of the invariance of I. The preceding examples may suffice to render the invariants I and ω familiar. We now sketch the proof of the invariance of I in the geometrical form given by Segre. The surface considered is supposed to have no multiple points beyond a double curve with triple points, which are triple for the surface, and, possibly, isolated nodal points of general character. On this surface we consider a linear pencil of curves, of which the general curve is supposed irreducible unless the contrary be stated, the curves of the pencil having base points only at simple points of the surface. These base points may be multiple, say s-fold, for curves of the pencil, but it is supposed in general that at such a point, the s tangents are variable from curve to curve of the pencil; we denote by σ the number of such base points, without regard to their multiplicity. The condition for a curve of the pencil to have a double point is a single condition; we suppose then that there is a finite number of curves of the pencil which have a double point. We expect to find a proper correction for the cases when a curve of the pencil has several double points, or a multiple point of higher order, or when a curve of the pencil becomes reducible. The whole number of double points arising, *at simple points of the surface which are not base points of the pencil*, we denote by δ. Denoting by p the effective genus of a general curve of the pencil, we seek to establish the invariance of the number $\delta - \sigma - 4p$, subject to a correction for the isolated nodes of the surface, when such exist, whatever be the pencil of curves by which this number is found; and the invariance of this number in the passage to another surface, in birational correspondence with the original, subject then to a correction for curves of the surface which are thereby transformed to single points.

On the original surface, consider two such linear pencils of curves as we have described, say γ and γ_1, which are independent of one another; and denote by m the number of intersections of a general curve of the pencil γ with a general curve of the pencil γ_1. That a curve of one pencil should, at an intersection with a curve of the other pencil, have three coincident intersections, requires two conditions; and there are two disposable parameters, one for each pencil. We assume therefore that there exists a finite number of such three-pointic intersections. This number we denote by τ; it is reckoned with the exclusion of possible intersections at the base points of the pencils γ or γ_1, and of points which are multiple points for curves of either pencil, or are multiple points of the surface. Now, on each curve of the pencil γ, there is a finite number of points at which there is coincidence of two points of a set, of the linear series $g_1{}^m$ determined by all the curves of γ_1 upon this curve γ, these points constituting the Jacobian set of this series $g_1{}^m$. There is therefore a curve C which is the locus of points at which two of the m intersections, of a curve of γ with a curve of γ_1, coincide with one another. This curve C contains every ordinary contact of a curve of γ with a curve of γ_1, not generally touching either of these curves at this point. The curve C also passes through every base point of the pencil γ_1 (supposed not also a base point of the pencil γ); for there is a definite curve of γ passing (simply) through this point, and a definite curve of γ_1 touching this curve of γ at this point. It can in fact be shewn that, if the point is s_1-fold for the curves of γ_1, then C has there a $(2s_1-1)$-ple point, its tangents consisting of the tangent of the curve of γ which passes through this point, together with the $(2s_1-2)$ double rays of the pencil of sets of s_1 tangent lines of curves γ_1 at this point; similarly for base points of the pencil γ. Apart from the base points, the curve C is met by any curve of the pencil γ in $2m+2p-2$ points, where p is the genus of a curve of γ; these $2m+2p-2$ points are in fact, on this curve γ, the Jacobian set of the linear series of sets of m points cut thereon by all curves γ_1. Thus, if $N=2m+2p-2$, we have, on C, a linear series $g_1{}^N$ cut by all curves γ. The curves γ_1 likewise cut on C a linear series $g_1{}^{N_1}$, where $N_1=2m+2p_1-2$, if p_1 be the genus of a curve of γ_1. Consider now the Jacobian set, on C, of the series $g_1{}^N$; this consists of $2N+2\pi-2$ points, if π be the genus of C. A point of this set arises at an ordinary contact of a curve γ with the curve C, and thus at a contact of a curve γ with a curve γ_1 if at this point the tangent of C coincides with the common tangent of the curves of γ and γ_1. Consider first a simple base point, O_1, of the pencil γ_1; there is one curve of γ through this point, and, of the m intersections of every curve of γ_1 with this curve γ, one falls at O_1; there are therefore only $2(m-1)+2p-2$, or $N-2$, other points of coincidence of pairs of intersections of curves γ_1 with this particular γ; thus, of the series $g_1{}^N$ upon C, by curves of γ, the set by this particular γ has two points at O_1; the Jacobian set of $g_1{}^N$ has thus one point at O_1; there is a definite curve γ_1 through O_1 which touches this particular γ at O_1, and the curve C touches both these at O_1. If, next, we consider a point O_1, which is an s_1-ple base point for curves γ_1, there will still be a particular curve γ through O_1, and a definite curve γ_1 of which one tangent coincides with that of this particular γ; hence also a single point of the Jacobian set of $g_1{}^N$ on C will fall at O_1. But, further, among the pencils, each of s_1 lines, formed by the tangents at O_1 of a curve γ_1, there are $2(s_1-1)$ pencils for each of which two of the s_1 tangents coincide; these give $2(s_1-1)$ tangents of C, as we have already indicated. And, reasoning as in the case when O_1 is a simple base point of curves γ_1, we can prove that there are no other than the $1+2(s_1-1)$, or $2s_1-1$, tangents of C at O_1 thus obtained; for, the particular curve γ through O_1 has $2(m-s_1)+2p-2$, or $N-2s_1$, contacts with curves γ_1, other than at O_1; and this is the number of points of the set of the series g_1^N on C determined by this particular curve γ; thus C meets this γ in $2s_1$ points at O_1; these are those accounted for by the $2s_1-1$ tangents of C we have spoken of. Thus whatever be the multiplicity of O_1 for curves γ_1, there is just one point of the Jacobian set of the series g_1^N on C which falls at O_1.

Take as illustration the case when $s_1=2$, the curves γ_1 having double points at O_1. There will then be two curves γ_1 which have a cusp at O_1; on the particular curve γ which passes through this point, the curve C has four points at O_1, two of them arising by the contact, with this particular curve γ, of a certain γ_1 through O_1; each cusp gives another, and hence a tangent for C at O_1; but there is no coincidence of pairs of otherwise distinct intersections of a curve γ with C arising therewith, and no point of the Jacobian set of g_1^N on C arising corresponding to the cusp.

If then σ_1 be the number of points at which the pencil γ_1 has base points, there are on C just σ_1 points of the Jacobian set of the g_1^N on C which fall at these base points. We prove further that, beside these, there are $\delta+\tau$ points in this Jacobian set, where, as stated, δ is the number of double points of curves γ, and τ is the number of points where a curve γ has three coincident intersections with a curve γ_1. Consider a curve γ with a double point (not at base points of the pencil, or at double points of the surface); this curve will then be of genus $p-1$. Curves γ_1 thus cut, on this γ, a series of sets of m points, among which there are $2m+2(p-1)-2$ coincidences of pairs of a set, other than at the double point, which number is $N-2$. The curve C contains the double point, since, on the curve γ_1 which passes through this point, there is here a coincidence of two points of a set of the series cut on this γ_1 by all curves γ. Thus, in the series g_1^N, cut on C by curves γ, the set cut by the curve γ with the double point has two intersections coinciding at this double point. This double point is thus a point of the Jacobian set, upon C, of the series g_1^N; and thus the Jacobian set contains the δ double points of the pencil γ. That the Jacobian set also contains the τ points of osculation of a curve γ with a curve γ_1 may be regarded as obvious. We assume now that all the points of this Jacobian set on C have been enumerated. Thus we obtain the equation $2N+2\pi-2=\sigma_1+\delta+\tau$. By a similar argument for the linear series $g_1^{N_1}$ cut on C by curves γ_1, we obtain

$$2N_1+2\pi-2=\sigma+\delta_1+\tau.$$

By subtracting these equations, replacing N, N_1 respectively by

$2m + 2p - 2$ and $2m + 2p_1 - 2$, we arrive at the equation we desired, namely $\delta - \sigma - 4p = \delta_1 - \sigma_1 - 4p_1$.

It may be worth while to illustrate the preceding proof of the number $\sigma_1 + \delta + \tau$, for the Jacobian points of the series g_1^N on C, by taking the case when both the pencils γ, γ_1 are formed by sections of the surface made by planes through fixed lines. Let the former pencil be by planes through the axis a, and the latter be by planes through the axis a_1, this not meeting a. A contact, or two-pointic intersection of a curve γ with a curve γ_1, will then arise at a point of the surface where there is a tangent line which meets both a and a_1; the curve C is the locus of such points. The number N, of intersections of a curve γ with the curve C, will be the order of C diminished by the number of base points of the pencil γ, which are the intersections of the line a with the surface (here supposed all simple). The number $2N + 2\pi - 2$ will be the number of contacts, or two-pointic intersections, of plane sections γ with the curve C. Such a point evidently arises at a point of contact, with the surface, of a tangent plane through the axis a, at which point the curve γ has a double point. This corresponds to the term δ in the number of Jacobian points. The term σ_1 arises from the fact that a Jacobian point arises at a point where the axis a_1 meets the surface. There is clearly a curve γ through such a point, and also a curve γ_1 which touches this curve at this point; it is necessary to see that the curve C, which evidently passes through this point, touches both the curves γ and γ_1; of this an algebraic verification may be given: Let the axis a be $z = 0$, $t = 0$, and the axis a_1 be $x = 0$, $y = 0$, the surface being $f = 0$; the curve C is then the locus of a point (x, y, z, t) of $f = 0$, at which the tangent plane passes through the line of intersection of the planes $x'y - y'x = 0$, $z't - t'z = 0$ (x', y', z', t' being current coordinates); the curve C is thus the intersection of $f = 0$ with either of the surfaces $xf_1 + yf_2 = 0$, $zf_3 + tf_4 = 0$, where now x, y, z, t are current coordinates, and $f_1 = \partial f/\partial x$, etc. Now, if $(0, 0, 0, 1)$ be a base point for the pencil γ_1, the equation $f = 0$ is of the form $(ax + by + cz)t^{n-1} + \dots = 0$, and the tangent line of the curve C at this point is the intersection of $ax + by + cz = 0$ with the tangent plane of $zf_3 + tf_4 = 0$ at this point, that is, with

$$zc + (n-1)(ax + by + cz) = 0;$$

or, the tangent line of the curve C at this point is $z = 0$, $ax + by = 0$. The curve γ through this point is on $z = 0$, and thus touches the curve C; as was to be shewn.

As has been said, the preceding investigation of the invariance of $\delta - \sigma - 4p$ assumes, not only that the multiple points of the curves γ are only double points, and that the tangents at a base point are all variable, but also that the curves γ are irreducible. Suppose, to indicate the necessary modifications when the last condition is not satisfied, that there is a curve γ which breaks up into a curve χ of genus ρ occurring doubly, and a curve ψ. Then the curve C, the locus of contacts of curves of the pencil γ with curves of the pencil γ_1, which we have supposed in general to be irreducible, contains the curve χ as part; let τ be its other part, which we suppose irreducible. We consider the series g_1^N cut upon the curve τ, by the curves γ, and the series $g_1^{N_1}$ cut upon τ by the curves γ_1. If μ be the number of intersections of the curve χ with any curve of the pencil

γ_1, it can then be proved that $N = 2m + 2p - 2$, as before; but $N_1 = 2m + 2p_1 - 2 - \mu$. Consider the Jacobian sets of these series on the curve τ; we have as before

$$2N + 2\pi - 2 = \delta' + \tau + \sigma, \quad 2N_1 + 2\pi - 2 = \delta_1 + \tau + \sigma,$$

where, however, δ' consists of two parts $\Delta + \xi$, in which Δ is the number of double points of the curves of the pencil γ other than on the particular degenerate curve, and $\xi = 2\mu + 2\rho - 2 + 2i$, where $2\mu + 2\rho - 2$ is the number of contacts with χ of curves of the pencil γ_1, and $2i$ is twice the number of intersections of the parts χ, ψ of the particular curve γ which degenerates. Thus the same formula holds as in the general case if for δ we put $\Delta + 2i + 2\rho - 2$ (cf. above p. 189, the calculation of I from an irrational pencil).

To illustrate the wide application of the formula for I, we may consider in a plane the pencil of curves of order kn, given by an equation $u^k + \lambda v^k = 0$, where $u = 0$, $v = 0$ are two general curves of order n. Each curve of the pencil is then an aggregate of k curves of the pencil $u + \mu v = 0$, among which there are $3(n-1)^2$ curves having a double point. The genus p of the curve $u^k + \lambda v^k = 0$, which may be computed from $\frac{1}{2}(kn-1)(kn-2) - \frac{1}{2}n^2 k(k-1)$, is given by $p = \frac{1}{2}kn(n-3) + 1$. Counting each of the curves $u^k = 0$, $v^k = 0$ as furnishing $(k-1)(2p_0 - 2)$ double points, where p_0 is $\frac{1}{2}(n-1)(n-2)$, we find for I the value

$$3(n-1)^2 + 2(k-1)n(n-3) - n^2 - 2kn(n-3) - 4,$$

which is -1, the proper value for a plane.

Having thus obtained a number I, $= \delta - \sigma - 4p$, for the given surface, which is independent of the pencil of curves by which it is defined, we consider the modification of this number in passing to another surface which is in $(1, 1)$ birational correspondence with the original surface. If the transformation be such that no point of either surface is changed into a curve of the other surface, and no simple point of either into a multiple point of the other, the value of I is the same for both surfaces; for, any pencil of curves of either surface, with base points only at simple points of the surface, is transformed into a pencil of like kind, with an equal number of base points, and the same genus; and there will be the same number of curves in both pencils having double points at simple points of the surfaces respectively. But, taking first the commonest and simplest case of exception to this, suppose that the transformation is such that there is one point, O, which we suppose to be a simple point, of the original surface, which is represented by a curve on the derived surface. Consider in turn the two possibilities: (a), that the value of I on the original surface is computed by a pencil of curves of which O is not a base point; (b), that this pencil has O for a base point. Call the curve, on the derived surface, which arises from O, the *exceptional* curve; denote the pencil of curves of the original surface by (ϕ), and the corresponding pencil on the derived surface by (ϕ'). In case (a), as the curves (ϕ) do not pass through O, the curves (ϕ') will not be met by the exceptional curve, except, possibly, at the base points of (ϕ'); the exceptional curve will be what, for this reason, is called a *fundamental* curve of the pencil (ϕ'). There is, however, one curve, say ϕ_0, of the pencil (ϕ), which passes through O, with a definite tangent at O; while, the various directions of approach to O, on the surface, correspond to the various points of the exceptional curve on the derived surface. Thus, the curve on the latter surface which corresponds to ϕ_0 will break up into the exceptional curve, and another curve intersecting the exceptional curve in a definite point. This result will be considered, in a more detailed algebraic manner, below (pp. 212, 218). The pencil (ϕ') will thus contain a curve which must be considered to have a double point, on the curve corresponding to ϕ_0; and this is in addition to those derived by transformation of the double points arising on curves (ϕ). Thus, if δ' denote the number of double points of curves of the pencil (ϕ'), and δ of curves of (ϕ), we have $\delta' = \delta + 1$. Next, consider the case (b), when O is a base point of the pencil (ϕ). Then all the curves (ϕ'), instead of having as a base point a point arising from O, will have the exceptional curve as component, but have as base points the transformations of the other $\sigma - 1$ base points of the pencil (ϕ). Thus, if σ' be the number of base points of the pencil (ϕ'), we have $\sigma' = \sigma - 1$. In both cases, therefore, if I, I' be the invariants of the original and derived surfaces, computed from the pencils (ϕ) and (ϕ'), we have $I' - 1 = I$. This argument, though only for the simplest case, suggests the conclusion which we have verified in many examples, that if there be e curves of the first surface which, by the transformation, become simple points of the derived surface, and e' curves of the latter arising from simple points of the original, then $I' - e' = I - e$. In both cases (a), (b), a similar argument may be employed when the exceptional curve arises by transformation of an isolated *multiple* point of the original surface.

The algebraic formulation of the matter is on lines which we now indicate. Let (x, y, z, t) be coordinates for the original surface $F = 0$, and (x', y', z', t') for the derived surface $F' = 0$; and let the formulae of transformation be $x'/f = y'/\phi = z'/\psi = t'/\chi$, where f, ϕ, ψ, χ are polynomials of the same order in x, y, z, t. Suppose that these polynomials all vanish at the same point O of the original surface, which we take to be $(0, 0, 0, 1)$. Consider the effect of this transformation upon any point of the surface $F = 0$ lying in the neighbourhood of O, say the point $(\xi, \eta, \zeta, 1)$, where ξ, η, ζ are small. Retaining, in f, ϕ, ψ, χ, only terms of the lowest order in ξ, η, ζ, these will be replaced by polynomials, of the same order, say s, in ξ, η, ζ. To the first approximation, F is likewise a polynomial in ξ, η, ζ, say of order k. In particular, when O is a simple point of the surface, $k = 1$; and,

if f, ϕ, ψ, χ vanish only to the first order at this point, also $s=1$.
Thus, for variation of $(\xi, \eta, \zeta, 1)$ on the surface, in the neighbourhood
of O, the point (x', y', z', t') describes a curve, of which the coordinates of a point are rational functions, of order s, of the
coordinates of a point, (ξ, η, ζ), which varies upon a plane curve of
order k. In particular when $(k=1)$ O is a simple point of $F=0$, the
point (x', y', z', t') varies on a rational curve, of order s, whose
points correspond to the directions on $F=0$ issuing from O; this
is the case we have described above. The reverse transformation
which by hypothesis exists and is expressible in a form

$$x/f' = y/\phi' = z/\psi' = t/\chi',$$

where f', ϕ', ψ', χ' are homogeneous polynomials of the same order
in x', y', z', t', must be such as to lead to the point $(0, 0, 0, 1)$ when
(x', y', z', t') is any point of the exceptional curve on the derived
surface $F'=0$; thus the surfaces $f'=0$, $\phi'=0$, $\psi'=0$ all pass through
the exceptional curve. We shall suppose that the *geometrical* relations
are sufficiently represented by the hypothesis that, by means of $F'=0$,
the equations are capable of the forms $x/u'f_1' = y/u'\phi_1' = z/u'\psi_1' = t/\chi'$,
where $u'=0$ is a surface passing through the exceptional curve on
$F'=0$ (cf., for further detail, Noether's fundamental paper, *Math.
Ann.* VIII, § 9). Now consider a linear system of curves, of freedom
r, on the surface $F=0$, given by surfaces of order m, whose equation,
with $x, y, z, 1$ as current coordinates, is of the form

$$\lambda(A+Bx+Cy+Dz+Ex^2+...)+...$$
$$+\lambda_r(A_r+B_rx+C_ry+D_rz+E_rx^2+...)=0;$$

and denote the constant $\lambda A + ... + \lambda_r A_r$ by Δ. The linear system of
curves on $F'=0$, derived by direct transformation with the equa-
tions we consider, is then given by an equation

$$\Delta\chi'^m + u'(\beta f_1' + \gamma\phi_1' + \delta\psi_1')\chi'^{m-1}$$
$$+ u'^2(\epsilon f_1'^2 + ...)\chi'^{m-2} + ... + u'^m(\quad) = 0,$$

where $\beta = \lambda B + ... + \lambda_r B_r$, etc.; and this equation is of the form

$$u'(\lambda V + ... + \lambda_r V_r) + \Delta\chi'^m = 0,$$

where $V, ..., V_r$ are certain polynomials in the coordinates x',
y', z', t', independent of $\lambda, ..., \lambda_r$. Thus, curves of the original
linear system which do *not* pass through the point O, or $(0, 0, 0, 1)$,
for which therefore $\lambda, ..., \lambda_r$ are such that Δ is not zero, become
curves on the derived surface for which the curve given by $u'=0$ is
fundamental; namely, this exceptional curve does not meet the
curves on the derived surface, except at points for which $\chi'=0$;
these are then base points common to all these curves. These points
are also base points of all the transforming curves of the reverse

transformation, namely $u'f_1' = 0$, $u'\phi_1' = 0$, $u'\psi_1' = 0$, $\chi' = 0$; thus, as we see next, they correspond to exceptional curves on the original surface. For, to curves of the linear system on $F = 0$ which are such that $\Delta = 0$, that is, which pass through O, correspond on the derived surface, by the direct transformation, the curves given by $u'(\lambda V + \ldots + \lambda_r V_r) = 0$; these break up into the curve given by $u' = 0$, with a linear system of freedom $r - 1$.

Ex. 1. The circumstances we have dwelt on are familiar for the transformation from a plane, with coordinates x, y, z, to a cubic surface, with coordinates x', y', z', t'. Let $\phi = 0$, $\psi = 0$ be two conics in the plane, containing respectively the points $(1, 0, 0)$ and $(0, 1, 0)$; for definiteness let $\phi = by^2 + cz^2 + fyz + gzx + hxy$ and $\psi = a_1 x^2 + c_1 z^2 + f_1 yz + g_1 zx + h_1 xy$. Take the transformation expressed by $x'/x\phi = y'/z\phi = z'/y\psi = t'/z\psi$, wherein the four cubic curves, $x\phi = 0$, $z\phi = 0$, $y\psi = 0$, $z\psi = 0$, have six points in common, namely the four intersections of $\phi = 0$, $\psi = 0$, and the two points $(1, 0, 0)$, $(0, 1, 0)$; the reverse transformation is given by $x/x't' = y/y'z' = z/y't'$. By this transformation the points of the plane correspond to the cubic surface whose equation is

$$x'\{z'(ht' - h_1 y') + t'(gt' - a_1 x' - g_1 y')\}$$
$$+ y'\{z'(bz' + ft' - f_1 y') + t'(ct' - c_1 y')\} = 0.$$

It is easy to see that the point $(1, 0, 0)$ of the plane changes into the line of the cubic surface given by $y' = 0$, $gt' + hz' - a_1 x' = 0$; and that this is fundamental for the curve of the surface corresponding to any curve of the plane which does not contain the point $(1, 0, 0)$. So, the line $t' = 0$, $f_1 y' + h_1 x' - bz' = 0$ of the surface corresponds to the point $(0, 1, 0)$ of the plane.

Ex. 2. Prove that a quartic surface having $t = 0$, $yz + zx + xy = 0$ as double conic, and also containing the line $y = 0$, $z = 0$, has an equation of the form

$$t^3(my + nz) + t^2(by^2 + cz^2 + 2fyz + 2gzx + 2hxy) + (yz + zx + xy)^2 = 0.$$

We may then obtain a transformation between this surface and the plane $x = 0$, by drawing the line through the point $(0, \eta^{-1}, \zeta^{-1}, \tau^{-1})$ to meet the line $y = z = 0$, and also the double conic, and considering the fourth intersection of this line with the surface. If this be (x, y, z, t), prove that $x/D = -y/\zeta N = -z/\eta N = t/K$, where

$$D = \tau^3 + \tau(b\zeta^2 + c\eta^2 + 2f\eta\zeta) + \eta\zeta(m\zeta + n\eta),$$
$$N = \tau(2g\eta + 2h\zeta) + (\eta + \zeta)(m\zeta + n\eta),$$

and $\quad K = (\tau^2 + b\zeta^2 + c\eta^2 + 2f\eta\zeta)(\eta + \zeta) - (2g\eta + 2h\zeta)\eta\zeta$,

so that $\tau K - (\eta + \zeta)D + \eta\zeta N = 0$. The reverse transformation is then $\eta/zt = \zeta/yt = \tau/(yz + zx + xy)$. The curves $D = 0$, $N = 0$, in the plane (η, ζ, τ), have 5 common points, beside the point $\tau = 0$, $m\zeta + n\eta = 0$, which does not lie on $K = 0$; the four cubic curves $D = 0$, $\zeta N = 0$, $\eta N = 0$, $K = 0$ have 5 points in common. These transform into lines of the quartic surface, and are the only exceptional curves arising in the transformation.

Since for a plane $I = -1$, this gives for the quartic surface $I_0 = 4$. We can verify indeed that $\mu_0 = 4$, $\mu_1 = 8$, $\mu_2 = 12$, $\nu_2 = 4$, and so obtain, from the early formulae of this chapter, both $I_0 = 4$, and $\omega_0 - 1 = 4$. The quartic surface is obtainable by projecting the intersection of two quadrics in space [4] from a point; and this surface in [4] can be projected on to a plane from a line lying thereon. The quartic surface considered in this example is obtained by projecting from the point $(0, 0, 0, 0, 1)$ in space

[4], where the coordinates are x, y, z, t, u, the surface which is the inter-section of the quadric $yz + zx + xy + ut = 0$ with the quadric

$$t(my + nz) + by^2 + cz^2 + 2fyz + 2gzx + 2hxy + u^2 = 0.$$

Remark. Before passing from the consideration of the in-variant I, it should be said that the equation $I = \delta - \sigma - 4p$ is re-garded by Segre (*Atti...Torino*, XXXI, 1896) as the first of a series of equations connecting the values of invariants I belonging to manifolds of successive dimensions, of the form $I_n = \delta - I_{n-2} - 2I_{n-1}$, for $n = 2, 3, \ldots$, with $I_2 = I$. If we put $J_2 = I + 4$, $J_n = (-1)^n I_n + 2n$, this general equation becomes $J_n - 2J_{n-1} + J_{n-2} = (-1)^n \delta$. In a remarkable paper, which the reader should consult, J. W. Alex-ander (*Rend. Lincei*, XXIII, 1914) has identified J_n with Poincaré's topological invariant of a manifold (*J. d. l'Éc. polyt.* 1895) which is expressed by $J_n = \sum_{i=0}^{n} (-1)^i \alpha_i$ (see the chapter on Riemann Surfaces, Vol. v, preceding, p. 135).

The invariance of ω in a birational transformation. Consider first the case when the surface for which ω is defined has no multiple points other than a double curve, with triple points. We have given a general definition of ω, and reached (p. 197) the equation $\omega - 1 = (C_J - 3C, C_J - 3C)$, where the system of curves $|C|$ consists of the plane sections of the surface, and $|C_J|$ is the complete system defined by the proper curves of contact of tangent planes drawn to the surface from an arbitrary point, or, is the complete Jacobian system of $|C|$. We consider this result now in more detail, with the object of shewing, first, that ω can be defined by this equation also when $|C|$ denotes any linear system of curves on the surface for which the general curve of the Jacobian system is irreducible, and, second, to find (p. 226) the modification in ω arising in any birational transformation of the surface. For the first purpose we need a result which, for clearness, we state without proof and apply to the consideration of ω. The proof of this result, given below, p. 222, requires discussions not immediately concerned with the invariant ω, which have application however to the consideration of the modification of ω in a birational transformation. The result referred to, when, as on p. 197, (P, Q) denotes the number of intersections of two curves P, Q of the surface, other than at base points pre-scribed for these curves, is, that if $|C|$ be *any* linear system of curves without prescribed base points, whose complete Jacobian system $|C_J|$ consists of generally irreducible curves, the number $(C_J, A) - 3(C, A)$, when A is any curve of the surface, is inde-pendent of the system $|C|$, and is thus a number characteristic of the curve A. Without any assumption as to the existence of the system $|C_J - 3C|$, we may denote this number by $(C_J - 3C, A)$; and, more generally, a symbol $(P - Q, R - S)$ may be used to denote

Invariants of surfaces 215

the number $(P, R) - (Q, R) - (P, S) + (Q, S)$. With the assumption of this result, if $|C|$, $|C'|$, $|B|$, $|D|$ be any systems of curves without prescribed base points, whose Jacobian systems $|C_J|$, $|C_J'|$, $|B_J|$, $|D_J|$ are irreducible, we have

$$(B_J - 3B, D_J - 3D) = (C_J - 3C, D_J - 3D) = (D_J - 3D, C_J - 3C)$$
$$= (C_J' - 3C', C_J - 3C);$$

if we suppose that both $|C|$ and $|C'|$ denote the system of plane sections of the surface, this shews that $\omega - 1$ can be computed from the first form. It is thus independent of the linear system, and in particular is independent of the net of curves, from which it is computed. We pass then to the discussion of the result assumed.

The canonical system of curves of a surface, and the definition of ω. Still supposing the surface to have no singularities except a double curve, with triple points, let $|A|$, $|B|$ denote any two linear systems of curves, whose base points, supposed to be independently assigned, are at simple points of the surface. Following a phraseology used by Enriques (*Memorie...Società Italiana d. Scienze*, x, 1896, pp. 1–81), we look upon an s-fold multiple point of a curve A, lying on the surface, as a break, or gap, in the curve; and regard a curve which has the same point as multiple point of multiplicity $s - 1$, as derivable from A by the addition of this point of the surface counted once. Then, if this point be P, we may denote by $|A + P|$ the complete system of curves which differs only from the complete system $|A|$ in the fact that the point P is prescribed to be multiple of one less multiplicity for $|A + P|$ than for $|A|$. With this notation, denote by α the aggregate of the prescribed base points of the system $|A|$, which are not prescribed base points of the system $|B|$, each to be taken simply; likewise, denote by β the base points of $|B|$ which are not base points prescribed for $|A|$, each taken simply; and denote by μ the aggregate of points, each taken simply, which are prescribed as base points, generally of different orders, for both $|A|$ and $|B|$. Now consider the two complete systems of curves $|(A + B)_J|$, $|A_J + 3B + \beta|$, of which the former is the Jacobian system of the system $|A + B|$, while $|A_J|$ is the complete Jacobian system of $|A|$. As already said, the only prescribed base points for $|A_J|$ are at the base points prescribed for $|A|$, and the prescription is that $|A_J|$ have a $(3s - 1)$-ple point at an s-ple point of $|A|$. The two systems named will have the same base points. For a point, simple on the surface itself, which is an s-fold base point for $|A|$, and a t-fold base point for $|B|$, will be $(s + t)$-fold for $|A + B|$, and therefore $[3(s + t) - 1]$-fold for $|(A + B)_J|$. On the other hand, a point which is s-fold for $|A|$, and not a base point for $|B|$, will be a base point of multiplicity $3s - 1$ for $|A_J + 3B + \beta|$; a point which is not a base point for $|A|$, but is

t-fold for $|B|$, will be $(3t-1)$-fold for $|A_J+3B+\beta|$, being $3t$-fold for $|3B|$; and a point which is s-fold for $|A|$, and t-fold for $|B|$, with $s>0$, $t\geqslant 0$, will be $(3s-1+3t)$-fold for $|A_J+3B+\beta|$. And we may notice that a point which is accidentally a base point of the same order for all Jacobian curves of a linear system, though not a base point of this system (and, therefore, we have agreed, is not a prescribed base point of the complete Jacobian system), enters equally in curves defining the two complete systems which we are comparing. To complete the proof that these two systems are the same it is only necessary to prove now that they consist of curves of the same order. But it may be more convincing to develop the algebraic basis of the theorem. For this, we enunciate an algebraic identity, which has other applications: let the given surface be $F=0$; let the Jacobian determinant, for a net of curves given by the variable part of the intersection of $F=0$ with surfaces $\lambda f+\mu\phi+\nu\psi=0$, in which λ, μ, ν are variable parameters, namely the determinant $\partial(f,\phi,\psi,F)/\partial(x,y,z,t)$, be denoted by $J(f,\phi,\psi,F)$, or simply by J. Also let f, ϕ, ψ be of order m; and let u, U be any two homogeneous polynomials of respective orders k and $k+m$; and let p, q, r be constants. The identity in question expresses the Jacobian determinant of the three forms $uf+pU$, $u\phi+qU$, $u\psi+rU$, which are all of order $k+m$, in a certain way; namely

$$J(uf+pU,\ u\phi+qU,\ u\psi+rU,\ F)=(1+km^{-1})(u^3J-uK),$$

where J denotes $J(f,\phi,\psi,F)$, and K is a determinant of 5 rows and columns formed by suitably "bordering" J; the five rows of K consist respectively of the elements f_1,f_2,f_3,f_4,p; ϕ_1, ϕ_2, ϕ_3, ϕ_4,q; $\psi_1,\psi_2,\psi_3,\psi_4,r$; $F_1,F_2,F_3,F_4,0$; and $w_1,w_2,w_3,w_4,0$, in which $f_1=\partial f/\partial x$, etc., and $w_i=uU_i-Uu_i$, for $i=1,2,3,4$. Assuming this, let the net of curves on $F=0$, given by $\lambda f+\mu\phi+\nu\psi=0$, define the system $|A|$. The Jacobian curve of this net satisfies the equation $J=0$, which is satisfied, however, also at all the multiple points and curves of the surface $F=0$, as well as at base curves, and base points thereon, of the surfaces $f=0$, $\phi=0$, $\psi=0$. Also let a definite single curve of another system $|B|$ be given by the equation $b=0$, parts of the intersection of this with $F=0$ which belong to the prescribed base system of $|B|$ being disregarded. We may then form the Jacobian determinant of the net formed from the three composite curves which are given by $fb=0$, $\phi b=0$, $\psi b=0$. Referring to the algebraic identity we have remarked, taking therein $p=q=r=0$, and hence $K=0$, we see that the Jacobian curve of this composite net $(\lambda f+\mu\phi+\nu\psi)b=0$, defined in this way, consists of the Jacobian curve of the net (f,ϕ,ψ) compounded with the relevant part of the curve $b=0$ taken three times over. And it is proper, and easy, to consider the geometrical meaning of this when the Jacobian curve is defined, as was originally done, by a locus of double points. This

Jacobian curve evidently belongs to the Jacobian system$|(A+B)_J|$, and also, by the identity employed, also to the system $|A_J+3B|$; but, it is special, as having, at any t-ple base point of the system $|B|$ which is not a base point of $|A|$, a $3t$-ple base point. As the system $|(A+B)_J|$ has $3t-1$ for its prescribed multiplicity at such a point, and a complete system, of prescribed base, is determined by one of its curves, we infer that the systems $|(A+B)_J|$, $|A_J+3B+\beta|$ are the same; β here denoting the aggregate of all the base points of $|B|$, which are not base points of $|A|$, each taken once.

From this result, with a like definition for α, we infer that the systems $|A_J+3B+\beta|$, $|B_J+3A+\alpha|$ are the same. From this, the systems $|A_J-2A+B+\beta|$, $|B_J-2B+A+\alpha|$, both supposed to exist, are also the same, both being identical with $|(A+B)_J-2A-2B|$. But, further, if μ be the aggregate of the common base points of $|A|$ and $|B|$, each taken once, so that $\alpha+\mu$ denotes all the base points of $|A|$, each simply, and $\beta+\mu$ those of $|B|$, we can infer that the three complete systems

$$|(A+B)_J-3A-3B-\alpha-\beta-\mu|, \quad |A_J-3A-\alpha-\mu|,$$
$$|B_J-3B-\beta-\mu|,$$

are the same, if they exist. Conventionally, we say that two linear systems of curves, $|P-Q|$, $|R-S|$ are the same, even when the subtractions are impossible, provided there exists a system $|U|$ for which $|P-Q+U|$ and $|R-S+U|$ exist and are the same. It can be shewn that this definition is independent of the system employed (cf. Severi, *Rend. Ist. Lombardo*, XXXVIII, 1905). Hence finally we have the conclusion; let $|C|$ be any complete linear system, with base points only at simple points of the surface, $|C_J|$ the complete Jacobian system of this, and γ the aggregate of the base points prescribed for $|C|$, each taken simply; then the system $|C_J-3C-\gamma|$ is independent of the system $|C|$, and has no base points at base points of $|C|$. It is this system, supposed existing, which is called the *canonical system* of the surface.

The modification of the canonical system of curves in a birational transformation. If a birational transformation of the given surface be made to another surface, also having no multiple points other than a double curve, with triple points, and, in this transformation, no exceptional curves of either surface arise from points of the other surface, it is clear, from the geometrical meaning of the Jacobian system of a given linear system, that the canonical system of curves of one surface changes into the canonical system of the other. But suppose, to take the simplest hypothesis, that a simple point, O, of the original surface, $F=0$, is replaced, on the derived surface, $F'=0$, by an exceptional curve, given thereon by a surface $u'=0$. The point O may, (a), *not* be a base point of the system $|C|$ used to define the canonical system on $F=0$; or (b), it may be. In the first case (*a*), the system $|C'|$, of curves on $F'=0$, arising by direct transformation of the system $|C|$, will, by what was seen earlier (p. 212), be given by surfaces having an equation of the form $\lambda_0(u'f'+pU')+\lambda_1(u'\phi'+qU')+\lambda_2(u'\psi'+rU')+\ldots=0$, in which U' denotes what was before written as χ'^m, and p, q, r are constants. Thus, by the algebraic identity given above, the Jacobian system of the transformed system $|C'|$ will be *defined by* curves all containing, as part, the curve given by $u'=0$; this is *fundamental* for the system $|C'|$, intersecting the curves of this system only at its base points, if at all. These conclusions are sufficiently obvious without appeal to the algebraic formulation; for, since the curves $|C|$ do not have O as a base point, the condition for such a curve to contain O is one-fold; thus the condition for a curve of the corresponding system $|C'|$ to contain the curve $u'=0$ is likewise one-fold, which shews that this curve is fundamental for this system. The curves of any net contained in $|C'|$ which pass through a point of $u'=0$ consist therefore of this curve together with a pencil of curves, of which one can be supposed to pass through this point; this shews that the curve $u'=0$ is part of the Jacobian curve of this net. The Jacobian system of $|C'|$ is thus *defined by* curves containing $u'=0$ as part. The canonical curves of the derived surface are thus defined by curves $|C_J'-3C'-\gamma'|$ containing $u'=0$ as part (cf. however the discussion, p. 194, above). Consider next the case, (*b*), when the point O is a base point of the system $|C|$ by which the canonical system for the surface $F=0$ is defined. Then, if, on the derived surface, we consider the system $|C'|$ which is obtained *by direct algebraical transformation* from $|C|$, every such curve will contain $u'=0$ as part; thus, by the algebraic lemma, the Jacobian system of $|C'|$ will be defined by curves which contain the curve $u'=0$ three times over, the Jacobian determinant containing the factor u'^3 (the constants p, q, r in this lemma are zero, and K vanishes). In this case the factor u'^3 enters both in C_J' and in $3C'$, and the canonical curves defined by $|C_J'-3C'-\gamma'|$ are not proved in this way to contain the curve $u'=0$. But we have seen that the canonical curves may be determined from any system of curves, and, therefore, from a system $|D'|$, on the new surface, of which $u'=0$ is not a part. Thus the conclusion must be: *The canonical system of the derived surface, formed from curves $|C'|$ which are the direct transformation of a system $|C|$ of the original surface, is defined by curves having as part all the exceptional curves of the new surface arising from (simple) points of the original surface.* This is proved as here when these points are not base points of the system $|C|$ considered. When, however, they are such base points, the curves $|C'|$ obtained by direct transformation are, as we have seen, *composite*; they consist of the curve $u'=0$, and another system, say $|D'|$, which, in the simplest case, will not contain $u'=0$. Thus we infer from the former case that, if we compute the canonical system from the system $|D'|$, as $|D_J' - 3D' - \delta|$, the resulting curves will contain $u' = 0$. But this result is also indicated geometrically; for a curve D', and hence also a curve D_J', will be in $(1, 1)$ correspondence respectively with a curve C and a curve C_J; if C have an s-ple point at O, C_J will have a $(3s-1)$-ple point, and thus D' and D_J' will meet $u' = 0$ in respectively s and $3s-1$ points; thus the curves $|D_J' - 3D' - \delta|$, if existing and meeting $u' = 0$ in points, would meet this curve in -1 points, which is impossible. The conclusion must therefore be that the curves, if existing, break up into the curve $u' = 0$ together with a part not intersecting $u' = 0$ (cf. Enriques-Campedelli, p. 83). We may compare a familiar result on a cubic surface: If $|C|$ denote the linear system of curves of order $3m$ thereon, obtained by its intersection with surfaces of order m, and $|D|$ the system of curves of order $3m-1$ obtained by its residual intersection with surfaces of order m which pass through a definite line lying on the cubic surface, the curves $|C|$ have m intersections with this line, and the curves $|D|$ have $m+1$ intersections with this line; the curves $|C-D|$ consist of the line only. Later (p. 222) we shall have occasion to consider the system $|C_J - 2C|$, and the corresponding system $|C_J' - 2C'|$ on the derived surface. When the point O is not a base point of $|C|$, the preceding argument shews that the curve $u' = 0$ is part of the curve defining the system $|C_J' - 2C'|$, as arising in the Jacobian curve of any net of curves contained in $|C'|$. When O is a base point of $|C|$, if $|C'|$, containing $u' = 0$ as part, be the direct transformation of $|C|$, the preceding argument would lead to $u' = 0$ as part of $|C_J' - 2C'|$, because u'^3 enters into C_J'; in this case, however, the curve D', which is the variable part of C', meets $u' = 0$ in s points (with the notation above), while D_J' meets $u' = 0$ in $3s-1$ points; thus a curve of $|D_J' - 2D'|$ meets $u' = 0$ in $s-1$ points, and does not usually contain $u' = 0$.

In the preceding sketch, the separation of the curves which we have called exceptional, from the canonical system of curves, has been reached in connexion with the theory of the transformation of the surface. But consider a surface, say of order n, possessing also, in order to make a general statement, multiple curves, say, typically, j-fold curves, and isolated nodal points, say, typically, k-fold points. Let $|C|$ denote the system of plane sections of this surface; the Jacobian curve of a net of such sections is formed by the first polar of a point in regard to the surface. Such a first polar will have a j-ple curve of the surface as a $(j-1)$-ple curve, and a k-ple node of the surface as a $(k-1)$-ple node. The system of curves $|C_J - 3C|$ will then lie on surfaces of order $n-4$, passing $(j-1)$ times through each j-ple curve of the surface. We have, so far, not dealt with the question of defining the canonical system in the case

of a surface with isolated nodes; we consider the surfaces of order
$n-4$, passing $j-1$ times through every j-ple curve, and $k-2$ times,
(not $k-1$ times), through each k-ple node of the surface (the reason
for this will be made clearer later; historically the behaviour first
presented itself in the transformation of a double integral associated
with the surface); the intersections of these surfaces with the original
surface are the canonical curves of the surface as defined by
Noether in his fundamental paper, "Zur Theorie des eindeutigen
Entsprechens, u.s.w.," *Math. Ann.* VIII, 1875. Such curves may
be composite, consisting of a variable, and a fixed part, common to
all. It is this fixed part which constitutes, in Noether's definition,
the aggregate of the exceptional curves of the surface. These curves
are then determined, for any given surface, by the surface itself,
without reference to any transformation of the surface into another.
Noether states, however, and indicates a proof of the conclusion,
that a transformation of the surface is possible, whereby these exceptional curves are changed to points (*loc. cit.* § 10). Without
considering this question*, we shall distinguish the exceptional
curves so arising by indicating their origin, and denote the number
of such curves on a given surface by e_N. Two examples may be
given. (1), Consider a sextic surface having a double conic θ, and
a double line l, which does not meet θ. If $|C|$ denote the system of
plane sections, the proper Jacobian system $|C_J|$ will be given by
quintic surfaces containing both θ and l. Thus the system $|C_J-2C|$
will be given by cubic surfaces, also containing θ and l. The system
$|C_J-3C|$ will be given by quadric surfaces, also through these; and
these surfaces, as meeting the plane of θ outside this conic, namely
on l, will contain this plane as part, and will thus consist, beside this
plane, of planes through the line l. The plane of θ, meeting the quintic
surface also in a conic with a double point on l, will contain two
lines of the surface intersecting on l. These two lines, being a fixed
component (beside the multiple curves) of the intersection of the
surface with all the Noether canonical surfaces, are the exceptional
curves of the surface in Noether's sense; the variable parts of the
canonical curves are ∞^1 plane quartic curves, in planes through l,
each having two double points, where its plane meets the conic θ.
The quintic surfaces by which the curves $|C_J|$ are determined have
here two intersections with each exceptional line, other than on the

* Enriques (*Mem. Soc. Ital. d. Scienze*, x, 1896, p. 72) states that it is
not always possible to make a transformation whereby a Noether exceptional
curve is changed to a simple point of the derived surface, quoting as an
example, suggested by Castelnuovo, the case of a quintic surface having
3 tacnodes, whose plane meets the surface in a cubic curve and a conic
touching this in three points. The case of a tacnode is examined in Picard-
Simart, *Fonct. algéb.* I, 1897, pp. 76, 185, with reference to Noether, *Gött.
Nachr.* 1871. Cf. Ex. 6, p. 241, below.

multiple curves of the surface; but the cubic surfaces determining the curves $|3C|$ have three such intersections; accordingly the curves $|C_J-3C|$ contain the line. As a second example, (2), consider a quintic surface with two triple nodes, of which the joining line, therefore, lies on the surface. For the system $|C|$, obtained by arbitrary plane sections, the Jacobian system $|C_J|$ is obtained by quartic surfaces having each of the triple points as a quadratic node; whence the system $|C_J-C|$ is obtained by cubic surfaces likewise having these quadratic nodes, and therefore containing their joining line. Thus the system $|C_J-3C|$ is obtained by planes through these nodes. These planes then contain the join of the nodes, which is an exceptional line in Noether's sense, and the reduced canonical system consists of quartic curves, in planes through the exceptional line, each having a node at both triple points of the surface. In both these examples, the exceptional curve may be said to arise from the presence of the term $3C$ in $|C_J-3C|$, and not from the term C_J, as in the preceding discussion. In the former example, the exceptional lines are not a necessary part of the curves $|C_J-2C|$, which are given by cubic surfaces containing the double line and double conic. These two examples are among those discussed in the note at the end of this chapter (p. 226).

Return to the definition of ω. From this necessary discussion of the canonical curves of a surface, we return now to the question of the invariance of ω, limiting ourselves in the first place, as before, to the case when the surface has no multiple points other than a double curve with triple points thereon. From the theorem obtained, that the linear system expressed by $|C_J-3C-\gamma|$ is the same whatever the system $|C|$ may be, it follows, as was stated above for the case when $|C|$ has no prescribed base points, that the number of intersections of a curve U, of the surface, with a curve of $|C_J-3C-\gamma|$, depends on U only. We proceed to express this number in terms of the characters of U. Suppose, first, that U is not an isolated curve, but, with prescribed base points, at simple points of the surface, determines a linear system $|U|$, of freedom at least 2. Let n be the effective grade of this linear system, namely the number of intersections of two curves of the system other than at the base points, and ν the *completed* grade (cf. p. 139), the number of intersections including those at the base points, so that if t be the effective multiplicity of all curves of the system at one such base point $\nu=n+\Sigma t^2$; similarly let p be the effective genus of the curves $|U|$, in Riemann's sense (which depends on the multiple base points of the curves), and call $p+\frac{1}{2}\Sigma t(t-1)$ the *completed* genus, momentarily, denoting this by π. Then we prove that the number of intersections of a curve of the canonical system with the curve U is $2\pi-2-\nu$; by prescription the canonical system has no base points, and we can suppose that none of the intersections in question fall at base points of the system $|U|$. We call this number the *canonical number** of the curve U, or of the system $|U|$; and we remark that, by the definition, the canonical number of a linear system $|U+V|$, formed by the sum of two systems $|U|$ and $|V|$, is the sum of the canonical numbers of these. To obtain the result stated, we suppose the canonical system to be computed from the system $|U|$; if

$$au + bv + cw = 0$$

determine a net of curves of this system, this net determines on the particular curve $u=0$, outside the base points of the system, a linear series of freedom 1, given by the pencil $bv+cw=0$; and it is well known that the Jacobian set of this series, where two points of a set coincide, is equivalent, on the curve $u=0$, to two sets of the linear series, taken with a set of the canonical series on this curve; the linear system $|U_J-2U|$, where $|U_J|$ denotes the Jacobian system of $|U|$, thus determines on $u=0$, outside the base points of the system $|U|$, a set of $2p-2$ points, this being an immediate consequence of the definition of the Jacobian curve of the net

$$au + bv + cw = 0.$$

If then, for two curves P, Q, the symbol (P, Q) denote the number of intersections other than at base points prescribed for either of these curves, and ω denote the aggregate of the base points of the system $|U|$, each taken simply, we have† $(U_J-2U, U)=2p-2$; and the number of intersections of U with the canonical system, not at the base points of $|U|$, or the number $(U_J-3U-\omega, U)$, is $2p-2-(U+\omega, U)$. Here $(U+\omega, U)$ will be the number of intersections, other than at the base points of $|U|$, of the curve U, with a curve of the same order having, at a t-ple base point of U, only a $(t-1)$-ple point, and will therefore be $n+\Sigma t^2-\Sigma t(t-1)$, or $n+\Sigma t$, the summation being extended to all the base points of $|U|$. Putting however $\nu=n+\Sigma t^2$, $\pi=p+\frac{1}{2}\Sigma t(t-1)$, we find at once that $2p-2-n-\Sigma t$ is the same as $2\pi-2-\nu$.

The system $|U_J-2U|$ is called the *adjoint* system of $|U|$ (cf. Vol. v, p. 213).

---

* Cf. Severi, "Il genere aritmetico ed il genere lineare", *Atti...Torino*, xxxvii, 1902, No. 1; also "Su alcune questioni", *Rend. Palermo*, xvii, 1903, No. 4; and "Sulle intersezioni", *Mem. Torino*, lii, 1903, No. 8. We define a "class of immersion" below (p. 225, Ex. 5, and Chap. vi, p. 251), and connect it with the canonical number.

† This remains true when the curves $|U|$ have variable double points on the double curve, p being the effective genus taking account of such. By a theorem, due to Bertini, a linear system of curves can have variable multiple points only on the multiple curves of the surface. But such points arise for instance when the system $|U|$ consists of plane sections of the surface, the system $|U_J|$ being then composed of proper curves of contact of enveloping cones to the surface.

Some particular remarks should be made in connexion with this result: (1), for the case when $|U|$ is the canonical system itself, supposed now to be of freedom 2 at least, if N denote the grade and Π denote the genus for this system, which has no base points, we infer $2\Pi - 2 - N = N$, or $N = \Pi - 1$ (cf. the result for the canonical system, as defined by Noether, given in the preceding chapter, pp. 162, 168); (2), If ν_1, π_1 be respectively the completed grade and genus of one linear system of curves $|A_1|$, and ν_2, π_2 the corresponding numbers for a system $|A_2|$, while ν_{12}, π_{12} similarly belong to the sum system $|A_1 + A_2|$, and i is the complete number of intersections of a curve of $|A_1|$ with a curve of $|A_2|$, since the complete number of intersections of two curves of $|A_1 + A_2|$, say $[A_1 + A_2, A_1 + A_2]$, is clearly equal to $[A_1, A_1] + 2[A_1, A_2] + [A_2, A_2]$, we have

$$\nu_{12} = \nu_1 + \nu_2 + 2i;$$

(3), We also have $2\pi_{12} - 2 = 2\pi_1 - 2 + 2\pi_2 - 2 + 2i$. For if, from an arbitrary point, a curve of each of $|A_1|$, $|A_2|$, $|A_1 + A_2|$ be projected, on to a plane, into curves of respective orders $m_1, m_2, m_1 + m_2$, and we denote by h_1, h_2 and k the numbers of lines from this arbitrary point which respectively meet A_1 in two points, A_2 in two points and meet A_1 and A_2 each once, we have the three equations

$$2\pi_1 - 2 + 2\delta_1 = m_1(m_1 - 3) - 2h_1, \quad 2\pi_2 - 2 + 2\delta_2 = m_2(m_2 - 3) - 2h_2$$

and $\quad 2\pi_{12} - 2 + 2\delta_1 + 2\delta_2 = (m_1 + m_2)(m_1 + m_2 - 3) - 2h_1 - 2h_2 - 2k$,

where δ_1, δ_2 represent the number of double points of $|A_1|$, $|A_2|$ on the double line of the surface (and no correction of π_{12} is necessary in the last equation owing to the double points of a degenerate plane curve $A_1 + A_2$ which arise from an intersection of A_1 and A_2 on the surface; for the sum system $|A_1 + A_2|$ on the surface has no variable double points not included in $\delta_1 + \delta_2$). Taking account then of $m_1 m_2 = k + i$, where i is the completed number of intersections of $|A_1|$ and $|A_2|$, we obtain the result in question. This proof, suggested in Picard-Simart, *Fonct. algéb.* (cf. Vol. v, p. 209), is valid when either $|A_1|$ or $|A_2|$ consists of an isolated curve, supposed irreducible. When both $|A_1|$ and $|A_2|$ have freedom 2 at least, another proof arises from an equation found above (p. 217), expressing that the system $|(A_1 + A_2)_J - 2A_1 - 2A_2|$ is equivalent either to $|A_{1J} - 2A_1 + A_2 + \alpha_2|$, or to $|A_{2J} - 2A_2 + A_1 + \alpha_1|$; this shews, putting down the number of intersections of a curve of the first system with a curve of $|A_1 + A_2|$, other than at base points, as equivalent to the sum of

$$(A_{1J} - 2A_1 + A_2 + \alpha_2, A_1) \text{ and } (A_{2J} - 2A_2 + A_1 + \alpha_1, A_2),$$

that $2\pi_{12} - 2 - \Sigma(s_1 + s_2)(s_1 + s_2 - 1)$ is equal to the sum of

$$2\pi_1 - 2 - \Sigma s_1(s_1 - 1) + i - \Sigma s_1 s_2 \text{ and } 2\pi_2 - 2 - \Sigma s_2(s_2 - 1) + i - \Sigma s_1 s_2;$$

and this gives the formula in question. See also Noether, *Acta*

Math. VIII, 1886, p. 182; (4), The formula in (2) enables us to give
a meaning to ν_2 when the system $|A_2|$ consists of an isolated curve;
for, if this be given, and $|A_1|$ be a system of freedom at least 2, it
is easy to see that the system $|A_1+A_2|$ is then of freedom at least
as great as that of $|A_1|$; and then, further, that the value of ν_2
thence defined is independent of the system $|A_1|$ adopted. Further
the formulae (2), (3) enable us to define ν_2 and π_2 when $|A_2|$ consists
of reducible curves, and it can be shewn that the definition leads
to definite values. It is then to be remarked that these formulae
lead to $2\pi_{12}-2-\nu_{12}=2\pi_1-2-\nu_1+2\pi_2-2-\nu_2$; this is an immediate
consequence of the result $(C_J-3C-\gamma,\ U)=2\pi-2-\nu$, proved above
for the case when $|U|$ is a system of freedom at least 2. But, by
what is now said, this result is now proved even when U is an isolated curve; (5), The formulae in (2) and (3), but not the formula
$(C_J-3C-\gamma,\ U)=2\pi-2-\nu$, continue to hold when the completed
grade, genus, and number of intersections, $\nu,\ \pi,\ i$ are respectively
replaced by their effective values $\nu-\Sigma t^2,\ \pi-\frac{1}{2}\Sigma t(t-1),\ i-\Sigma t_1t_2$,
where t is an actual multiplicity; and also when these are replaced
by their *prescribed*, or *virtual*, values $\nu-\Sigma s^2,\ \pi-\frac{1}{2}\Sigma s(s-1),\ i-\Sigma s_1s_2$,
where s is a prescribed multiplicity, at a prescribed base point. (The
actual multiplicity may exceed the prescribed multiplicity.) This
is obvious because both the functions $x^2,\ x(x-1)$ satisfy the equation $\phi(x_1+x_2)=\phi(x_1)+\phi(x_2)+2x_1x_2$.

After this long discussion it is clear that *we may define the number*
$\omega-1$ *as the grade of the canonical system* of the surface, itself defined
from any linear system $|C|$ as the system $|C_J-3C-\gamma|$. And as we
have considered the modification of the canonical system in the case
of a birational transformation when a simple point of the original
surface is replaced by a curve on the derived surface, we are in a
position to find the modification of $\omega-1$ due to the transformation.

Ex. 1. The equation for the canonical number

$$(C_J-3C-\gamma,\ U)=2\pi-2-\nu,$$

if modified to the form $(3C+\gamma-C_J,\ U)=\nu-(2\pi-2)$, will be found to
hold in cases when the canonical system consists of ideal curves, as for
instance in the case when the surface is a plane or a rational surface.
For example, if we consider a system of curves of order k in a plane
with prescribed s-ple points, the system $|C_J|$ is that of curves of order
$3k-3$ with prescribed $(3s-1)$-ple points, and the system $|3C+\gamma-C_J|$
is that of the cubic curves of a plane. These meet the curves of order n
in the plane in $n^2-n(n-3)$, or $3n$ points.

Ex. 2. As stated above (p. 222), the equation $(U_J-2U,\ U)=2p-2$ is
unaffected by the presence of a double curve on the surface. For example,
if $|U|$ be the system of plane sections of a Steiner quartic surface ($p=0$),
the system $|U_J|$ consists of curves of order 6, and $|2U-U_J|$ of conics,
meeting a plane in $2-2p$ points. Or, if $|U|$ be the system of plane
sections of a quartic surface which has a double line, the system $|U_J|$
consists of curves of order 10, and the system $|U_J-2U|$ consists of

conics, which meet a plane in $2p-2$ or 2 points. Or again the example (p. 229) of a sextic surface with a double conic and a double line may be quoted.

Ex. 3. If the canonical system $|C_J-3C-\gamma|$ break up into a system of irreducible curves $|K|$, together with other isolated curves $\epsilon_1, \epsilon_2, \ldots,$ of grade $\nu_1, \nu_2, \ldots$ and genera $\pi_1, \pi_2, \ldots$, the equation $N=\Pi-1$ for the unreduced canonical system leads to the corresponding equation $N_0=\Pi_0-1$ for the system $|K|$, if, and only if,

$$\Sigma(\pi_i-1-\nu_i)=\Sigma(K, \epsilon_i)+\Sigma(\epsilon_i, \epsilon_j).$$

This equation is true if every π_i be zero and every ν_i be -1, and if the isolated curves do not intersect the curve K, nor one another (intersections on the double curve of the surface being neglected).

Ex. 4. If N, Π be the complete grade and genus of the system $|C_J-3C-\gamma|$, and ν, π of the system $|C|$, while ν', π' are the complete grade and genus of $|C_J-2C|$, we have from $C_J-2C=C_J-3C-\gamma+C+\gamma$, the results $\nu'=N+\nu+2(2\pi-2-\nu)$, and $\pi'=\Pi+\pi+2\pi-2-\nu-1$; hence as $N=\Pi-1$, we have $\nu'-\pi'=\pi-2$. And since

$$(t-1)^2-\tfrac{1}{2}(t-1)(t-2)=\tfrac{1}{2}t(t-1),$$

this remains true when the grades and genera involved are the actual, or are the prescribed (or virtual) values.

Ex. 5. We have considered, in the last chapter, the curve, of order μ_1, on a surface in space [4], which is the locus of points whereat the tangent plane of the surface meets an arbitrary general line, l. This is the Jacobian C_J, of the net of curves $|C|$, on the surface, determined by primes through the line l, all such prime sections through a point of this curve having a common tangent line which meets the line l. We shall have occasion, in the next chapter, to consider the number of points of a given curve U of the surface whereat the tangent plane of the surface meets an arbitrary line l; it will be called the *class of immersion* of the curve U on the surface, and usually denoted by z. From what we have said, z is equal to the number of points of intersection of U with the curve C_J arising from the line l. The canonical number of the curve U has, however, been defined as the number of its intersections with a curve of the canonical system of the surface, when this exists; which is given by $|C_J-3C|$. Thus if ϵ_0 be the order of U, its canonical number is equal to $z-3\epsilon_0$. This then is equal to $2\pi-2-\nu$, where π, ν are the completed genus and grade of U. As a simple example, consider a line on the Segre quartic surface of intersection of two quadrics in [4]. Here z is at once seen to be 2; thus $\nu=-1$. More generally, for an isolated curve, of order s, on the surface, for which $\pi=0$, $\nu=-1$, we infer $z=3s-1$. This is in agreement with the fact that for a linear system of curves on a surface, with a common s-fold base point at a simple point of the surface, the Jacobian system has a $(3s-1)$-fold point at this point (cf. p. 254, Ex. 2).

The computed characters of an exceptional curve. Suppose that in the birational transformation of a surface a simple point of the original surface becomes an exceptional curve on the derived surface. Let the point be O, and the curve into which it is transformed be ϵ'. Take two curves A, B on the original surface, without prescribed base points, intersecting at O and in n other points. On the derived surface these become, by *direct* transformation, as we have seen (p. 218, above), two composite curves $A'+\epsilon'$, $B'+\epsilon'$, whose total number of intersections is the same as of A and B on the original surface. Thus we have

$$n+1=(A, B)=(A'+\epsilon', B'+\epsilon')=(A', B')+(A', \epsilon')+(B', \epsilon')+(\epsilon', \epsilon');$$

here (A',B') is equal to the number of intersections of A and B which are not at O, namely n, and (A', ϵ'), (B', ϵ') are both equal to 1 because A and B pass simply through O. Thus we deduce $(\epsilon', \epsilon')=-1$. The canonical number of the curve ϵ' is also -1; for let K denote a canonical curve of the original surface, which we suppose to be irreducible, and not to pass through O; we have proved above (p. 218) that the direct transformation of this, on the derived surface, is a curve $K'+\epsilon'$, whereof K' will be in $(1, 1)$ correspondence with K, so that $(K', \epsilon')=0$, as K does not contain O; thus $(K'+\epsilon', \epsilon')=-1$. Thus if π' be the complete genus of ϵ', we have $2\pi'-2-\nu'=-1$, wherein ν' has been shewn equal to -1. Whence also $\pi'=0$. As ϵ' is the representative of a simple point of the original surface, its effective genus is 0; thus it has no multiple points. A particular consequence (see Ex. 3 preceding) is that, if N',Π' denote the grade and genus of the curves of the reduced canonical system K', on the derived surface, we have $N'=\Pi'-1$. More generally this is true for the reduced canonical system on any surface if this is formed from the system $|C_J-3C-\gamma|$ by the omission of curves which can all be regarded as arising by transformation from simple points of another surface (see Ex. 3, preceding).

The modification of ω in a birational transformation. In the simplest case, just considered, where one simple point O of the original surface becomes an exceptional curve ϵ' of the derived surface, we have, with the same notation,

$$\omega'-1=(K'+\epsilon', K'+\epsilon')=(K', K')+2(K', \epsilon')+(\epsilon', \epsilon'),$$

wherein $(K', \epsilon')=0$, $(\epsilon', \epsilon')=-1$, and (K', K'), the number of intersections of two curves which are in $(1, 1)$ correspondence with the canonical curves of the original surface, is equal to $\omega-1$. Thus we have $\omega=\omega'+1$. Thus, if, on the first surface, there be e curves which change into simple points of the second surface, and, on this surface, there be e' curves arising by the transformation from simple points of the original surface, we may infer that $\omega+e=\omega'+e'$.

Note I. Examination of examples cited by Noether. It will be interesting, before pursuing the general theory in more detail (see Note II, p. 232, below), to examine the application of what has been said to a set of examples quoted by Noether (*Math. Ann.* VIII, 1875, § 11), *and from his point of view.* Of his sixteen examples one involves intersecting multiple curves on the surface (a septimic surface with a triple line, and two double conics in planes through this line); we consider the other fifteen, numbering these as in his paper. He computes in each case two numbers: (i), The number p_g of actually existing surfaces of order $\mu_0 - 4$, where μ_0 is the order of the given surface ψ^{μ_0}, which pass $(j-1)$ times through every j-ple curve of the surface, and have a $(k-2)$-ple node at every k-ple node (supposed general) of the surface. Such so-called *adjoint surfaces* meet ψ^{μ_0} in the curves called by Noether the *canonical* curves; it may happen however that all these adjoint surfaces have common an undesigned curve, or curves, lying on ψ^{μ_0}; such curves we shall call Noether's *exceptional* curves, and the number of irreducible curves of this character will be denoted by e_N. The variable parts of the canonical curves constitute a linear system which we call the *reduced* canonical system. Noether also computes (ii), the effective grade of this reduced system, namely the number of variable intersections of two curves of the system other than at the multiple points of the surface, denoting this by $p^{(2)}$; and he computes the effective genus of a curve of this system (which takes account of multiple points of curves of this system, if any, which lie at the multiple points of ψ^{μ_0}). It appears in all cases that $p^{(2)} = p^{(1)} - 1$; this is a consequence of what has been given above (see p. 223) when there are no exceptional curves, and the surface has no isolated nodes, and is inferred to hold in general from an indicated proof that the exceptional curves can be transformed to simple points of another surface. Beside p_g and $p^{(1)}$ we shall, in the following examples, compute p_n and I, ω, or, in case the surface have isolated nodes, the numbers I_0, ω_0 which we have used above (p. 199) to mark the presence of such nodes. That surfaces exist for which $p_g > p_n$ is a discovery, not taken account of in Noether's paper, dating from Cayley's remark (1871, *Papers*, VIII, p. 394) that, for a non-rational ruled surface, p_n is negative, while $p_g = 0$ (see preceding chapter, Ex. 9, p. 167). One of the earliest examples of a surface with $p_g > p_n$ which is not ruled and not transformable to a ruled surface, appears to be the surface of order $2k$ with k-ple nodes at 8 associated points, considered by Castelnuovo (1891; see p. 203, above). But other cases are the hyperelliptic surfaces, representing the pairs of points of a curve of genus 2, for which $p_g = 1$, $p_n = -1$ (Picard, *Liouville's Jour.* I, 1885 and V, 1889; Humbert, *ibid.* IX, 1893); and, more generally, the surfaces (considered in some detail below, in Chap. VII) which represent the pairs of points of two curves, which may coincide (De Franchis, *Rend. Palermo*, XVII, 1903; Severi, *Atti...Torino*, XXXVIII, 1902-3). In all Noether's cases the so-called *irregularity* $p_g - p_n$ is zero. It is apposite to quote here the theorem (of Castelnuovo-Enriques, p. 140, above), that a surface on which there is a linear series of curves (of freedom $r \geqslant 0$), with complete genus $\pi > 0$, having a negative canonical number, $2\pi - 2 - \nu$, can be birationally transformed to a ruled surface.

In the first seven of the following examples (3), (6), (9), (10), (11), (13), (15), there is an ordinary double curve, and no isolated node, or exceptional curve; in the next four cases (1), (5), (7), (14) there is no isolated node, but exceptional curves arise; in the last four cases (2), (4), (8), (12) there are nodes, and, except in (8), also exceptional curves. When there are e_N exceptional curves, and κ nodes (of any orders), it will be verified in every case that $\omega_0 - \kappa + e_N = p^{(1)}$ (see Note II, p. 232, below. Above, pp. 215–24, nodes were excluded).

(3) For a sextic surface, ψ^6, with a double elliptic quartic curve, using the formulae which have been previously given

$$\mu_1 = \mu_0(\mu_0 - 1) - 2\Sigma\epsilon_0, \quad \nu_2 = \Sigma\left[(2\mu_0 - 4)\epsilon_0 - 2\epsilon_1 - 6t\right],$$

$$\mu_2 = \mu_0(\mu_0 - 1)^2 - \Sigma\left[(7\mu_0 - 12)\epsilon_0 - 4\epsilon_1 - 15t\right],$$

$$I + 4 = \mu_2 - 2\mu_1 + 3\mu_0, \quad \omega - 1 = \mu_2 + \nu_2 - 6\mu_1 + 9\mu_0,$$

$$12(p_n + 1) = I + 4 + \omega - 1,$$

we find

$$\mu_0 = 6, \quad \mu_1 = 22, \quad \mu_2 = 62, \quad \nu_2 = 16, \quad I + 4 = 36, \quad \omega - 1 = 0, \quad p_n + 1 = 3.$$

There are two quadric surfaces through the double curve, or $p_g = 2, = p_n$; two such quadric surfaces have no further intersection, so that $p^{(2)} = 0$; and $p^{(1)} = 1, = \omega$.

(6) A ψ^6 with a double rational cubic curve. There are three quadric surfaces through the double curve, or $p_g = 3$; any two of these quadric surfaces meet in a chord of the cubic curve, which has two further intersections with the surface, so that $p^{(2)} = 2$. A quadric surface through the cubic curve meets ψ^6 further in a sextic curve, for which (Ex. 10, p. 168, Chap. IV) the genus is $p^{(1)} = 3$, so that $p^{(2)} = p^{(1)} - 1$. Moreover, the formulae quoted give $\mu_0 = 6, \mu_1 = 24,$ $\mu_2 = 76, \nu_2 = 16, I + 4 = 46, \omega - 1 = 2, p_n + 1 = 4$, so that $\omega = p^{(1)}$, and $p_g = p_n$.

(9) A ψ^6 with two skew double lines. There are four quadric surfaces through these lines, or $p_g = 4$; moreover, two such quadric surfaces meet further in two skew lines, each meeting both the given lines and hence meeting ψ^6 residually in two points; thus $p^{(2)} = 4$. A quadric surface through the double lines meets the surface further in an octavic curve for which (Ex. 10, p. 168, Chap. IV) the genus $p^{(1)} = 5$, so that $p^{(2)} = p^{(1)} - 1$. The formulae give $\mu_0 = 6, \mu_1 = 26, \mu_2 = 90,$ $\nu_2 = 16, I + 4 = 56, \omega - 1 = 4, p_n + 1 = 5$, so that $\omega = p^{(1)}, p_g = p_n$.

(10) A ψ^7 with a double elliptic quartic curve, and a double line not meeting this. Cubic surfaces through the quartic curve and the line are, in tale, $20 - 12 - 4$, or 4, so that $p_g = 4$; two such cubic surfaces meet, beside, in a curve of order 4, which meets the quartic curve in 8 points, and the line in 4 points (as we may see, for example, by taking for such a cubic surface the aggregate of a quadric

surface through the quartic curve and a plane through the line); such cubic surfaces have therefore $28-24$, or 4, further intersections with the ψ^7; thus $p^{(2)}=4$. The formulae give $\mu_0=7$, $\mu_1=32$, $\mu_2=99$, $\nu_2=34$, $I+4=56$, $\omega-1=4$, $p_n+1=5$. Thus $p_n=p_g$ and $\omega-1=p^{(2)}$.

(11) A general ψ^5. The tale of all planes is 4; and two planes meet in a line having 5 intersections with the surface; thus $p_g=4$, $p^{(2)}=5=p^{(1)}-1$. The formulae give $\mu_0=5$, $\mu_1=20$, $\mu_2=80$, $\nu_2=0$, $I=51$, $\omega-1=5$, $p_n=4$. Whence $p_n=p_g$ and $\omega=p^{(1)}$. Cf. No. (12) below.

(13) A ψ^7 with a double sextic curve of genus 3. The tale of cubic surfaces through the sextic curve is $20-(18-3+1)$, $=4$; so that $p_g=4$. Two such cubic surfaces meet further in a rational cubic curve having 8 intersections with the sextic curve, and hence $21-16$, or 5, other intersections with the surface. Thus $p^{(2)}=5$. The formulae give $\mu_0=7$, $\mu_1=30$, $\mu_2=94$, $\nu_2=28$, $I=51$, $\omega-1=5$, $p_n=4$; so that $p_n=p_g$ and $\omega-1=p^{(2)}$. Evidently, if $U=0$, $V=0$, $W=0$, $K=0$ be independent cubic surfaces through the sextic curve, the equations $x'/U=y'/V=z'/W=t'/K$ transform the surface ψ^7 into a ψ^5; which explains the equality of the values of I and ω in this case with those in the preceding No. (11).

(15) A ψ^6 with a plane cubic double curve. The quadric surfaces through this curve break up into the plane of the curve and a further arbitrary plane. Thus $p_g=4$ and $p^{(2)}=6=p^{(1)}-1$. The formulae give $\mu_0=6$, $\mu_1=24$, $\mu_2=84$, $\nu_2=12$, $I=50$, $\omega-1=6$, $p_n=4$, so that $p_n=p_g$ and $\omega-1=p^{(2)}$.

(1) A ψ^6 with a double conic, and a double line not meeting this. The plane of the conic meets the surface again in a curve of order 2 with a double point on the double line, that is, in a pair of lines. These are exceptional lines according to Noether's definition, since a quadric surface through the double conic and the double line must break up into the plane of the conic and a plane through the line. The reduced canonical curves of the surface are then quartic curves in planes through the double line, each with two double points on the double conic. Thus $p_g=2$, $p^{(1)}=1$. The formulae give $\mu_0=6$, $\mu_1=24$, $\mu_2=68$, $\nu_2=20$, $I=34$, $\omega-1=-2$, $p_n=2$. Thus $p_n=p_g$ and $\omega+e_N=p^{(2)}$, where e_N, $=2$, is the number of exceptional lines. As was remarked earlier (p. 225), the exceptional lines do not lie on the adjoint cubic surfaces, $|C_J-2C|$, drawn through the double conic and the double line, which meet the surface residually in curves of order 12. Of such cubic surfaces the tale is 9, and they may be used to transform the given surface (as in No. (13) above) into a surface lying in space [8]; of these 9 surfaces there are 8 which contain a particular one of the two exceptional lines, so that this line becomes a point of the transformed surface—as, similarly, does the other exceptional line.

(5) A ψ^7 with a double sextic curve of genus 1. Such a curve has 3 quadrisecant chords; these chords therefore lie entirely on the surface ψ^7, and also on any cubic surface drawn through the curve. Thus, two cubic surfaces through the double curve have no further intersection with the surface than these three lines, and $p^{(2)}=0$. Through the double curve there are $20-18$, or 2, cubic surfaces, and $p_g=2$. The formulae give $\mu_0=7$, $\mu_1=30$, $\mu_2=78$, $\nu_2=36$, $I=35$, $\omega-1=-3$, $p_n=2$. Thus $p_n=p_g$ and $\omega+e_N=p^{(2)}+1$, where ϵ_N, $=3$, is the number of exceptional lines. As in the preceding case, No. (1), the adjoint quartic surfaces $|C_J-2C|$ through the sextic curve are such that each of the quadrisecants imposes a single condition for such a quartic surface to contain this line, which can then be transformed to a point.

(7) A ψ^7 with a double sextic curve of genus 2. In this case the curve has one quadrisecant which lies on the surface and on all cubic surfaces through the double curve. There are 3 such cubic surfaces, or $p_g=3$. Further, two such cubic surfaces meet again in a conic, beside the quadrisecant, and this conic meets the double curve in 6 points; whence $p^{(2)}=14-12$, $=2$. The formulae give $\mu_0=7$, $\mu_1=30$, $\mu_2=86$, $\nu_2=32$, $I=43$, $\omega-1=1$, $p_n=3$, so that $p_n=p_g$ and $\omega-1+e_N=p^{(2)}$, where $e_N,=1$, is the number of exceptional lines. And, as in the last example, the quadrisecant lies on all but one of the independent quartic surfaces containing the sextic curve.

(14) A ψ^7 with a rational quintic double curve. This curve has one quadrisecant line, which lies on the surface, and on all cubic surfaces through the quintic curve. Such surfaces are, in tale, $20-(15+1)=4$, so that $p_g=4$. Two such cubic surfaces meet further in the quadrisecant line and in a cubic curve having 8 intersections with the double curve; this cubic curve thus meets the surface, not on the double curve, in $21-16$ points, or $p^{(2)}=5$. The formulae give $\mu_0=7$, $\mu_1=32$, $\mu_2=99$, $\nu_2=34$, $I=52$, $\omega-1=4$, $p_n=4$. Whence $p_n=p_g$ and $\omega-1+e_N=p^{(2)}$, where $e_N=1$.

(2) A ψ^5 with two triple points. The join of these points lies on the surface; it also lies on any surface of order μ_0-4 (or plane) passing $k-2$ times (or once) through the triple points, and is an exceptional line. The canonical curves are the plane quartic curves with a double point at each triple point, so that

$$p_g=2, \quad p^{(2)}=p^{(1)}-1=0.$$

Assuming, as in the earlier part of this chapter, that an isolated k-ple point diminishes the class of the surface by $k(k-1)^2$, we have here, for $k=3$ twice over, $\mu_2=5.4^2-2.3.2^2=56$; while $\mu_1=20$. Whence, from the definitions (p. 200, preceding),

$$I_0+4=\mu_2-2\mu_1+3\mu_0+\Sigma(k-1),$$
$$\omega_0-1=\mu_2+\nu_2-6\mu_1+9\mu_0+\Sigma(k-1)(2k-1),$$

with $\nu_2 = 0$, using κ, $= 2$, for the number of isolated nodes, we have $I_0 + \kappa = 33$, $\omega_0 - \kappa - 1 = -1$, $p_n = 2$. Whence $p_n = p_g$ and

$$\omega_0 - \kappa + e_N = p^{(1)},$$

where e_N is the number of exceptional lines.

(4) A ψ^6 with a rational non-singular cubic as double curve, and also a cubic node. Quadric surfaces through the cubic curve and the node are two in all, and these surfaces contain the chord of the cubic curve which can be drawn from the node, which lies on the surface and is thus an exceptional line. Thus $p_g = 2$, $p^{(2)} = 0$. With the formulae of p. 162 of the preceding chapter, and those in the preceding example, we find $\mu_0 = 6$, $\mu_1 = 24$, $\mu_2 = 64$, $\nu_2 = 16$, $I_0 = 32$, $\omega_0 - 1 = 0$, $p_n = 2$. Thus $p_n = p_g$, and $\omega_0 - 1 - \kappa + e_N = p^{(2)}$, where $\kappa = 1$, $e_N = 1$. In this case also, the exceptional line is one condition for cubic surfaces containing the cubic curve and the node.

(8) A ψ^6 with a double plane cubic curve, and a cubic node. Here, quadric surfaces through the plane cubic curve contain the plane, and the canonical curves are given by arbitrary planes through the node. Whence $p_g = 3$, $p^{(2)} = p^{(1)} - 1 = 3$. Here there is no exceptional line. The formulae referred to in the preceding example lead to $\mu_0 = 6$, $\mu_1 = 24$, $\mu_2 = 72$, $\nu_2 = 12$, $I_0 = 40$, $\omega_0 - 1 = 4$, $p_n = 3$. Whence $p_n = p_g$ and $\omega_0 - \kappa = p^{(1)}$, where κ, $= 1$, is the number of nodes.

(12) A ψ^7 with a triple conic, and one cubic node. In this case, cubic surfaces having the conic as a double curve must contain the plane of the conic; thus the reduced canonical curves are given by quadric surfaces containing the conic and the node, and $p_g = 4$. Two such quadric surfaces meet further in a conic, through the node, having two intersections with the given conic; this conic then has $14 - 9$, or 5, intersections with the surface, other than at multiple points of the surface, so that $p^{(2)} = 5$. By the formulae referred to in the last example we find $\mu_0 = 7$, $\mu_1 = 30$, $\mu_2 = 92$, $\nu_2 = 20$, $I_0 = 51$, $\omega_0 - 1 = 5$, $p_n = 4$. The plane of the conic meets the surface further in an exceptional line. Thus, with $\kappa = 1$, $e_N = 1$, we have

$$\omega_0 - 1 - \kappa + e_N = p^{(2)}.$$

In this case also, the exceptional line is one condition for the adjoint quartic surfaces $|C_J - 2C|$ which pass through the triple point and have the conic as double curve. But the ψ^7 can be transformed to a quintic surface by considering the ∞^3 quadrics through the triple point and the double conic, these becoming planes in the new space; a particular plane section of the quintic surface breaks up into a conic, and a cubic curve, which is the transformation of the triple point of the ψ^7, but is not an exceptional curve of the quintic surface in Noether's sense. The values of the invariants for the quintic surface are found in No. (11) above, say $I' = 51$, $\omega' - 1 = 5$. Thus, in particular, $I_0 + \kappa - e_N = I'$, which can also be considered in the form $I_0 - e_N = I' - \kappa$, where κ would then refer to the cubic curve on the quintic surface, as on pp. 201 ff. See Note II following.

This completes the set of examples we quote from Noether. For these, see also D. W. Babbage, *Proc. Camb. Phi!. Soc.* xxix, 1933, p. 319. The following case by E. A. Maxwell may also be adduced: A surface ψ^9, of order 9, having an isolated 4-ple node and a rational quartic curve as triple curve. This quartic curve lies on only one quadric surface, and the residual intersection of this with the ψ^9 consists of six lines; the three chords of the quartic curve which can be drawn from the 4-ple node also lie on the ψ^9. The adjoint surfaces of order 5, passing doubly through the quartic curve, which have the node as a quadratic node, break up into the unique quadric surface through the quartic curve, together with cubic surfaces, passing through the quartic curve, which have a quadratic node at the 4-ple node of ψ^9. Thus we can prove that $p_g = 3$ and $p^{(2)} = 2$. The formulae give, in this case, $\mu_0 = 9$, $\mu_1 = 48$, $\mu_2 = 120$, $\nu_2 = 60$, $I_0 = 50$, $\omega_0 - 1 = -6$, $p_n = 3$. Thus again we have $p_n = p_g$ and

$$\omega_0 - 1 - \kappa + e_N = p^{(2)},$$

since there is one node ($\kappa = 1$), and 9 exceptional lines ($e_N = 9$).

Note II. The adjoint surfaces of a given surface; the exceptional curves. In the present state of the theory of the resolution of the multiple elements of a surface (cf. for example, Enriques-Chisini, *Teoria geometrica*, ii, pp. 649–58; and the references there given) it seems desirable to approach the theory of the invariants of a surface from the historical side, following Noether, as we have done. The conclusion that the adjoint surfaces of the given surface should be taken to have a $(k-2)$-ple point at an isolated k-ple point, and not a $(k-1)$-ple point as have the first polars, was first suggested, for the case of the canonical surfaces, by the conditions of finiteness of Clebsch's double integral, by which the invariant p_g was defined. But a geometrical aspect can be given to this conclusion, at least in simple cases, in two ways, which we now consider in turn (cf. Enriques-Campedelli, *Superficie algebriche*, 1932, p. 129). Consider a general isolated k-ple node of the surface, O; denote by D any section of the surface by a plane through O. The first polar of the point O, in regard to the surface, is a surface with a k-ple node at O, whose asymptotic cone coincides with that of the given surface; the intersection of this first polar with the given surface is therefore effectively that of a surface with a $(k+1)$-ple node at O. Hence, if $|D_J|$ denote the Jacobian curve of the net of curves $|D|$, the system $|D_J - 3D|$ on the given surface will be given by surfaces having at O a multiple point of multiplicity $[k(k+1)-3k]/k$, or $k-2$. This is for the case when the k-ple point is quite general, but the same suggestion avails in other cases.

Consider, for example, first, the case of a uniplanar double point; for this, the equation of the given surface, in non-homogeneous co-ordinates, for the neighbourhood of O, is of the form

$$z^2 + \phi(x, y, z) + \ldots = 0,$$

where ϕ is homogeneously of order 3 in x, y, z. The first polar of the point O intersects the surface, near O, as does a surface of equation $\phi(x, y, z) + \ldots = 0$; the behaviour of $D_J - 3D$ near the origin is then that of a fraction $\phi(x, y, z)/\psi(x, y, z)$, where ψ is also homogeneously of order 3 in x, y, z, and hence, that of a fraction $u(x, y)/v(x, y)$, where u, v are homogeneously of order 3 in x and y only. Thus the uniplanar point imposes no condition for the adjoint surface. If, however, we consider a tacnode, for which the equation of the surface is $z^2 + z\phi(x, y, z) + \ldots = 0$, in which ϕ is homogeneously quadratic in x, y, z, we shall be similarly led to consider the fraction $z\phi(x, y, z)/\psi(x, y, z)$, where ψ is of order 3, and hence a fraction zu/v, where u, v are homogeneously of orders 2 and 3, respectively, in x and y only; in this case, however, z is of an order of smallness comparable with that of a *quadratic* polynomial in x and y, and the fraction therefore vanishes to the first order. Thus the canonical curves of the surface are given by adjoint surfaces subject to the condition of passing through the tacnode. That the adjoint surfaces should be given near O by considering the system $|D_J - 3D|$ is of course suggested by the general theory given above for the case when the surface has no isolated multiple point. In Noether's theory ("Theorie des eindeutigen Entsprechens, u.s.w." § 3, *Math. Ann.* II, 1870) the adjoint surfaces determining the canonical curves, for the surface, of order n, expressed by $f(x, y, z, t) = 0$, are those surfaces, $\phi = 0$, of order $n - 4$, for which the double integral $\iint \phi \Delta \, du \, dv$ remains finite, where u, v are two parameters in terms of which the coordinates (x, y, z, t), of a point of the surface, are supposed to be expressible in the neighbourhood of a point considered, and Δ is given by $(a, y, z_1, t_2)/(a\partial f/\partial x)$, wherein (a, y, z_1, t_2) means the determinant of four rows, namely a, b, c, d; x, y, z, t; x_1, y_1, z_1, t_1; x_2, y_2, z_2, t_2, with $x_1 = \partial x/\partial u$, $x_2 = \partial x/\partial v$, etc., and $(a\partial f/\partial x)$ means $a\partial f/\partial x + b\partial f/\partial y + c\partial f/\partial z + d\partial f/\partial t$; and it is shewn that Δ is independent of a, b, c, d. In particular, if $a = b = c = 0$, Δ reduces to $(x, y_1, z_2)/\partial f/\partial t$, and, speaking in general terms, the determinant (x, y_1, z_2) here plays the part denoted by the term $3D$ in $|D_J - 3D|$. A similar formulation is possible for a $(r-1)$-ple multiple integral attached to a primal $f(x, y, \ldots) = 0$, of order n, in space of r dimensions, the ϕ then being a polynomial of order $n - r - 1$; and the condition of finiteness of the integral, when there is on $f = 0$ a k-fold multiple locus of dimension s, *in the most general case*, is that $\phi = 0$ should have this locus to multiplicity $k - (r - s - 1)$; for example, for a multiple k-fold point $(s=0)$, to multiplicity $k-r+1$; for a multiple k-fold curve $(s=1)$ to multiplicity $k-r+2$. When the multiple locus is of special character, these numbers may be altered; but it is possible to find them by considering the behaviour of Δ; for instance, for a uniplanar or tacnodal point of a surface in three dimensions, see Picard-Simart, *Fonctions algébriques*, I, 1897, pp. 77, 185. Another way of reaching the necessary behaviour of an adjoint surface at a multiple point (Enriques, *Soc. Ital. d. Sc.* x, 1896, p. 10; Enriques-Campedelli, *Superficie algebriche*, 1932, pp. 17, 66, 129) is based on the fact that it is possible to transform the surface birationally so that the neighbourhood of the multiple point of the surface is changed into a curve, ϵ, on the new surface; the sections $|C|$ of the surface by arbitrary planes are thereby changed into a system $|C'|$ on the new surface, for which the curve ϵ is fundamental, not intersecting the curves $|C'|$, except at base points of these. Such a curve ϵ will then be a fixed part of the Jacobian curve of any net of curves contained in the system $|C'|$, as explained above (p. 216). We may therefore regard the Jacobian system $|C_J|$, of the system $|C|$, as being of the form $|\Gamma+\Sigma O|$, where, under the hypothesis that the surface is transformed to one upon which all the isolated multiple points, O, of the original surface, are represented by curves, ϵ, the curve Γ is the curve of the original surface which transforms into the part of the Jacobian system $|C_J'|$ other than these curves ϵ. If D denote the intersection of the original surface with a plane through a general k-ple point O, it is natural to write the number of intersections (D, O) as k; and, therefore, as a general plane section, equivalent to $D+O$, does not pass through O, to write $(D+O, O)=$ zero, and hence $(O, O)=-k$. We also suppose two curves ϵ, ϵ', arising by transformation from two multiple points O, O', not to intersect. Further, as the first polar of a point, in regard to the original surface, has a $(k-1)$-ple point at O, there are $k(k-1)$ points of its intersection which are in the immediate neighbourhood of O; thus we write $(\Gamma, O)=k(k-1)$. Hence we have

$$(C_J, O)=(\Gamma+\Sigma O, O)=(\Gamma+O, O)=k(k-1)-k;$$

or $(C_J, O)=k(k-2)$. Thus, identifying Γ with the proper curve of contact of tangent planes to the surface from an external point, with a recognised (or prescribed) $k(k-1)$-ple point at the k-ple point of the surface, and identifying $|C_J-3C|$ with the canonical system of the surface, we reach the conclusion that the adjoint surfaces determining these canonical curves have only a $(k-2)$-ple point at the multiple point. In this point of view we assign a grade $-k$ to the isolated multiple point O; if further, since this point transforms in general to a curve of order k, we assign to this point a genus $\frac{1}{2}(k-1)(k-2)$, we can reach the corrected value of the invariant $\omega - 1$ which we have adopted above (p. 200) when the surface has isolated multiple points*. For let π be the prescribed genus of the proper curve of contact Γ; then, by the formula $p_1 + p_2 + i - 1$ developed in the text (p. 223), the prescribed (or virtual) genus of C_J, that is of $\Gamma + \Sigma O$, is

$$\pi + \Sigma[\tfrac{1}{2}(k-1)(k-2) + k(k-1) - 1],$$

which is the same as $\pi + \Sigma[(k-1)(2k-1) - \frac{1}{2}k(k-1) - 1]$. Whence, with $(C, C) = \mu_0$, $(C_J, C) = (\Gamma, C) = \mu_1$, $(\Gamma, \Gamma) = \mu_2 + \nu_2$, we have

$$(C_J, C_J) = (\Gamma + \Sigma O, \Gamma + \Sigma O) = \mu_2 + \nu_2 + \Sigma[2k(k-1) - k]$$
$$= \mu_2 + \nu_2 + \Sigma[(2k-1)(k-1) - 1],$$

and, therefore,

$$(C_J - 3C, C_J - 3C) = \mu_2 + \nu_2 - 6\mu_1 + 9\mu_0 + \Sigma[(2k-1)(k-1) - 1].$$

While, if ϖ be the prescribed genus of C_J,

$$\varpi - 1 - 9(p-1) = \pi - 1 - 9(p-1) + \Sigma[(2k-1)(k-1) - \tfrac{1}{2}k(k-1) - 1].$$

These are the values found (p. 200) for $\omega_0 - 1 - \kappa$, if κ denote the number of isolated multiple points. The use of ω, $= \omega_0 - \kappa$, in place of ω_0, would involve the use of I, $= I_0 + \kappa$, in place of I_0. It is this value of I, in the case of a surface with isolated quadratic nodes, which is used by Severi, *Ann. Sc. d. l'École norm. sup.* xxv, 1908, No. 8, and *Acta Math.* xxxii, 1907, p. 351 (cf. Zeuthen, *Lehrbuch d. abz. Meth.* 1914, p. 167). This I would, for instance, be 5, not only for a general cubic surface, but also for cubic surfaces with nodes. There seems convenience in making the distinction, but the essential facts are not modified whichever convention be employed.

In regard to the exceptional curves of a surface, which arise from or can be transformed solely to simple points of another surface, we may accept it as proved that these are part of the system $|C_J - 3C|$, since such a curve would otherwise have $3s - 1$ intersections with C_J and $3s$ intersections with $3C$ (see above, pp. 219, 226). But, pending a more complete examination of Noether's suggested proof that all curves breaking off as fixed parts from the system $|C_J - 3C|$ are such as can arise by transformation from simple points of another surface, and of other connected questions, we have preferred to indicate, by the notation (e_N), the definition of Noether's curves. Cf. p. 220, above, footnote; p. 241, below, Ex. 6; and *Ann. d. Mat.* vi, 1901, p. 28.

---

* Although the fundamental ideas were subsequently found in the Enriques-Campedelli volume, referred to above, it should be stated that this was pointed out independently by D. W. Babbage; to him, and to E. A. Maxwell, the author is much indebted for criticism of the theory as first given, for the case of only double or triple points. The term $(k-1)(2k-1)$ was obtained, and the equation $\omega_0 - \kappa + e_N = p^{(1)}$ was verified in many particular cases, by Mr Maxwell. See also Mr Babbage's paper, *Proc. Camb. Phil. Soc.* xxix, 1933, p. 212.

Ex. 1. We may formulate in general terms the proper genus and grade to be assigned to an isolated multiple point, which we have exemplified above for the case when this is of quite general character (cf. Enriques-Campedelli, as above, p. 65). Let n, π be the prescribed (or virtual) grade and genus of an arbitrary plane section C of the surface (n being the order of the surface). Denote the multiple point to be considered by O, and a section through this point by C_1, writing $C \equiv C_1 + O$. Let the grade and genus to be assigned to O be n_0 and π_0, the grade and genus of C_1 being n_1 and π_1. Let d be the loss of genus of C by passing through O, or $\pi_1 = \pi - d$; let the corresponding loss of n be k, so that k is the number of intersections of the surface with an arbitrary line through O which are absorbed at O, or k is the multiplicity of O, and $n_1 = n - k$. Let it be assumed that we can assign a definite meaning to the number, i, of intersections of the section of the surface by an arbitrary plane through O, with O itself regarded as a curve, that is, of the number of points of the surface in the neighbourhood of O, which lie on the arbitrary plane section. Then, from $C \equiv C_1 + O$, we have $n = n_1 + n_0 + 2i$, $= n - k + n_0 + 2i$, and $\pi = \pi_1 + \pi_0 + i - 1$, $= \pi - d + \pi_0 + i - 1$. Thus $n_0 = k - 2i$, $\pi_0 = d - i + 1$. Particular cases are: (1), A general k-ple point, for which we may assume $i = k$, $d = \frac{1}{2}k(k-1)$. Thence $n_0 = -k$, $\pi_0 = \frac{1}{2}(k-1)(k-2)$, as above, in the text; (2), A double point for which the asymptotic cone breaks up into two planes. For this, similarly, $n_0 = -2$, $\pi_0 = 0$; (3), A uniplanar double point, whereat the asymptotic cone consists of a repeated plane. Here $d = 1$, $k = 2$ and hence $n_0 = 2(1-i)$, $\pi_0 = 2 - i$. In this case, the terms of lowest order in the equation of the surface may be supposed to be of the form $z^2 + \phi + \ldots = 0$, where ϕ is of order 3 in x, y, z; of a surface passing through the point, which is one of a system which may be used to transform the uniplanar point to a curve, the lowest terms, though linear in x, y, z, are effectively linear in x and y only; thus the uniplanar point is transformed to a line, meeting the transformation of the plane section through the point in one point. Thus we see that $i = 1$, and hence $n_0 = 0$, $\pi_0 = 1$; (4), For a tacnodal double point, with $k = 2$, $d = 2$, the tacnode is likewise transformable to a line, which, however, is double on the new surface. Thus $i = 2$, and $n_0 = -2$, $\pi_0 = 1$ (cf. p. 270, Ex. 11); (5), For a triple point on a double curve of the surface, triple for the curve and the surface, we have $k = 3$, $d = 2$, $i = 3$, and $n_0 = -3$, $\pi_0 = 0$.

Ex. 2. In this connexion reference may be made to the relation of an exceptional curve on the surface to the adjoint system (p. 222) of a given linear system of curves thereon. If, in the birational transformation from one surface to another, a system $|\,C\,|$ be changed to a system $|\,C'\,|$, and a point O, which is not a base point for $|\,C\,|$, become a curve ϵ on the new surface, the curve ϵ is fundamental for $|\,C'\,|$; hence, taking a net of curves from $|\,C'\,|$, given, suppose, by an equation $u'(\lambda\phi + \lambda_1\phi_1) + \lambda_2\psi = 0$, where $u' = 0$ on the curve ϵ, there is a curve of $\lambda\phi + \lambda_1\phi_1 = 0$ passing through an arbitrary point of the curve ϵ; this point is then a double point of a (degenerate) curve of the net; hence ϵ forms part of the Jacobian of any net contained in $|\,C'\,|$; thus, if the complete Jacobian system $|\,C_J'\,|$, of $|\,C'\,|$, contain only curves of which ϵ forms a part, or even, this not being so, if the complete adjoint system $|\,C_J' - 2C'\,|$ contains only curves of which ϵ forms a part, we can infer that the exceptional curve ϵ, arising by transformation from a simple point which is not a base point of the original system $|\,C\,|$, is a fixed part of the adjoint system of the transformed system $|\,C'\,|$ (cf. Enriques, "Intorno ai fondamenti...", *Atti... Torino*, xxxvii, 1901–2, No. 21; Severi, in Pascal's *Repertorium*, ii, 2, p. 749).

Ex. 3. A sextic surface having a double line, and a double conic which does not meet this, contains two exceptional lines, in the plane of the conic, meeting on the double line. These arise from simple points of a surface obtainable by transformation of the sextic surface by cubic surfaces containing the double conic and the double line; as a general plane section, C', of the sextic surface meets either of these lines, the simple point from which this line arises is a base point of the system $|C|$ which transforms into the plane sections $|C'|$ of the sextic surface; and the adjoint system, $|C_{J}' - 2C'|$, given by cubic surfaces through the double conic and the double line, does not contain the exceptional lines. These lines form part, however, of the system of adjoint curves defined by the plane of the conic and a variable quadric surface through the double line.

Ex. 4. For the surface $f = 0$, given by $z^2 t^{2n-2} - (x, y, t)_{2n} = 0$, with a particular singularity at $x = y = t = 0$, prove that for the integral $\iint \phi \, dx \, dy / \partial f / \partial z$ to be finite, it is necessary and sufficient that $\phi = (x, y, t)_{n-3} t^{n-1}$; so that the adjoint surfaces of order $2n - 4$ contain the plane, $t = 0$, $(n-1)$ times, and are otherwise cones with vertex at $x = y = t = 0$. Hence also $p_g = \frac{1}{2}(n-1)(n-2)$. It is assumed that the curve $(x, y, t)_{2n} = 0$ is without singular points. See Hodge, *Proc. Lond. Math. Soc.* xxx, 1930, pp. 133–40. A geometrical theory is given in the recent volume, Enriques-Campedelli, *Superficie algebriche*, 1932, p. 370.

Note III. Details of the algebra of a particular Cremona transformation of the Weddle surface and the Kummer surface. That there is a (1, 1) birational transformation between a Kummer surface and a Weddle surface is very well known (Cayley, *Papers*, VII, p. 133, Feb. 1870; Darboux, *Bull. Math. Astr.* I, 1870, p. 348; and other papers by Reye, R. de Paolis, Schottky); and this is immediately clear from Segre's generation of a Kummer surface (cf. *Proc. Lond. Math. Soc.* I, 1904, p. 249). Detailed algebra in close connexion with the fundamental binary sextic is given in the writer's *Mult.-periodic functions* (Cambridge, 1907), p. 39; an account is also given in Hudson's *Kummer's quartic surface* (Cambridge, 1905), pp. 160–8. The following is a self-contained simple account of a Cremona transformation between the surfaces.

Before proceeding to the formulae, it may be well to describe in general terms what the transformation is. In general, cubic surfaces through a sextic curve of genus 3 are ∞^3 in aggregate; and, as the residual curve intersection of two of these meets the sextic curve in 8 points, three of these cubic surfaces have a single free intersection; thus these cubic surfaces form a so-called homaloidal system, which, being taken to correspond to the planes of another space, determine a birational transformation; and it is easy to see that the reverse transformation is of the same character, determined by cubic surfaces through a sextic curve of genus 3. If we form a matrix, of three rows and four columns, of which each element is linear in the coordinates, the determinants of three rows and columns, in this matrix, determine four cubic surfaces having common such a cubic curve; and new coordinates, taken proportional to these four determinants, give such a transformation, expressible by three bilinear relations connecting the coordinates of the two spaces.

Thus, if a sextic of genus 3 can be taken on the Kummer surface, containing ten nodes of the surface, cubic surfaces through this sextic determine a transformation of the Kummer surface into another surface whose order is the number (12) of intersections of the Kummer surface with the variable cubic curve intersection of two of these cubic surfaces, less the number (8) of intersections of this cubic curve with the sextic curve; namely the new surface is of order 4. By considering the equations of the cubic surfaces for the neighbourhood of one of the ten nodes, supposed to be a simple point on these surfaces, we see that the node transforms into a line of the new surface; this new surface therefore has only six nodes.

In the actual transformation taken here, the base sextic curve on the Kummer surface consists of three conics, with one point (D) in common, a node of the surface, any two of these having a further common point, also a node of the surface (at A, or B, or C). The ten nodes of the surface which are transformed to lines of the Weddle surface are the nodes on these three conics other than A, B, C. The behaviour at A, B, C is exceptional, in that the cubic surfaces have a common tangent plane at each; and the behaviour at D is exceptional, in that the cubic surfaces have a quadric node, whose cones have three common generators. On the Weddle surface, the base sextic curve degenerates into a rational cubic curve and three coplanar lines, each meeting this curve once. That, in both cases, the composite sextic curve has genus 3, follows by applying the formula $p_1 + p_2 + i - 1$ for the genus of a curve which consists of two curves of genera p_1 and p_2 with i common points.

In the space (ξ, η, ζ, τ), with $PQRO$ as tetrahedron of reference, O being $(0, 0, 0, 1)$, we consider the *base-cubic-curve* which is given by taking ξ, η, ζ, τ respectively proportional to $p^2(qc-rb)$, $q^2(ra-pc)$, $r^2(pb-qa)$, pqr, wherein a, b, c are constants but p, q, r are any parameters subject to $p+q+r=0$. This curve contains the points $(0, 1, 1, 0)$, $(1, 0, 1, 0)$, $(1, 1, 0, 0)$, lying respectively on the edges QR, RP, PQ. Through this curve, and these three edges, forming a composite sextic curve, pass ∞^3 cubic surfaces, whose system may be taken to have the equation

$$\lambda U\tau + \mu V\tau + \nu W\tau + \rho\Psi = 0,$$

where λ, μ, ν, ρ are variable parameters, $U=0$ is the quadric surface, through the cubic curve, given by

$$U = a(a+b+c)\tau^2 + [a(\eta-\zeta)+(b-c)\xi]\tau + \xi(-\xi+\eta+\zeta),$$

V, W being formed from this by cyclical changes, and $\Psi = 0$ is the cubic surface given by $\Psi = (bc\xi + ca\eta + ab\zeta)\tau^2 + \xi\eta\zeta$. This composite sextic base lies then also on the quartic surface $\Omega = 0$, where

$$\Omega = (\eta c - \zeta b + 6bc\tau) U\tau + (\zeta a - \xi c + 6ca\tau) V\tau$$
$$+ (\xi b - \eta a + 6ab\tau) W\tau + [4e\tau - 2(\xi + \eta + \zeta)]\Psi,$$

in which e is a further constant. The surface $\Omega = 0$ is in fact a Weddle surface, being the locus of the vertices of the quadric cones containing the three points P, Q, R and three other points which we now define. Consider the *node-cubic-curve*, given by taking ξ, η, ζ, τ respectively proportional to $aq'r'(q' - r')$, $br'p'(r' - p')$, $cp'q'(p' - q')$, $p'q'r'$, in which p', q', r' are any parameters subject to $p' + q' + r' = 0$. This curve contains P, Q, R, and can be shewn to meet the base cubic curve above in 4 points. Beside P, Q, R we consider the three points in which this node cubic curve is met by the plane $e\tau = \xi + \eta + \zeta$. It can be shewn that the equation of the general quadric surface through these six points is given by $xf_1 + yf_2 + zf_3 + tf_4 = 0$, where x, y, z, t are variable parameters, $f_4 = 2\tau(e\tau - \xi - \eta - \zeta)$ and $f_1 = 3bc\tau^2 + (\eta c - \zeta b)\tau + \eta\zeta$, while f_2, f_3 are obtained from this by cyclical changes. The conditions for this quadric surface to be a cone are four in number, and are linear both in ξ, η, ζ, τ and x, y, z, t. The equation $\Omega = 0$ may easily be shewn to be the result of eliminating x, y, z, t from these four equations. Solving the first three of the four bilinear equations for x, y, z, t, however, we find $x/U\tau = y/V\tau = z/W\tau = t/\Psi$; and these equations are those of a Cremona transformation, leading, without reference to the equation $\Omega = 0$, to $\xi/xL = \eta/yM = \zeta/zN = \tau/xyz$, where $L = \omega_2 - \omega_3 + xt$, etc., $\omega_1 = (y - z)t + ayz$, etc. Thus, if x, y, z, t be coordinates in another space, with A, B, C, D as points of reference, D being $(0, 0, 0, 1)$, the Weddle surface is transformed by this Cremona transformation into another surface, whose equation is obtainable by eliminating ξ, η, ζ, τ from the four bilinear equations. The result is the symmetrical determinant whose rows consist of 0, z, y, u; z, 0, x, v; y, x, 0, w; u, v, w, ϖ, in which $u = bz - cy - 2t$, $v = cx - az - 2t$, $w = ay - bx - 2t$, $\varpi = 6(bcx + cay + abz) + 4et$. When expanded this is the same as $t^2X + 2tY + Z^2 = 0$, in which

$$X = x^2 + y^2 + z^2 - 2yz - 2zx - 2xy,$$
$$Y = ayz(y - z) + bzx(z - x) + cxy(x - y) + exyz,$$
$$Z = ayz + bzx + cxy.$$

The order, 4, of this surface corresponds to the fact that the variable residual cubic intersection of two cubic surfaces of the family $\lambda U\tau + \mu V\tau + \nu W\tau + \rho\Psi = 0$ meets the Weddle surface in 4 points not on the composite sextic base of the family. This surface is the Kummer surface, having 16 trope conics, of which 4 consist of $t = 0$, $ayz + bzx + cxy = 0$ together with 3 such as $x = 0$, $\omega_1 = 0$.

Through the composite sextic base formed by the last 3 of these conics, there pass 4 cubic surfaces, belonging to the family given by $lxL + myM + nzN + kxyz = 0$, where l, m, n, k are variable, and $L = \omega_2 - \omega_3 + xt$, as above, etc. In the transformation, the three nodes of the Kummer surface at A, B, C, correspond to the nodes P, Q, R of the Weddle surface; the three other nodes of the Kummer surface which lie on the trope conic $t = 0$, $ayz + bzx + cxy = 0$, correspond to the nodes of the Weddle surface which are the intersections of the plane $e\tau = \xi + \eta + \zeta$ with the node cubic curve. The cubic surfaces $lxL + \ldots + kxyz = 0$ have a node at the point D, whose asymptotic cone contains the lines DA, DB, DC, and the residual variable sextic intersections of these surfaces with the Kummer surface all pass once through the 10 nodes of the Kummer surface other than those on $t = 0$. Each of these 10 nodes corresponds to a line on the Weddle surface (the intersection of a plane through 3 nodes of this surface with the plane through the other three nodes). The Weddle surface contains 15 other lines, each a join of two of its nodes; these correspond to the 15 trope conics of the Kummer surface other than that in $t = 0$; in particular, the lines QR, RP, PQ correspond to the trope conics in $x = 0, y = 0, z = 0$.

Ex. Verify that the 27 lines of the cubic surface $\Psi = 0$ are made up of three, say l, m, n, such as $\xi = 0 = \tau$, each counting four times; of three, say a_1, b_1, c_1, such as $\eta/b = -\zeta/c = \tau$, each counting twice; of three, say a_2, b_2, c_2, such as $\eta/b = -\zeta/c = -\tau$, each counting twice; and of three, say u, v, w, such as $\xi = 0$, $\eta/b = -\zeta/c$, each counting once. And specify the complete intersection of $U = 0, \Psi = 0$; and of $V = 0, W = 0, \Psi = 0$.

Note IV. Miscellaneous Examples.

1. For the incidental consequence of the theory of the invariant I, that, in a pencil of curves in a plane, of genus p, with base points at σ points of the plane, the number of curves of the pencil which have double points is $\sigma + 4p - 1$, compare Cremona, *Ann. d. Mat.* VI, 1864; Caporali, "Sopra i sistemi lineari triplamente infiniti di curve algebriche piane", *Collect. math. in mem. Chelini*, 1881, or *Mem. di Geom.*, 1888; also Guccia, *Rend. Palermo*, IX, 1895 (Mem. II).

2. For two pencils of plane curves, of orders n, n', the curve which is the locus of points of contact of a curve of one pencil with a curve of the other is of order $2n + 2n' - 3$, has a $(2s - 1)$-ple point at an s-ple base point of either pencil, and is of genus $4nn' + 6(p + p') - \sigma - \sigma' - 2$, where σ, σ' are the numbers of points of the plane at which the pencils have base points. The number of 3-pointic contacts of curves of the two pencils is $12(nn' + p + p') - 3(\sigma + \sigma') - 9$. See Steiner (for $\sigma = \sigma' = 0$), *Werke*, II, p. 500 (1848); Berzolari, *Atti...Torino*, XXXI, 1895–6. The number of double contacts of curves of the two pencils is $4N$, where

$$N = n^2 n'^2 + (nn' - 6)(p + p') + pp' - 7nn' + \sigma + \sigma' + 5.$$

3. For illustrations of the equation $I - e = I' - e'$ in the transformation of one plane into another, compare Cremona, *Mem. Accad. Bologna*, V, 1865, 2ᵃ, § 5; Jung, *Ann. d. Mat.* XV, XVI, 1888, 1889 and *Rend. Ist. Lombardo*, XXI, 1888.

4. On an algebraic manifold of 3 dimensions, consider a linear pencil of surfaces, with base curves (simple or multiple) of genus p. Let δ be the number of these surfaces which have double points, and I the invariant for any one of these surfaces. Then the number $\delta - 2p - 2I$ is independent of the linear pencil (p. 214, above). For the case when the 3-fold manifold is ordinary space, this number is 2. Thus, in ordinary space, a pencil of surfaces of invariant I, with a base curve of genus p, has, in general, $2(I+p+1)$ surfaces with a double point outside the base curve. In particular, for general surfaces of order n, whose base is the complete intersection of two such surfaces, we have $I = (n-2)(n^2-2n+2)$, $p = n^2(n-2)+1$, and hence $\delta = 4(n-1)^3$. For such a pencil of surfaces, with a single isolated k-ple point, Cremona gives (*Grundzüge einer allgem. Theorie der Oberflächen*, No. 119) $\delta = 4(n-1)^3 - 2(k-1)^2(2k+1)$. Cf. Guccia, *Compt. rend.* cxx, 1895, p. 896; Pannelli, *Gior. d. Mat.* xli, 1903, p. 97; xlii, 1904, p. 197; Pieri, *Gior. d. Mat.* xxiv, 1882.

5. For a net of irreducible curves, whose Jacobian curve is irreducible, on a surface of invariants I, p_n, the following are given by Severi, *Atti... Torino*, xxxvii, 1901–2. Let p be the genus of a curve of the net, n the grade, β the number of points of the surface at which the curves have base points. Then (a), in the net there are $24(p+p_n)$ curves which have cusps; (b), the number of curves of the net which have two double points is $\frac{1}{2}M$, where

$$M = (n+\beta+4p+I)^2 - 3n - \beta - 78p - I - 72p_n + 2.$$

For example, for the number of planes from an arbitrary point which cut a general cubic surface in sections with 2 double points, we find 27 ($\beta = 0 = p_n$, $n = 3$, $p = 1$, $I = 5$); (c), the number of pencils of curves of the net having two curves with 3-pointic contact is $3P$, where

$$P = n + 6p - \beta - 1 + 8p_n - 2;$$

(d), the number of pencils having two curves with double contact is $2D$, where
$$D = (n+p)^2 - 17p - 5n + 2\beta + 2I - 18p_n + 6.$$

Cf. Zeuthen, *Compt. rend.* lxxxix, 1879 (2me sem.), pp. 899–901.

6. The canonical system of a surface has been taken here to have no *prescribed* simple base points. But, for example (Castelnuovo-Enriques, *Math. Ann.* xlviii, 1897, p. 279; see also p. 281), on a quintic surface with two tacnodes, the canonical curves (by planes through the tacnodes) all have common the simple point of the surface on the join of the tacnodes; to preserve the equation $p^{(2)} = p^{(1)} - 1$, this point must be counted in $p^{(2)}$. If the surface be transformed so that this point becomes a curve, the *reduced* canonical system of the new surface will have this as common part. Another case of this behaviour is a double plane of which the branch curve has the singularity described as two consecutive triple points (three ordinary branches with a common tangent); see Enriques-Campedelli, *Superf. alg.* pp. 389, 399. (Cf. also p. 220, above, footnote.)

CHAPTER VI

SURFACES AND PRIMALS IN FOUR DIMENSIONS.
FORMULAE FOR INTERSECTIONS

The chord curve and trisecant curve for a surface in four dimensions. Rational surfaces. For surfaces ψ, in space of four dimensions, with no multiple points other than accidental double points, we have, in Chap. IV, found formulae in terms of four characteristics μ_0, μ_1, μ_2, ν_2, and, in Chap. V, we have added the formulae $I+4=\mu_2-2\mu_1+3\mu_0$, $12(p_n+1)=2\mu_2-8\mu_1+12\mu_0+\nu_2$. Conversely we can express the various numbers in terms of the four μ_0, p, I, p_n, where p is the genus of the prime section. Then we have

$$\mu_1=2\mu_0+2p-2, \quad \mu_2=\mu_0+4p+I, \quad \nu_2=2\mu_0+8p-4-2I+12p_n,$$
$$d=\tfrac{1}{2}(\zeta_0-\nu_2)=\tfrac{1}{2}(\mu_0-2)(\mu_0-3)-5p+I-6p_n,$$
$$\zeta_0=(\mu_0-1)(\mu_0-2)-2p,$$
$$t=\tfrac{1}{6}(\mu_0-2)(\mu_0-3)(\mu_0-4)-(\mu_0-8)p-I+8p_n,$$
$$2Q-2=2(\mu_0-2)(\mu_0-4)+4(\mu_0-10)p+6I-48p_n,$$
$$2P-2=(\mu_0-2)(\mu_0-5)+2(\mu_0-12)p+4I-30p_n;$$

here d is the number of double points, ζ_0 is the order of the curve in which the surface is met by the chords from an arbitrary point (*the chord curve*), t is the number of trisecant chords of the surface from an arbitrary point, Q is the genus of the chord curve, and P is the genus of the double curve, with t triple points, on the surface in space [3] into which the surface ψ projects. We have also denoted the order of this double curve by ϵ_0, and its rank by ϵ_1, using ζ_1 for the rank of the chord curve on the surface ψ, and have remarked that $\zeta_0=2\epsilon_0$, $\zeta_1=2\epsilon_1+\nu_2$, $2Q-2=2(2P-2)+\nu_2$. For the projected surface in [3] we also used three numbers, i, κ, ρ. Of these, i was the class of the developable formed by the stationary tangent planes of the surface, in general the same as the number of inflexional generators of the proper enveloping cone of the surface from an arbitrary point; it was given by $i=4(\mu_2-\mu_1)+2\nu_2$. The number κ was the number of inflexional tangents of the surface passing through an arbitrary point, in general the same as the number of cuspidal generators of the proper enveloping cone. The number ρ was the class of the developable formed by the tangent planes of the surface at points of the double curve. Expressed as above these are respectively

$$i=24(p_n+p), \quad \tfrac{1}{3}\kappa=\mu_0+6p-2-I+8p_n,$$
$$\rho=(\mu_0-2)(2\mu_0-5)+2(\mu_0-11)p+3I-24p_n.$$

For a ruled surface the formulae can be simplified by $p_n = -p$, $I = -4p$ (p. 206).

One incidental consequence of the equations, following from the facts that neither d nor t can be negative, is that

$$p \leqslant \tfrac{1}{6}(\mu_0 - 1)(\mu_0 - 2) + 2p_n/(\mu_0 - 3),$$

as we see by adding the forms for d and t. For a ruled surface this leads to $p \leqslant \tfrac{1}{6}(\mu_0 - 2)(\mu_0 - 3)$.

Another consequence of greater interest is that

$$2Q - 2 + 6t = \zeta_0(\mu_0 - 4).$$

This result we can now obtain in another way, by proving that canonical sets on the chord curve are obtained by primals of order $\mu_0 - 4$ passing through the $3t$ double points which this curve possesses (see Chap. IV, p. 172). We have in fact, in the preceding chapter (pp. 217, 222), by reasoning founded on the consideration of the Jacobian curve of a net of curves, which is equally applicable to a surface in space [4], defined the canonical system of curves, $|K|$, on a surface; and shewn that, with a linear system of curves $|C|$ without base, it gives a system $|K + C|$ defining canonical sets on any curve of $|C|$; the system $|K + C|$ is that called the adjoint system of $|C|$. Now, in the case in hand, if the surface ψ have a canonical system of curves, these will be of order $2p - 2 - \mu_0$; for, when added to the system of prime sections of the surface, they give canonical sets on a prime section. The adjoint system, of the system of curves on ψ defined by the chord curve, is obtained by adding the canonical system of the surface to curves of the system defined by the chord curve, and thus consists of curves of order $2p - 2 - \mu_0 + \zeta_0$. By the formula for ζ_0 this is the same as $\mu_0(\mu_0 - 4)$. Thus, sets of the canonical series on the chord curve are obtainable by its intersections with primals of order $\mu_0 - 4$, account being taken of the multiple points of the chord curve. This curve has $3t$ double points, intersections of branches of the curve which lie in the same sheet of the surface. It has also d double points, at the accidental double points of the surface, where branches of the curve, which lie in different sheets of the surface, cross one another; these latter points, on projection into space of three dimensions, give ordinary points of the double curve, not affecting the genus P of this, nor, therefore, the genus Q of the chord curve on ψ. Hence we have $2Q - 2 = (\mu_0 - 4)\zeta_0 - 6t$. The reader may compare the somewhat different interpretation of what is essentially the same relation, for the case of the double curve of the surface in [3], given in Chapter IV, p. 168, Ex. 11. The argument does not apply to surfaces without a canonical system of curves, in particular not to rational surfaces. But, with suitable modifications, numerical results such as this remain valid in many cases, with use of a fictitious (virtual) canonical system; and we have in fact obtained the present formula independently, from the preceding numerical results only.

We may also consider the trisecants of the surface which pass through a point P of the surface, each meeting the surface in two points beside P; the locus of these points is a curve on the surface which we may here call the *trisecant curve*. Denote the order of this curve by ζ_0', its genus by Q', and the number of quadrisecants of the surface which pass through P by t', these giving rise, as for the chord curve, to $3t'$ double points on the trisecant curve; also, let ν_2' be the number of lines from P which touch the surface elsewhere. By projecting from P, on to a space [3], we obtain from ψ a surface of order $\mu_0' = \mu_0 - 1$, with a double curve, the projection of the trisecant curve, of order ϵ_0', equal to the order of the conical sheet formed by the trisecants, and of genus P', with ν_2' pinch points. The trisecant curve has a multiple point at the point of ψ from which the trisecants are drawn, say of order k. To state the values of these numbers it is convenient to treat the case when the surface ψ is ruled apart from the general case. In the general case $k = \mu_0 - 4$ and

$$\zeta_0' = \zeta_0 - \mu_0, \quad \epsilon_0' = \epsilon_0 - (\mu_0 - 2), \quad t' = t - (\epsilon_0' - k), \quad \nu_2' = \nu_2 - 4,$$
$$2Q' - 2 + 6t' = 2(\mu_0 - 5)\epsilon_0', \quad 2(2P' - 2) + \nu_2' = 2Q' - 2;$$

but when ψ is a ruled surface, $k = \mu_0 - 3$, and, using p for the genus of a prime section,

$$\zeta_0' = \zeta_0 - \mu_0 + 1, \quad \epsilon_0' = \epsilon_0 - (\mu_0 - 2), \quad t' = t - (\epsilon_0' - k), \quad \nu_2' = \nu_2 - 2,$$
$$2Q' - 2 = 2(\mu_0 - 5)(\mu_0 - 3 + 2p), \quad 2P' - 2 = (\mu_0 - 6)(\mu_0 - 3 + 2p).$$

It seems unnecessary to give proofs of these formulae (a detailed consideration will be found in the *Proc. Camb. Phil. Soc.* xxviii, 1932, p. 62); but some remarks may be made. In the case when ψ is not ruled, for ϵ_0', t', P', ν_2' we may consider the surface in space [3], obtained by projection from the point P, and apply thereto the Salmon formulae, with $p_n' = p_n$ and $I' - 1 = I$, because the point P becomes a line on the projected surface. The same is true when ψ is a ruled surface, except that $I' - 1 = I - 1$, because the generator through P is an exceptional line of the surface ψ. Then Q' may be found by the Zeuthen correspondence formula. In the general case, sets of the canonical series on the trisecant curve are given by primals of order $\mu_0 - 5$, cutting the surface so as to have a $(\mu_0 - 5)$-ple point at the point P from which the trisecants are drawn, and also passing through the $3t'$ double points of the trisecant curve. The same is true for the case when ψ is a ruled surface, only then the generator through P is a fixed component of the complete adjoint system of the trisecant curve. In both cases, the proper adjoint system is given by primals of order $\mu_0 - 5$, but there are absorbed at P $k(k-1)$ or $k(k-1-1)$ intersections, with k respectively μ_0-4 or μ_0-3. The value of Q' for the case when ψ is a ruled surface forms an interesting application of a formula for a curve on a ruled surface which was given by Segre: For a curve of order ν and genus π, which is a simple curve on a ruled surface of order n whose prime section is of genus p, this curve meeting every generator in k points ($k>1$) and having the equivalent of d double points (neglecting those which may exist at accidental double points of the surface, arising from the nature of these points), we have

$$k(2p-2)=2\pi-2-\{(k-1)(2\nu-nk)-2d\}.$$

Now, it can be shewn that the trisecant curve of ψ, when this is a ruled surface, meets every generator in μ_0-3 points; and, as has been said, it has a point of multiplicity μ_0-3 at the point from which the trisecants are drawn. We can then put, in Segre's formula,

$$n=\mu_0, \quad k=\mu_0-3, \quad \pi=Q', \quad d=\tfrac{1}{2}(\mu_0-3)(\mu_0-4)+3t', \quad \nu=\zeta_0-\mu_0+1;$$

the result will be $2Q'-2+6t'=(\mu_0-5)(\zeta_0-2\mu_0+4)$, which leads to the value independently obtained for Q'.

Ex. 1. When the surface ψ is rational, its prime sections being represented on a plane by curves of given order with given multiple points, the formulae put down enable us to find the characters of the surface. For the value of p_n, equal to that of the plane, will be zero, and the value of I will be $-1+\sigma-e$, in which σ is the number of points of the plane which are base points of the plane system of curves, and e is the number of points of the surface (supposed simple) arising from exceptional curves of the plane. The formulae also enable us to obtain the curve in the plane which represents the chord curve of the surface, or that representing the trisecant curve. To obtain, for instance, the representation of the chord curve, when ψ is not ruled, supposing, for simplicity, that the system of representing plane curves has no fundamental curve, we may argue thus: Let the plane curves representing the prime sections be of order m, and their base points be of multiplicities denoted by k_i, so that the system of curves may be denoted by $c^m(k_i)$. Let the plane curve representing the chord curve be of order M, and have multiplicity K_i at the (k_i) base point of the system $c^m(k_i)$. An accidental double point of the surface ψ, which is a point intersection of two sheets, will correspond to two generally distinct points of the plane; these will lie on the curve representing the chord curve, and be such that every curve of the system $c^m(k_i)$ which passes through one of them also passes through the other. The three double points of the chord curve which lie on a trisecant of the surface through the vertex of the chord cone, will be represented by three double points of the plane curve which represents the chord curve, and will be such that every curve of $c^m(k_i)$ which passes through two of them will also contain the third. We may suppose that the plane curve representing the chord curve has no multiple points other than the $3t$ double points spoken of, and the multiple points (K_i) at the base points of the system $c^m(k_i)$. Canonical sets on the plane curve representing the chord curve correspond to canonical sets on the chord curve; and these, we have seen, are determined by primals of order μ_0-4 passing through the $3t$ double points of the chord curve. But, canonical sets on the plane curve representing the chord curve are obtained by curves of order $M-3$ passing K_i-1 times through a base point (k_i), and once through each of the $3t$ double points; a base point (k_i) corresponds, generally, to a curve of order k_i on the surface. Thus we infer, for each point (k_i), that $K_i-1=(\mu_0-4)k_i$. Further, as the curves of the surface which correspond to the lines of the plane are of order m, we also infer that $M-3=(\mu_0-4)m$. The curve in the plane which represents the chord curve of the surface ψ is thus one of a system which may be represented by $c^{3+(\mu_0-4)m}[1+(\mu_0-4)k_i,\,(2^3)_t]$, where the entry $(2^3)_t$ is to recall the three double points which correspond to each of the t trisecants of the chord curve. This result, for the representation of the double curve of a rational surface in ordinary space [3], was obtained by Clebsch, *Math. Ann.* I, 1869, p. 253. See also Zeuthen, *ibid.* IV, 1871, p. 26.

The representation of the trisecant curve in the plane is similarly given by a curve belonging to the system which may be denoted by

$$c^{3+(\mu_0-5)m}[1+(\mu_0-5)k_i,\,\mu_0-4,\,(2^3)_{t'}],$$

the trisecant curve having a multiple point of order μ_0-4.

These results assume that, for the plane curves representing the prime sections of the surface, the adjoint system of curves, which determine the canonical series on a curve of this representative system $c^m(k_i)$, are not reducible. But, if this system $c^m(k_i)$ have a fundamental curve, with base points furnishing independent conditions for the fundamental curve, then the adjoint system of $c^m(k_i)$ contains this fundamental curve as fixed part, and the canonical series on a curve of the system $c^m(k_i)$ is determined by the variable residual curves of this adjoint system. We have already illustrated this fact (Chap. V, p. 194), referring for the general proof to Castelnuovo, *Mem....Torino*, XLII, 1892. If the fundamental curve be of character $c^\mu(\kappa_i)$, and its base points, supposed to be only at the points (k_i), be so independent, the preceding equations for the chord curve must evidently be modified to

$$M-3-\mu=(\mu_0-4)m,\quad K_i-1-\kappa_i=(\mu_0-4)k_i.$$

In particular this is true when ψ is a ruled surface. For the representation of the trisecant curve of the ruled surface, another correction arises: in this case, the generator of the ruled surface ψ, through the point from which the trisecants are drawn, is represented in the plane by a fundamental line of the curve representing the trisecant curve; this is additional to the fundamental lines of the representative system $c^m(k_i)$. For details we refer to the Note, *Camb. Phil. Soc. Proc.* XXVIII, 1932, p. 62. The result given in the following Ex. 2, is also proved in this Note.

Ex. 2. If the system of plane curves, representing the prime sections of a rational surface in space $[r]$, have an equation of the form

$$\phi(\lambda_0 Q_0+\ldots+\lambda_{r-2}Q_{r-2})+\lambda_{r-1}U+\lambda_r V=0,$$

and the curve $\phi=0$ be of genus q, then the surface contains a line corresponding to this curve, and this line is $(q+1)$-fold on the surface. A simple example is the quartic surface with a double line in ordinary space, whose prime sections are represented by plane quartic curves having one double base point and 8 simple base points. Another example is the quartic intersection of two quadrics in space of four dimensions, whose prime sections are represented by the conics in a plane which are outpolar to two flat pencils of lines. Examples of octavic surfaces with a double line, in space [5], are considered by D. W. Babbage, *Camb. Phil. Soc. Proc.* XXIX, 1933, pp. 95, 405; cf. Roth, *ibid.* p. 186. The five surfaces considered are those represented by the respective systems of plane curves $c^5(2^2, 1^9)$, $c^6(4, 1^{12})$, $c^6(2^6, 1^4)$, $c^7(3, 2^8)$, $c^9(3^8, 1)$, all the base points being taken on a plane cubic curve.

Ex. 3. We have found, in Chap. IV, p. 175, the values of μ_0, μ_1, μ_2, ν_2 for the surface which is the complete intersection of two primals of orders ρ, ρ', in space of four dimensions, namely $\mu_0 = \rho\rho'$, $\mu_1 = \mu_0(k+k')$, $\mu_2 = \mu_0(k^2 + kk' + k'^2)$, $\nu_2 = \mu_0 kk'$, where $k = \rho - 1$, $k' = \rho' - 1$. We have given the values of I and p_n in terms of these characters at the beginning of this chapter, and we have $I + 4 + \omega - 1 = 12(p_n + 1)$. Hence prove that, for this complete intersection,

$$I + 4 = \mu_0[k^2 + kk' + k'^2 - 2k - 2k' + 3],$$

$$12(p_n + 1) = \mu_0(2k^2 + 3kk' + 2k'^2 - 8k - 8k' + 12),$$

and $\omega - 1 = \mu_0(k + k' - 3)^2$. This last is in accord with the fact that $\omega - 1$ is the grade of the canonical system, of which the curves are (as remarked above) of order $2p - 2 - \mu_0$, where p is the genus of the prime section, that is, of order $\rho\rho'(\rho + \rho' - 4) - \mu_0$, or $\mu_0(k + k' - 3)$. They are in fact given by primals of order $k + k' - 3$.

Intersections of loci in space of four dimensions. We now pass to some formulae* for the intersections of surfaces and primals, mainly, but not exclusively, in space [4]. These results may be provisionally classified as relating either to usual or to exceptional intersections. Thus, in space [r], two algebraic manifolds of dimensions k and $r - k + s$ usually meet in a manifold of dimension s; and we may be required, given a manifold of dimension s, through which two manifolds of dimensions k and $r - k + s$ both pass, to find the characters of the remaining intersection of these, supposed to be of dimension s, and to find, when $2s > r$, the points of this lying on their given partial intersection. This is a problem of usual intersection. A simple case, already considered (Vol. V, Chap. VIII), is that of two surfaces in space [3] passing through a curve, when we are required to characterise the remaining curve common to these surfaces, and the points of this which lie on the former curve. But it may happen that two manifolds of dimensions k and $r - k + s$ have common a manifold of dimension greater than s, and we may be required to find the character of their remaining intersection, supposed to be of dimension s, and the points of this, if any, which lie on their given partial intersection. For instance, two surfaces in space [4] may have a common *curve*, and we may be required to find the number of common points of the surfaces which do not lie on this curve. Or, it may happen that several manifolds, in space [r], of dimensions k_1, k_2, ..., have common a manifold of dimension greater than $k_1 + k_2 + ... - r$, and we may be required to find their

* For the more difficult of the results special acknowledgment is due to Severi's paper, "Sulle intersezioni delle varietà algebriche e sopra loro caratteri e singolarità projettive", *Mem....Torino,* LII, 1903—who refers to Caporali's previous results for space [4], *Mem. di Geom.,* Napoli (Pellerano), 1888.

remaining intersection of this dimension. We may provisionally classify such cases as relating to exceptional intersections.

In the first place, we deal with the cases which arise in space [4], though not entirely in logical order. In two respects the problems are more difficult than for higher space; first because the surfaces arising may have (accidental) double points, which we do not regard as usual in higher space; second because a primal, a manifold of dimension 3, passing through a surface, has generally isolated multiple points at ordinary points of the surface. This latter may be illustrated by the case when the given surface lies on two primals $U = 0$, $V = 0$, and the primal, through the surface, which is considered, has an equation of the form $uU + vV = 0$; the first polars of this primal vanish, on the surface, where the four equations $uU_i + vV_i = 0$ are satisfied ($U_i = \partial U/\partial x_i$ etc.); and the four equations $U = 0$, $V = 0$, $u = 0$, $v = 0$ have common solutions. We may also refer at once to a character of a curve which lies on a surface in space [4], of which generalisation is possible to higher cases, namely *the class of immersion* of the curve on the surface; this is the number of points of the curve at which the tangent plane of the surface meets an arbitrary line; it is also the order of the manifold, of three dimensions, described by the tangent planes of the surface at the points of the curve. By use of this character we may distinguish between curves of the same order lying on the surface. (Cf. p. 225, Ex. 5.)

Ex. On the ruled cubic surface of space [4], prove that a generator is of class of immersion 1, but the common transversal of the generators is of class of immersion 2.

I. Residual intersection of three primals with a common curve. Let three primals, of orders ρ, ρ', ρ'', have in common a curve of order ϵ_0 and class ϵ_1. Let their remaining intersection be a curve of order ϵ_0' and class ϵ_1'; it will be proved that this meets the former curve, say in i points, and that

$$\epsilon_0 + \epsilon_0' = \rho\rho'\rho'', \quad \epsilon_1 + i = \epsilon_0\sigma, \quad \epsilon_1' + i = \epsilon_0'\sigma,$$

where $\sigma = \rho + \rho' + \rho'' - 3$. From these ϵ_0', ϵ_1' and i can be found when, beside ρ, ρ', ρ'', we are given ϵ_0 and ϵ_1. The first of these equations is obvious. For the other two, consider the Jacobian primal of the three given primals and two arbitrary primes. This is expressed by the vanishing of a determinant of five rows and columns, of which the elements of any one of the three first rows are the derivatives, in regard to the coordinates, formed from the equation of one of the three primals, and the last two rows are similarly formed from the equations of the two arbitrary primes; these primes meet in a plane, say ϖ. This Jacobian is of order σ; evidently it vanishes at a point, common to the three given primals, at which the tangent line of their curve of intersection is such as to meet the plane ϖ; also,

Intersections of loci 249

evidently, this Jacobian vanishes at a point which is common to the two curves into which the intersection of the three primals breaks up—for, at such a point, the tangent primes of the three primals, having common the plane of the tangent lines of the two curves, belong to a pencil of primes, and the minor determinants formed from the first three rows of the Jacobian all vanish. We assume that the intersections of the two curves are of general character, and that the Jacobian does not vanish on either at other points than those described. By considering the intersections of the Jacobian primal with each of the two curves in turn, we then obtain the second and third of the equations put down.

Cor. It follows that if, exceptionally, four primals, of orders $\rho, \rho', \rho'', \rho'''$, have common a curve (ϵ_0, ϵ_1), then they intersect further, in general, in a number of points given by $\epsilon_0'\rho''' - i$, that is, in $\rho\rho'\rho''\rho''' - \epsilon_0(\rho+\rho'+\rho''+\rho'''-3)+\epsilon_1$ points.

More generally, if three primals (ρ), (ρ'), (ρ''') have common an aggregate of m curves $(\epsilon_0^{(r)}, \epsilon_1^{(r)})$, of which the two curves $(\epsilon_0^{(r)}, \epsilon_1^{(r)})$, $(\epsilon_0^{(s)}, \epsilon_1^{(s)})$ have $i_{r,s}$ intersections, then the three primals have further common a curve (ζ_0, ζ_1), having j_r intersections with the curve $(\epsilon_0^{(r)}, \epsilon_1^{(r)})$, where, writing ϵ_0 for $\overset{m}{\underset{1}{\Sigma}} \epsilon_0^{(r)}$ and ϵ_1 for $\overset{m}{\underset{1}{\Sigma}} \epsilon_1^{(r)}$, with

$$\sigma = \rho+\rho'+\rho''-3, \quad \zeta_0+\epsilon_0=\rho\rho'\rho'', \quad \zeta_1 = \sigma(\rho\rho'\rho''-2\epsilon_0)+\epsilon_1+2\underset{r,s}{\Sigma} i_{r,s},$$
$$j_r+\epsilon_1^{(r)}+(\underset{s}{\Sigma} i_{r.s}-i_{r,r})=\sigma\epsilon_0^{(r)}.$$

And, if four primals, (ρ), (ρ'), (ρ''), (ρ''') have in common an aggregate of m curves $(\epsilon_0^{(r)}, \epsilon_1^{(r)})$, they meet further, in general, in a number of points specified by saying that the *point-equivalence* of their common curve is $(\rho+\rho'+\rho''+\rho'''-3)\epsilon_0-\epsilon_1-2\underset{r,s}{\Sigma} i_{r,s}$.

Ex. 1. Three quadrics drawn through a rational quartic curve in space [4] meet further in a rational quartic curve, having 6 intersections with the former curve. Four quadrics through the original curve have two further common points. Conversely, through two rational quartic curves which have 6 common points, there can be drawn three linearly independent quadrics.

Through a rational quartic curve, in space [4], there can be drawn in all six linearly independent quadrics; if the quadratic functions giving the equations of these quadrics be supposed proportional to the coordinates in a space of five dimensions, the original space [4] is thereby represented by a primal, which, since two quadrics through the quartic curve have two further intersections, is a quadric, Ω, in this space [5]; a plane conic section of this quadric Ω, given by three primes of the space [5], represents then a quartic curve in the original space [4], having 6 points common with the fundamental quartic curve therein. It can be shewn that, in this representation, the neighbourhood of any point of this quartic curve, consisting of points lying in the ∞^2 planes through the tangent line of the curve at this point, is represented, in the space [5], by a plane, which then will lie on the quadric Ω; the aggregate of such planes, corresponding to all the points of the quartic curve, defines a manifold on Ω, of dimension 3, whose order, equal to the number of the component planes which meet three arbitrary primes of the space [5], is equal to the number of intersections of the fundamental quartic of the original space with the residual curve of intersection of three arbitrary quadrics through this fundamental curve; namely this order is 6. But this manifold V_3^6, of order 6 and dimension 3, on Ω, is in fact the intersection of Ω with a primal M_4^3, in the space [5], of order 3 and dimension 4. To see this we consider the representation in the space [5] of the chords of the quartic curve in [4]; any one of these chords lies on five of the quadrics containing the curve, and is represented by a single point of Ω; the surface formed by the ∞^2 points so arising is of order 4 because, as we may prove, two quadrics containing the quartic curve in [4] contain four chords of this curve; the ∞^1 chords of this curve which pass through one of its points are represented in the space [5] by a conic, there being two chords of the quartic curve through any one of its points which lie on a particular quadric containing the curve. The surface of order 4 representing the chords of the quartic curve is in fact a Veronese surface lying on Ω; and the M_4^3 spoken of contains the planes of conics thereon (of which we have spoken only of ∞^1). In the reverse transformation, any prime of the original space [4] is represented by an ∞^3 locus lying on Ω, whose order, equal to the number of points of this prime which lie on the residual curve of intersection of three quadrics through the original quartic curve in [4], is four. This locus is again the intersection of Ω with a primal, a quadric. And as the prime contains a point of any chord of the original quartic curve, we see that the primes of the original space [4] are represented by the intersections of Ω with the five quadrics, independent of Ω, which contain the Veronese surface. It should be added, however, that the transformation we have referred to, is derivable from the symmetric transformation between two spaces [5], in which the primes of either space represent the quadrics of the other space which contain a Veronese surface lying therein. It can be shewn that three quadrics through a Veronese surface meet in another Veronese surface, having an elliptic sextic curve common with the former Veronese surface. The neighbourhood of a point of the fundamental Veronese surface of one of the two spaces, in this transformation, corresponds to a· plane of the other space; and this is the plane of a conic lying on the fundamental Veronese surface of this other space; etc. Cf. Severi, "Intersezioni...", *Mem. Torino*, LII, 1903, No. 24; also Semple, *Trans. Roy. Soc.* A, CCXXVIII, 1929, p. 351.*

---

* The theorem here twice exemplified, that on a quadric in space [r], an algebraic manifold of $r-2$ dimensions is the complete intersection with a single primal, was recognised and proved by Klein (1872), *Ges. Abh.* I, 153, with reference to Cayley, *Papers*, IV, 1860, p. 455; see also Noether, *Berl. Abh.* 1882, §§ 11, 12, for the curves on a general surface, of ordinary space, of order > 3. In Klein's theorem it is necessary that $r > 3$, and the quadric, if a cone, must be general enough to require at least 5 homogeneous coordinates to express its equation. For more general cases of primals in space [r] whereon any manifold of dimension $r-2$ is a complete intersection, see III, p. 295, below. For a surface on a general quadric of space [4], Klein's argument may be summarised as follows: The surface must meet the tangent prime T at any point O of the quadric Ω in a curve; and every point of this curve, lying on Ω and on T, is on the quadric conical sheet Σ in which T meets Ω. Let m be the number of points

Intersections of loci 251

Ex. 2. If three quadrics be drawn through three skew lines in space [4] —and therefore also through the common transversal line of these—they will meet further in a rational quartic curve, having 6 intersections in all with the four lines common to the quadrics. If four quadrics be drawn through the 3 skew lines (and their common transversal) they have 2 intersections common, not on any of the 4 given lines. In accordance with the formulae $p_1 + p_2 + i - 1$ for the genus of a composite curve, the 3 skew lines and their common transversal make a degenerate quartic curve of genus zero, and the results may be regarded as cases of Ex. 1.

Ex. 3. Four quadrics through a conic in space [4], and through four lines each meeting the conic once, the lines consisting of two pairs each of intersecting lines, meet further in $16 - 5.6 + 2.6 + 2$ points, that is, not at all.

Ex. 4. Consider 10 lines in [4], of which each meets 3 others, there being 15 intersections in all. Then 4 cubic primals through the 10 lines have $81 - 9.10 + 30$, or 21 external intersections. Cf. Salmon, *Higher Algebra*, 1885, p. 298. Find the number of quadric surfaces in ordinary space which can be drawn through 5 given points to touch 4 given planes. Cf. Schubert, *Abzähl. Geom.* 1879, p. 105; and Ursell, *Proc. Lond. Math. Soc.* xxx, 1930, p. 322.

Ex. 5. In space [r], if $r-1$ primals, (ρ_i), meet in two curves (ϵ_0, ϵ_1) and $(\epsilon_0', \epsilon_1')$, we have, as for the case $r=4$, for the number i of the common points of these two curves $\epsilon_1 + i = \epsilon_0 \sigma$, $\epsilon_1' + i = \epsilon_0' \sigma$, where $\sigma = \Sigma(\rho_i - 1)$. Cf. Veronese, "Behandlung, u.s.w.", *Math. Ann.* xix, 1882, No. 37.

II. Residual intersection of two surfaces with a common curve. To find the point equivalence of the curve in which two surfaces in space [4] exceptionally intersect, namely the number of external points common to these surfaces.

Let the surfaces be of respective orders μ_0, μ_0'; and the curve in which they intersect be of order ϵ_0 and class ϵ_1. Suppose that the tangent plane of either surface at a point of their common curve does not touch this surface at any other point of this curve; and that the tangent planes of the two surfaces at a point of this curve define a definite prime. Let the order of the three-fold constituted by the tangent planes of one surface at the points of the common curve be z, the corresponding number for the other surface being z'; also denote by X the number of the primes, each defined by the tangent planes of the two surfaces at a point of the common curve, which pass through an arbitrary point; we may call this the class of the manifold constituted by these primes. We suppose z, z' to be given, as well as μ_0, μ_0', ϵ_0, ϵ_1; and prove that the point equivalence of the

in which this curve meets any generator of Σ. The order of the surface, equal to the number of its intersections with a plane of the space [4], is equal to the number of its intersections with a plane through O lying in T; and such plane contains two generating lines of Σ; thus the order of the surface is $2m$. Projected from O we thus obtain, in an arbitrary three-fold space Π, a surface of order $2m$ having the conic ω, in which Σ meets Π, as an m-fold curve; such a surface is entirely defined by a single equation. Expressing that this surface has ω as an m-fold curve, this equation is at once seen to be that of the intersection of Ω with a primal.

common curve is $X + \epsilon_0$, and that $X = z + z' - \epsilon_1$. Thus the number N, of external points common to the two surfaces, is given by $\mu_0 \mu_0' = N + X + \epsilon_0$.

We first prove, in two ways not essentially independent, that $X + \epsilon_1 = z + z'$. For the former proof, consider the dual figure in which the tangent planes of the first surface, ψ, at the points of the common curve, c, are replaced by the generating lines of a ruled surface, ρ; and, likewise, the tangent planes of the second surface, ψ', at the points of this common curve, c, are replaced by the generating lines of a second ruled surface, ρ'; then any generating line of ρ will have a point of intersection with a proper corresponding line of ρ'. Of the first surface ρ, there will be z generating lines meeting an arbitrary plane, namely ρ is of order z; and, similarly, ρ' is of order z'. Of the curve locus of the intersection of a generator of ρ with the corresponding generator of ρ', there will be X points in an arbitrary prime, namely this curve is of order X. Further, as there are ϵ_1 tangent lines of the curve c, or (ψ, ψ'), which meet an arbitrary plane, there will be ϵ_1 of the planes, each defined as containing a generator of ρ and the corresponding intersecting generator of ρ', which meet an arbitrary line in the dual figure; namely, the planes of these pairs of intersecting generators, of ρ and ρ', constitute a three-fold of order ϵ_1. Now take a section of this dual figure by an arbitrary prime; in this space we shall then have two curves, of orders z and z', the sections of ρ and ρ', these curves being in $(1, 1)$ correspondence, with X coincidences of corresponding points; and the joins of corresponding points of these curves will give a ruled surface of order ϵ_1. Whence, by a well-known theorem of space [3], we have $\epsilon_1 = z + z' - X$, as was stated. For the alternative proof spoken of, effectively the dual of the former, take an arbitrary plane, ϖ, and therein an arbitrary point, O; any line through O, in the plane ϖ, is met by z tangent planes of the surface ψ, whose points of contact are on the curve c; and the tangent planes of ψ' at these points meet the plane ϖ in z points, which can be joined to O; in this way to every line through O in the plane ϖ correspond z other lines; reversely, any one of these lines is met by z' tangent planes of ψ' whose points of contact are on c, and the tangent planes of ψ at these points give z' points of ϖ which can be joined to O. Thereby we define a (z', z) correspondence between rays of the pencil of lines through O in the plane ϖ, in which therefore there are $z + z'$ coincidences of corresponding rays. Such a coincidence can arise in two ways; either because the tangent planes of ψ, ψ' at a point of the curve c meet ϖ in the same point, namely for the ϵ_1 cases in which the tangent line of the curve c meets the plane ϖ; or because the prime determined by the tangent planes of ψ and ψ' at a point of the curve c contains the point O, in which case the line in which this prime meets ϖ intersects two corresponding tangent planes of ψ and ψ'; and this occurs in X cases. Thus as before we have the result $z + z' = \epsilon_1 + X$. We next prove the formula

$$\mu_0 \mu_0' = N + X + \epsilon_0.$$

For this we recur to the consideration, given in a preceding chapter (II, p. 92), of the correspondence between two surfaces, in which to any point of one surface, say ψ, of order μ_0, corresponds every point of another surface, say ψ', of order μ_0', and conversely. For this we have proved the formula $\mu_0 \mu_0' = \epsilon(0, 4)(4) + \epsilon(1, 4)(3)$, where, it will be recalled, the symbol ϵ denotes a coincidence of corresponding points, namely, here, refers to a common point of the surfaces ψ, ψ'; and, in $\epsilon(0, 4)(4)$, the (4) indicates that we consider such a common point lying anywhere in the space [4], while the (0, 4) indicates that the ultimate join of two corresponding points, which are to coincide at the common point of the two surfaces, passes through an arbitrary point of the space [4]. Such a coincidence of corresponding points arises at an isolated common point of the two surfaces, any line through this point being a possible position for the ultimate join of the two points which coincide there; it also arises at a point of the common curve c of the two surfaces, if the prime, determined by their tangent planes at this point, passes through the arbitrary point of the space [4]. Thus the term $\epsilon(0, 4)(4)$ has the value $N + X$. In $\epsilon(1, 4)(3)$, the (3) indicates coincidences lying on an arbitrary prime of the space [4], and the (1, 4) indicates that the ultimate join of the two coinciding points meets an arbitrary line of the space [4]; now an arbitrary prime will contain a common point of ψ and ψ' only at the ϵ_0 intersections of this prime with the curve c; ultimate joins of coinciding points, for such a point, are all the lines through this point which lie in the prime determined by the tangent planes of ψ and ψ' at this point; and there is one such line passing to the point in which this prime is met by the arbitrary line of the space [4], and no other such meets this last line. Thus $\epsilon(1, 4)(3)$ is ϵ_0. On the whole then we have, as stated, $\mu_0 \mu_0' = N + X + \epsilon_0$ (cf. p. 102, Ex. 7).

Ex. 1. As a simple example, suppose ψ is the surface of order 4 obtained by projecting the Veronese surface into space [4], and ψ' is a plane meeting this surface in a conic. Here z', the number of tangent planes of ψ' which meet an arbitrary line, is zero; to obtain z, we may assume that the tangent planes of the projected Veronese surface are duals of the chords of a rational quartic curve in [4]; then, the tangent planes of this surface ψ at points, of a conic section, are duals of chords of the quartic curve which meet a (properly taken) line; such chords are known to determine an involution on the curve, and to generate a ruled surface of order 3. Whence $z = 3$. These are in accord with the remark made in Ex. 5 (p. 225) of Chap. v, the grade of the conic being respectively 4 and 1 on the plane and on the surface. This leads to $X = 3 + 0 - 2$ or $X = 1$. This is equivalent to saying that the primes, through the plane of the conic section ψ, each containing the tangent plane of the surface at one point of the conic, consist of all the primes through this plane. From $X = 1$, we have $N = \mu_0 \mu_0' - X - \epsilon_0 = 4 - 1 - 2 = 1$. Thus the plane of a conic section of the projected Veronese surface meets the surface in one point not on the conic. This follows also from the fact that a solid, or space [3], drawn through the plane of a conic section of the Veronese surface in space [5], has one further intersection with the surface; by such solids the surface is projected on to a plane.

Ex. 2. Another simple example is given by taking ψ to be the Segre quartic surface which is the intersection of two quadrics in [4], and ψ' to be a plane containing a line of the surface ψ. Here, as before, $z' = 0$; but z, the number of tangent planes of ψ, at points of the line, which meet an arbitrary line, is 2; for the tangent primes, of the two quadrics which meet in ψ, at points of the line on ψ, meet an arbitrary line in related ranges. In fact the line is of grade 1 on the plane, but of grade -1 on the surface. Wherefore X, $= z + z' - \epsilon_1 = 2$, and $N = \mu_0 \mu_0' - X - \epsilon_0 = 1$. Thus the plane meets the surface in one point not on the line. By such planes the surface is projected on to a plane.

Ex. 3. A plane through a generator of a ruled surface in [4], of order μ_0, meets the surface again in $\mu_0 - 2$ points.

Ex. 4. The formula $\mu_0 \mu_0' = N + X + \epsilon_0$ may be employed to obtain the formula (III), p. 159, of Chap. IV. For this, we take the surface ψ' to be the cone of chords, drawn from an arbitrary point O, to the surface ψ, of order μ_0, so that μ_0' is what is called ϵ_0 in Chap. IV, and the ϵ_0 of the formula here is what is called ζ_0 in Chap. IV, the order of the chord curve. Thus the formula is $\mu_0 \epsilon_0 = N + X + \zeta_0$ in the notation of Chap. IV. There is an isolated intersection of the surface ψ with the cone of chords at each of the intersections with ψ of the trisecants from O, as we have seen, Chap. IV, p. 172; or $N = 3t$. To find X, the number, through an arbitrary point, P, of primes which are each defined by the tangent plane of ψ at a point of the chord curve, and the tangent plane of the chord cone at this point, we notice that, as O, on all the tangent planes of the chord cone, lies in all such primes, when P is in such a prime, the line OP will meet the tangent plane of ψ by which the prime is defined; and conversely. Thus, on projection from O into space [3], X is the number of tangent planes, of the surface obtained by projection, at a point of the double curve of this, which pass through an arbitrary point of the space [3], namely the point where OP meets this space. Thus X is the number denoted in Chap. IV by ρ. Thus we have $\mu_0 \epsilon_0 = 3t + \rho + \zeta_0$, or, as $\zeta_0 = 2\epsilon_0$, we have $\mu_0(\epsilon_0 - 2) = \rho + 3t$, which is the formula (III) in question. We may note, however, that, when we consider the two surfaces consisting of the surface ψ and the chord cone, the formula $X + \epsilon_1 = z + z'$ is to be replaced by $X + \zeta_1 = z + 2z'$, since a tangent plane of the chord cone has *two* contacts with the chord curve. Recurring to the argument given above, the correspondence in the pencil of rays in the plane ϖ is of indices z and $2z'$. Here $z = \rho + \nu_2$, and z' is $2\epsilon_0 + 2P - 2$, or ϵ_1, where ϵ_0, ϵ_1, P refer to the double curve of the projected surface. Thus we have $X + \zeta_1 = \rho + \nu_2 + 2\epsilon_1$, in agreement with $X = \rho$, $\zeta_1 = 2\epsilon_1 + \nu_2$ (IV, p. 174).

Ex. 5. If a curve (ϵ_0) lie on a surface (μ_0) in space [4] which has d accidental double points, at each of which the curve has a four-fold point, then the conical sheet of lines, through an arbitrary point O of the space, to points of the curve, has a number, N, of isolated intersections with the surface, other than on the curve, given by $N = \epsilon_0(\mu_0 - 1) - z - 4d$, where z is the class of immersion of the curve on the surface, namely the number of tangent planes of the surface, at points of the curve, which meet an arbitrary line. This follows as in the last example, from $N = \mu_0\mu_0' - \epsilon_0 - X$, replacing μ_0' by ϵ_0 (the order of the cone) and X by z. Or, if we project from O on to a space [3], we are required to find the number of intersections, N, of the double curve of the surface in [3] with the curve, thereon, obtained by projection of the given curve (ϵ_0). This latter curve is such that z of the tangent planes of the projected surface at points of this curve, pass through an arbitrary point, say P', of the space [3]; namely, z is the number of intersections of this curve with the first polar of P', of order $\mu_0 - 1$, other than those (N) on the double curve, and those ($4d$) on the double curve, where the projected curve has a four-fold point.

III. **Residual intersection of a primal and surface having a common curve.** Now consider the residual curve intersection of a surface ψ and a primal Π, which intersect in a given curve, sup-posing that this curve has no multiple point. For ψ the usual characters are μ_0, μ_1; for Π we have the characters ρ_0, ρ_1, of which ρ_0 is the order, and ρ_1 is the order of the surface formed by points of contact of tangent primes of Π which pass through an arbitrary point, so that $\rho_1 = \rho_0(\rho_0 - 1)$. The given curve of intersection, c, of ψ and Π, is of order ϵ_0, and rank ϵ_1, and has a class of immersion z in ψ, and a class of immersion Z in Π, the latter being the number of tangent primes of Π at points of c which pass through an arbi-trary point. Thus, when the surface ψ, and the curve therein, are given, ϵ_0, ϵ_1 and z are definite; and Z is the number of intersections of the curve with the first polar of an arbitrary point taken in regard to Π. For the residual curve of intersection c', of ψ and Π, we denote the characters by ϵ_0', ϵ_1', z', Z'; and the number of its intersections with c by i. These unknown characters are to be found from the formulae

$$\epsilon_0 + \epsilon_0' = \mu_0\rho_0, \quad \epsilon_1 + i = z + Z, \quad \epsilon_1' + i = z' + Z',$$

$$z + z' = \mu_1\rho_0, \quad Z + Z' = \mu_0\rho_1,$$

of which one can be replaced, if desired, by

$$\epsilon_1 + \epsilon_1' + 2i = \mu_0\rho_1 + \mu_1\rho_0.$$

The formula $\epsilon_0 + \epsilon_0' = \mu_0\rho_0$ is obvious. The formula $z + z' = \mu_1\rho_0$ follows because the locus of points of contact of tangent planes of ψ which meet an arbitrary line is, by definition, a curve on ψ of order μ_1; and this meets Π in $\mu_1\rho_0$ points; while all points of the complete intersection (ψ, Π), at which the tangent plane of ψ meets the arbi-trary line, are so obtainable. Similarly for $Z + Z' = \mu_0\rho_1$, the points of Π, whereat the tangent prime passes through an arbitrary point form a surface of order ρ_1, meeting ψ in $\mu_0\rho_1$ points. To obtain the formula $\epsilon_1 + i = z + Z$, we use formulae of correspondence developed in an earlier chapter (Chap. i). Let ϖ be an arbitrary plane. This is met, by a tangent plane of ψ at a point of the curve c, in a point, say F, and, by the tangent prime of Π at the same point of c, in a line, say l; let L denote any point of l. We consider, in ϖ, the ∞^2 pairs of corresponding points in which to a point F corresponds any point L of the line l; every point F lies on a definite curve, γ, of the plane ϖ, that in which this plane ϖ is met by the tangent planes of ψ at the points of the curve c on ψ; and we assume, ϖ being an arbitrary plane, that, to every point of γ corresponds just one point of c, and hence a definite line l of points L; conversely, every point L in ϖ lies on the tangent prime at one of Z points of c, and each of these gives a definite point F. Coincidences in this correspondence arise when F lies on the associated line l. This may happen in two ways: Either, because the tangent line of the curve c, which is generally the intersection of the tangent plane of ψ with the tangent prime of Π at a point of c, meets the plane ϖ; and the number of such cases is ϵ_1; or, because the tangent prime of Π, at a point of c, contains the tangent prime of ψ at this point; this happens at one of the i intersections of the curves c, c', and does not happen otherwise if c have no double point. Thus the total number of coincidences in the correspondence is, in general, $\epsilon_1 + i$. On the other hand, it was shewn, in the chapter referred to, that the number of coincidences is $(0)(2)' + (1)(1)' + (2)(0)'$, where $(m)(n)'$ means the number of existing pairs of corresponding points in which F lies in an arbitrary space $[m]$, of the plane ϖ, and L lies at the same time in another arbitrary space $[n]$. Thus, here, the term $(0)(2)'$ vanishes, there being positions of F only on the curve γ of the plane ϖ. The term $(1)(1)'$ has a contribution when F is on an arbitrary line, and thus at one of the z intersections of γ with this line, while L lies on an arbitrary line; this latter will meet the line l, associated with F, in one point; whence $(1)(1)'$ is z. The term $(2)(0)'$ has a contribution when L is at an arbitrary point, the corresponding point F being anywhere in the plane ϖ; wherefore $(2)(0)' = Z$. Hence we have the result $\epsilon_1 + i = z + Z$; and the proof of $\epsilon_1' + i = z' + Z'$ is similar.

Ex. 1. For a curve which is the complete intersection of a surface (μ_0, μ_1) and a primal (ρ_0, ρ_1), the characters are given by $\epsilon_0 = \mu_0 \rho_0$, $\epsilon_1 = \mu_1 \rho_0 + \mu_0 \rho_1$, $z = \mu_1 \rho_0$, $Z = \mu_0 \rho_1$, where $\rho_1 = \rho_0(\rho_0 - 1)$.

Ex. 2. If a primal, of order $\rho + 1$, be put through the curve which is a prime section of a surface ψ, or (μ_0, μ_1), we may find the characters of the residual curve of intersection of the primal with the surface. The order and rank of the prime section, and the class of its immersion in ψ are given by $\epsilon_0 = \mu_0$, $\epsilon_1 = \mu_1$, $z = \mu_1$; the class of immersion, of the prime section of ψ, in the primal of order $\rho + 1$, is the number of intersections of this curve with the first polar of a point in regard to this primal. Hence, from $\epsilon_0' + \epsilon_0 = \mu_0(\rho + 1)$, $z + z' = \mu_1(\rho + 1)$, $Z + Z' = \mu_0(\rho + 1)\rho$, we have $\epsilon_0' = \mu_0 \rho$, $z' = \mu_1 \rho$, $Z' = \mu_0 \rho^2$; while, from $\epsilon_1 + i = z + Z$, $\epsilon_1' + i = z' + Z'$, we have $i = \mu_0 \rho$, $\epsilon_1' = \mu_1 \rho + \mu_0 \rho(\rho - 1)$. Thus, as was to be anticipated, ϵ_0', ϵ_1', z' are the same as for a complete intersection of the surface with

Intersections of loci 257

a primal of order ρ. Test the formulae similarly, for the residual intersection with ψ, of a primal of order $\rho + \rho'$, put through the complete intersection of ψ with a primal of order ρ.

Ex. 3. Consider the modifications arising in the formulae when the given curve common to ψ and Π has multiple points.

IV. Residual intersection of two primals having a surface common. We consider now the residual intersection of two primals passing through the same surface. This is a more laborious task than the preceding, and the results obtained will be stated at the conclusion, when the notation has become familiar.

We consider a general surface ψ, in space [4], with characters $\mu_0, \mu_1, \mu_2, \nu_2$ and d accidental double points. Through this surface we put two primals Π, Π', of orders ρ, ρ', which intersect further in a surface ψ'. We shew that ψ' has double points, in general, only at the d double points of ψ, these being equally accidental for ψ'; also that the surfaces ψ, ψ' have no isolated intersections, their common points lying on a curve, c; of this curve the characters are denoted by ϵ_0, ϵ_1. We find these, as also the characters $\mu_0', \mu_1', \mu_2', \nu_2'$ of ψ'; and the numbers δ, δ' of double points of Π and Π' which occur at simple points of ψ (and therefore of double points of Π, Π' at simple points of ψ').

That an accidental double point of ψ is an accidental double point of ψ' is clear enough when the tangent planes of the sheets of ψ at this point have only a point intersection and do not meet in a line. A general line through this point meets ψ in two points coincident thereat (is the limit of a chord of the surface), and so meets Π in two coincident points; or Π has a double point at this point; but a line through the point in one of the two tangent planes of ψ has a higher intersection with Π, and so lies on the osculating cone of Π at this point. Thus the tangent planes of ψ are planes, of the same system, of this osculating cone of Π; and the same may be said for Π'. The two osculating cones have therefore in common the two tangent planes of ψ at the double point; thus, being quadric cones, as is supposed, they also have common two other planes, both of the opposite system, likewise meeting only in this double point. These are then the tangent planes of the residual intersection, ψ', of Π and Π', which therefore has also a double point, of accidental character. But, to meet the possibility of the tangent planes of ψ at the double point having a line in common, we may also proceed differently, and shew that the rank of the curve section of ψ', by a prime through the double point of ψ, is equal to the rank of the section of ψ' by any other general prime. From this, as, by the preceding argument, ψ' certainly has a double point at the point, it will follow that this is an accidental double point for ψ'. Denote the double point of ψ by D; denote by k the curve in which

ψ is met by a general prime through D; let the surfaces in which this prime meets the primals Π, Π' be called ϕ and ϕ'. Form the Jacobian, in this prime, of these two surfaces ϕ, ϕ', and of two arbitrary planes lying in this prime. The curve k has D for a double point, and (because ψ lies on both Π and Π') lies on both ϕ and ϕ'; we can shew that the Jacobian, say J, has D for a double point. Thus J has four intersections with k at D. Its other intersections with k are at the points of this where the tangent line meets the line of intersection of the two arbitrary planes taken in the prime, and at the points of this where the surfaces ϕ, ϕ' touch one another. The number of intersections of the former kind, by the property of an accidental double point, is the same as would be the case for an arbitrary prime section of ψ, and is μ_1. As the Jacobian is of order $\rho + \rho' - 2$, we infer that the surfaces ϕ, ϕ', and hence also the primals Π, Π', touch, in this prime, other than at D, in

$$\mu_0(\rho + \rho' - 2) - \mu_1 - 4$$

points. Now, when two surfaces of orders m, m', in space [3], intersect in two curves of orders n_1, n_2, and ranks r_1, r_2, with i common points, we have

$$n_1 + n_2 = mm', \quad i + r_1 = n_1(m + m' - 2), \quad r_1 r_2 = (n_1 - n_2)(m + m' - 2);$$

hence, considering the curves in which ψ and ψ' are met by a quite general prime, we have $\mu_1' - \mu_1 = (\mu_0' - \mu_0)(\rho + \rho' - 2)$, and these curves meet in $\mu_0(\rho + \rho' - 2) - \mu_1$ points. This number, say ϵ_0, is the order of a curve, say c, of intersection of the surfaces ψ, ψ'; at points of this curve the primals Π, Π' have a common tangent prime, determined by the tangent planes of ψ, ψ'. Thus, the result obtained above may be expressed by saying that, in any prime through D, the primals Π, Π' touch, outside D, in $\epsilon_0 - 4$ points. Such points are the intersections, not at D, of the curve c, common to ψ, ψ', with this prime. Thus, this curve, of order ϵ_0, passes through D, and has thereat a four-fold point. Now consider, in the same prime through D, the curve k' in which this prime meets the surface ψ', and the intersections of k' with the same Jacobian of ϕ, ϕ' and two planes. We have seen that the common curve of ψ and ψ' is met in four points, coinciding at D, by any prime though this point; thence, as k has there a double point, it follows that k' has also a double point at D; thus the Jacobian, which has a double point at D, meets k' in four points at D. It follows, then, as before, that the primals Π, Π' have contacts, not at D, in this prime, whose number is $\mu_0'(\rho + \rho' - 2) - (\mu_1') - 4$, where (μ_1') denotes the rank of the curve in which this prime meets ψ'. This number is then equal to the former, $\mu_0(\rho + \rho' - 2) - \mu_1 - 4$. Whence, from the equation $\mu_1' - \mu_1 = (\mu_0' - \mu_0)(\rho + \rho' - 2)$, we infer that the rank (μ_1'), of the section of ψ' by a prime through D, is the same as the rank μ_1 of a section by an arbitrary prime. Hence D is an *accidental* double point of ψ'. A similar argument to the foregoing shews that the surfaces ψ, ψ' have no point in common which is not on the curve c. For, if D be a point common to ψ and ψ', which is an ordinary point of both, and is either simple for both Π and Π', or simple for one of these and double for the other, we prove in the same way that an arbitrary prime through this point D contains $\epsilon_0 - 1$ other points where Π and Π' touch; thus the curve c, common to ψ and ψ', passes once through this point.

As the surfaces ψ, ψ' are thus proved to have the same accidental double points, it follows, from the formulae for the number of these, such as $\nu_2 + 2d = \mu_0(\mu_0 - 1) - \mu_1$, and the formula for $\mu_1' - \mu_1$ given above, that $\nu_2' - \nu_2 = (\mu_0' - \mu_0)(\rho - 1)(\rho' - 1)$. Thus μ_0', μ_1', ν_2' are found. Next, if z, z' be the classes of immersion of the curve c, common to ψ, ψ', in these, respectively, and X the class of immersion of c in Π and Π', which have the same tangent prime at a point of c, determined by the tangent planes of ψ and ψ' at this point, it follows, as in a preceding case (pp. 252, 253, above) that*
$\mu_0\mu_0' = X + \epsilon_0$, $z + z' = X + \epsilon_1$. Further, we have

$$\mu_2 + \nu_2 + z = \mu_1(\rho + \rho' - 2).$$

For consider, on the surface Ψ, the curve of order μ_1, say θ, which is the locus of the points of contact of tangent planes of ψ which meet an arbitrary line, say l. This defines two series of primes, each ∞^1 in aggregate, in $(1,1)$ correspondence, namely the tangent primes of Π and Π' at points of this curve θ. Of these series of primes, respectively $\mu_1(\rho - 1)$ and $\mu_1(\rho' - 1)$ pass through an arbitrary point, as we see by taking the first polars of this point in regard to Π and Π'. Also, there are $\mu_2 + \nu_2$ tangent planes of ψ, at points of the curve θ, which meet a further arbitrary line, say m; this we may see by taking the line m so as to intersect l; then there are μ_2 tangent planes of ψ which meet the plane (l, m) in a line, and so meet l, so that their points of contact are on θ; and there are ν_2

---

* The former of these gives an independent proof of Severi's formula for the number d of double points of the surface ψ. For, consider a surface which is the intersection of the first polars of an arbitrary point O in regard to the primals Π and Π'. This surface, of order $(\rho - 1)(\rho' - 1)$, meets ψ in $\mu_0(\rho - 1)(\rho' - 1)$ points, consisting of (i), the X points of the curve c whereat the common tangent prime of Π and Π' is such as to pass through O; (ii), the ν_2 points of ψ whereat the tangent plane of ψ, common to the tangent primes of Π and Π', passes through O; (iii), the accidental double points of ψ, which, we have seen, are double points for Π and Π', and lie in the first polars. Thus we have

$$X + \nu_2 + 2d = \mu_0(\rho - 1)(\rho' - 1);$$

this gives $\nu_2 + 2d = \mu_0(\rho - 1)(\rho' - 1) - \mu_0\mu_0' + \epsilon_0$; but we have $\mu_0' = \rho\rho' - \mu_0$, and, by an arbitrary prime section, we have $\epsilon_0 = \mu_0(\rho + \rho' - 2) - \mu_1$. Whence

$$\nu_2 + 2d = \mu_0(\mu_0 - 1) - \mu_1.$$

tangent planes of ψ which pass through the point (l, m); thus there are $\mu_2 + \nu_2$ tangent planes of ψ which meet two arbitrary lines. Further, among the pairs of tangent primes of Π and Π' at points of the curve θ, there are z coincidences; for, we assume, coincidences of tangent primes of Π and Π' arise only at points of the curve c, or (ψ, ψ'); and the number of points of this curve at which the tangent plane of ψ meets the line l is z; these points are then on the curve θ. Now consider the dual figure; dual to the two series of primes, we shall have two curves, in $(1, 1)$ correspondence, of respective orders $\mu_1(\rho - 1)$ and $\mu_1(\rho' - 1)$, with z common corresponding points; and the ruled surface formed by joins of corresponding points is of order $\mu_2 + \nu_2$. Hence the formula $\mu_2 + \nu_2 + z = \mu_1(\rho + \rho' - 2)$ follows, as an application of a well-known result. From this z can be found. The corresponding formula for ψ' is $\mu_2' + \nu_2' + z' = \mu_1'(\rho + \rho' - 2)$, from which μ_2' could be found if z' were known; we have as yet, however, only obtained $z + z' = X + \epsilon_1$, $= \mu_0 \mu_0' - \epsilon_0 + \epsilon_1$. We proceed then further, and first prove that $X + 4d + \delta = \epsilon_0(\rho - 1)$, where δ is the number of double points of the primal Π which arise at simple points of ψ. Such a point, as lying on ψ, lies on Π', and counts doubly among the points common to Π and Π', which are the points of ψ and ψ'; such a point, simple on ψ, must then be also on ψ', or it would count only singly among the intersections of Π and Π'; being on ψ and ψ', which have no isolated common point, it is thus on the curve c. Thus δ is also the number of double points of Π which occur at a common simple point of ψ and ψ' lying on the curve c. We can then prove the result $X + 4d + \delta = \epsilon_0(\rho - 1)$ by considering the intersections of c with the first polar of Π in regard to an arbitrary point; these intersections consist of the X points of c for which the common tangent prime of Π and Π' passes through the arbitrary point, then of the double points of ψ and ψ', each a four-fold point for c, but simple on the first polar, and lastly of the δ double points of Π spoken of, each likewise simple on the first polar. Lastly we prove the simple result,

$$\epsilon_1 + 4d = \epsilon_0(\rho + \rho' - 8);$$

but the proof is rather intricate. Consider the cone Γ, with vertex at an arbitrary point O, which passes through the curve c; let N, N' be the numbers of isolated intersections of this cone respectively with ψ and ψ'; also let γ be the curve, of order $\epsilon_0\rho - \epsilon_0$, which is the remaining intersection of this cone with the primal Π. We first prove that $N + N' + 2X + 8d + \delta = \epsilon_0(\rho - 1)\rho'$. For this we consider the intersections of the curve γ with the primal Π'. The $N + N'$ isolated intersections of the cone Γ with ψ and ψ' are on Π, and hence on the curve γ, and on Π'; at one of the X points of c where the common tangent prime of Π and Π' passes through O, the curve γ touches Π'; at a common double point of ψ and ψ', where, by what has preceded, both c and γ have a four-fold point and Π' has a double point, there are 8 intersections of γ with Π'; lastly, at a simple common point of ψ and ψ' where Π has a double and Π' has a simple point, both c and γ have simple points, and there is one intersection of γ with Π'; there are δ such points. By considering all these intersections of γ with Π', the formula is proved. From this then, by the result last proved, namely $X + 4d + \delta = \epsilon_0(\rho - 1)$, we have $N + N' + X + 4d = \epsilon_0(\rho - 1)(\rho' - 1)$. We have, however (see Ex. 5, p. 254, above), $N + z + 4d = \epsilon_0(\mu_0' - 1)$; and

$$N' + z' + 4d = \epsilon_0(\mu_0' - 1);$$

and, as $\mu_0 + \mu_0' = \rho\rho'$, these give $N + N' = -z - z' - 8d + \epsilon_0(\rho\rho' - 2)$. Thus, eliminating $N + N'$ by means of

$$N + N' + X + 4d = \epsilon_0(\rho - 1)(\rho' - 1),$$

we obtain $z + z' + 4d = \epsilon_0(\rho + \rho' - 3) + X$. It was seen above that $z + z' = X + \epsilon_1$. Thus finally we reach the simple result

$$\epsilon_1 + 4d = \epsilon_0(\rho + \rho' - 3).$$

The equations we have found are just sufficient to give the characters of the residual surface ψ', and of the curve c in which this meets ψ. Beside $\mu_0 + \mu_0' = \rho\rho'$, they give

$$\mu_1' - \mu_1 = (\mu_0' - \mu_0)(\rho + \rho' - 2), \quad \nu_2' - \nu_2 = (\mu_0' - \mu_0)(\rho - 1)(\rho' - 1),$$

of which a particular consequence is that

$$\mu_0(\rho - 1)^2 - \mu_1(\rho - 1) + \nu_2 = \mu_0'(\rho - 1)^2 - \mu_1'(\rho - 1) + \nu_2'.$$

The two equations such as $\mu_2 + \nu_2 + z = \mu_1(\rho + \rho' - 2)$, using

$$z + z' = X + \epsilon_1, \quad \mu_0\mu_0' = X + \epsilon_0, \quad 2d = 2d' = \mu_0(\mu_0 - 1) - \mu_1 - \nu_2$$

and $\epsilon_1 + 4d = \epsilon_0(\rho + \rho' - 3)$, lead to

$$\mu_2' + \mu_2 = \mu_0'\mu_0 + 8d + \rho\rho'(k^2 + kk' + k'^2) - 3\epsilon_0(k + k'),$$

where $k = \rho - 1$, $k' = \rho' - 1$. Herein d, or d', is given by the formula just quoted, and $\epsilon_0 = \mu_0(\rho + \rho' - 2) - \mu_1$; also, we may notice, the equation $X + \epsilon_0 = \mu_0\mu_0'$ may be replaced by the formula

$$X + \nu_2 + 2d = \mu_0(\rho - 1)(\rho' - 1),$$

which is proved in the footnote to p. 259, above. If we denote by p the genus of the prime section of the surface ψ, so that $\mu_1 = 2\mu_0 + 2p - 2$, the equation $\epsilon_0 = \mu_0(\rho + \rho' - 2) - \mu_1$ is the same as $\epsilon_0 + 2p - 2 = \mu_0(\rho + \rho' - 4)$; this expresses that a primal of order $\rho + \rho' - 4$, passing through the curve c, would meet a prime section of ψ in a number of points equal to that of a canonical set on this prime section. Put into the form $2p-2-\mu_0=\mu_0(\rho+\rho'-5)-\epsilon_0$ it expresses that a canonical curve on ψ, including exceptional curves, and also including the curve c, has the order of the intersection of ψ with a primal of order $\rho+\rho'-5$ containing the surface ψ' (cf. p. 269 below; Severi, *Rend....Palermo*, XVII, 1903, No. 18). The equation $\epsilon_1+4d=\epsilon_0(\rho+\rho'-3)$, which we have found more difficult to establish, being equivalent, in terms of the genus, q, of the curve c, to $2q-2+4d=\epsilon_0(\rho+\rho'-5)$, expresses that a primal of order $\rho+\rho'-5$, put through the common accidental double points of ψ and ψ', meets the curve c in a residual set of points whose number is that of a canonical set* of c. Lastly, from the equations $\mu_0\mu_0'=X+\epsilon_0$, $X+4d+\delta=\epsilon_0(\rho-1)$, we see that the number of double points, δ, of Π, at simple points of ψ, on c, and the corresponding number δ' for Π', are given by $\mu_0'\mu_0+4d=\epsilon_0\rho-\delta=\epsilon_0\rho'-\delta'$ or by $\delta+2d=\mu_0(\rho-1)^2-\mu_1(\rho-1)+\nu_2$; and we have had the auxiliary formulae $X=\mu_0'\mu_0-\epsilon_0=z+z'-\epsilon_1$, and $\mu_2+\nu_2+z=\mu_1(\rho+\rho'-2)$, and also $N+z+4d=\epsilon_0(\mu_0-1)$. We notice that, from the formulae for the number of double points of a primal of order ρ passing through a surface (μ_0, μ_1, ν_2), with d double points, and through a second surface (μ_0', μ_1', ν_2'), with d' double points, namely $\delta=\mu_0(\rho-1)^2-\mu_1(\rho-1)+\nu_2-2d$ and

$$\delta'=\mu_0'(\rho-1)^2-\mu_1'(\rho-1)+\nu_2'-2d',$$

we have

$$\delta+\delta'-2i=(\mu_0+\mu_0')(\rho-1)^2-(\mu_1+\mu_1')(\rho-1)+\nu_2+\nu_2'-2(d+d'+i),$$

where i is the number of intersections of the two surfaces, assuming that the common points are not among the double points of either component surface. This gives the number of double points for a primal containing a composite surface.

Ex. 1. Denoting $\rho-1$, $\rho'-1$ by k and k', and denoting by p_n, ω the invariants of ψ and by p_n', ω', those of ψ', prove from the formulae found that (cf. Ex. 3, p. 247, preceding)

$$p_n+1+p_n'+1+\tfrac{1}{2}\epsilon_0(k+k'-3)-d$$
$$=\tfrac{1}{12}(\mu_0+\mu_0')(2k^2+2k'^2+3kk'-8k-8k'+12),$$

and $\omega-1+\omega'-1+2\epsilon_0(k+k'-3)+z+z'-6\epsilon_0=(\mu_0+\mu_0')(k+k'-3)^2$,

and in fact that $\omega-1+z-3\epsilon_0=(k+k'-3)(2p-2-\mu_0)$.

Ex. 2. If it be possible to put a primal of order $k+k'-3$ through the surface ψ' which does not contain ψ, its curve of intersection with ψ will meet the curve c in $z-3\epsilon_0$ points. For (by p. 255 preceding, under III), this number of intersections is given by $z+Z-\epsilon_1$, where Z is the number of variable intersections of the first polar of this primal with the curve c;

---

* If ϵ be the deficiency of the series, and $d-\eta$ be the number of independent conditions which the d accidental double points impose for a primal of order $\rho+\rho'-5$ put through them, it is proved by Severi (*loc. cit.*) that $\epsilon+\eta$ is equal to the sum of the irregularities (p. 227) of the surfaces ψ and ψ'.

this first polar, of a primal passing through the surface ψ' with d accidental double points, passes simply through each of these, which are four-fold points for the curve c. Thus $Z = (k + k' - 4)\,\epsilon_0 - 4d$; we have, however, $\epsilon_1 = \epsilon_0(k + k' - 1) - 4d$. It has already been seen (Chap. v, p. 225) that $z - 3\epsilon_0$ is the number of intersections of the curve c with the canonical curves of the surface ψ, when these exist.

Ex. 3. If we now further assume that the primals Π of order $k + k' - 3$ through ψ' cut canonical curves on ψ, and that two such canonical curves K_0, K cut in $\omega - 1$ points (see Chap. v, p. 197), and also that canonical curves are of order $2p - 2 - \mu_0$ (since such a curve, added to a prime section, gives a curve belonging to the adjoint system of the prime sections); then, taking intersections on ψ, we have $(\Pi, K) = (K_0 + c, K) = \omega - 1 + i$; which is therefore equal to $(k + k' - 3)(2p - 2 - \mu_0)$. We thus have an alternative proof of the last formula of Ex. 1.

Several applications of the formulae for intersections here given will be found in the following chapter.

The postulation of a surface for primals containing it. It is convenient to add here a proof, of summary and preliminary character, of the formula expressing the number of conditions which a surface, without multiple points, imposes upon a primal that is to contain the surface, when the order, ρ, of the primal is sufficiently high. Let the surface be of order μ_0, of numerical genus p_n, and its prime sections be of genus p. Then the number of conditions is

$$\tfrac{1}{2}\mu_0\rho(\rho + 1) - \rho(p - 1) + p_n + 1 - k,$$

where k is some number $\geqslant 0$. It appears that the formula is exact when the surface has accidental double points, the number of these being k. In general k is zero if the primals of order ρ determine a complete system of curves on the surface, supposed without multiple points, and, when this is not so, is the deficiency of this system. In terms of the characters of the surface this postulation is $\tfrac{1}{2}\mu_0(\rho + 1)(\rho + 2) - \tfrac{1}{6}\mu_1(3\rho + 4) + \tfrac{1}{6}\mu_2 + \tfrac{1}{12}\nu_2 - k$. To prove this, we assume a result due to Castelnuovo, *Ann. d. Mat.* xxv, 1897, p. 83: *The characteristic series* (p. 124), *for the complete system of curves determined on a surface without singular points, lying in space of any dimensions, by the primals of any order sufficiently high, is not complete, but of deficiency* $p_g - p_n$, *where* $p_g - 1$ *is the freedom of the canonical system of curves on the surface.* We also assume that the canonical series, on a curve of a linear system of curves on the surface, which has no base points, is the sum of the characteristic series on this curve, and the series cut on this curve by the canonical system of curves of the surface (above, p. 222). From this it follows that any set of the characteristic series on the curve is contained in a number equal to p_g of canonical sets on the curve, diminished by the number of canonical curves of the surface containing the curve in question. Thus, when this curve is of sufficiently high order, the index of specialness of its characteristic series is p_g.

(Cf. *Proc. Lond. Math. Soc.* XII, 1913, p. 19.) Now consider a linear
system of curves on a surface, cut by all the primals of sufficiently
high order; let n be the grade of this system, and r the freedom of
the complete system to which this system belongs; thence the
characteristic series is of sets of n points, and is of freedom $r-1$.
This series belongs then, by the Castelnuovo result above quoted,
to a complete series on the curve of freedom $r-1+p_g-p_n$. Whence,
by the Riemann-Roch theorem on the curve, we have

$$r-1+p_g-p_n=n-\pi+p_g,$$

where π is the genus of the curve. Wherefore $r=n-\pi+1+p_n$. Now
let the primals be of order ρ, so that $n=\mu_0\rho^2$, if μ_0 be the order of
the surface. We can obtain the genus π by a theorem for a composite curve, $\pi_1+\pi_2+i-1$, previously obtained (Chap. V, p. 223);
for let the genus of a prime section of the surface be p; then the
section by a primal of order ρ is equivalent to the section by a
primal of order $\rho-1$, together with the section by a prime; thus, if
π_ρ, $\pi_{\rho-1}$ be the genera of these sections, we have

$$\pi_\rho=\pi_{\rho-1}+p+\mu_0(\rho-1),$$

leading to π, or π_ρ, $=\tfrac{1}{2}\mu_0\rho(\rho-1)+\rho(p-1)+1$. Applying then the
preceding result, for the freedom of the complete system of curves
on the surface determined by its sections with primals of order ρ,
we have $r=\mu_0\rho^2-\pi_\rho+1+p_n$, or $r=\tfrac{1}{2}\mu_0\rho(\rho+1)-\rho(p-1)+p_n$. The
freedom of the system of curves, possibly not complete, cut by the
primals, is this number less a correction, say k. In order that such
a primal should contain the surface it is necessary and sufficient
that the primal should contain one more point of the surface than
is necessary to determine a curve of the system. Thus the postulation of the surface, for primals of order ρ, sufficiently great, is
$\tfrac{1}{2}\mu_0(\rho+1)\rho-\rho(p-1)+1+p_n-k$.

The formula $\tfrac{1}{2}\mu_0\rho(\rho+1)-\rho(p-1)+p_n$, for the freedom of the
complete system of curves on the surface determined by primals of
order ρ, is the same as that for the freedom of the system determined by primals of order $\rho+1$ put through a prime section of the
surface, if we assume the usual value, $\mu_0(\rho+1)-(p-1)$, of the
postulation of a curve of order μ_0 and genus p for primals of order
$\rho+1$, sufficiently great; such primals of order $\rho+1$ give a system of
variable curves of the same characters as do the unrestricted primals
of order ρ (cf. Ex. 2, p. 256). But suppose the surface has a single
accidental double point; then the postulation, for primals of order
$\rho+1$, of a prime section of the surface through the double point,
which has a double point thereat, is one less than before, two intersections with the prime section at the double point being obtained
by one condition for the primal. Hence we infer that, for a surface
with such a double point, the system of sections by unrestricted

Intersections of loci 265

primals of order ρ, has a freedom less by one than the system of sections by primals of order $\rho+1$, put through a fixed prime section containing the double point. This suggests that the formula above obtained for the postulation of a surface, when ρ is sufficiently great, which is obtained under the hypothesis that the surface has no double points, may be exact for a surface with d improper double points, if k be replaced by d; that is, that the only correction on account of the possible incompleteness of the system cut by primals of order ρ arises in virtue of the double points, when ρ is sufficiently great, and the surface has no other singularities. Examples in verification of this result arise below; for a complete result see Severi, *Rend....Palermo*, XVII, 1903, No. 32. We may remark that, for a composite surface, consisting of two surfaces with respectively d and d' double points, having i intersections, we have, denoting

$$\tfrac{1}{2}\rho(\rho+3)\mu_0 - \tfrac{1}{2}\rho\mu_1 + p_n + 1 - d$$

by $F(\mu_0, \mu_1, p_n, d)$, the equation

$$F(\mu_0, \mu_1, p_n, d) + F(\mu_0', \mu_1', p_n', d'+i)$$
$$= F(\mu_0+\mu_0', \mu_1+\mu_1', P_n, d+d'+i),$$

where $P_n+1 = p_n+1+p_n'+1$. Cf. Severi, *Rend....Palermo*, XXVIII, 1909, No. 5.

Ex. 1. If two primals (ρ), (ρ') have a plane in common, their remaining surface of intersection has the characters

$$\mu_0 = \rho\rho' - 1, \quad \mu_1 = (\rho\rho'-2)(\rho+\rho'-2), \quad \nu_2 = (\rho\rho'-2)(\rho-1)(\rho'-1),$$

while, with $k = \rho-1$, $k' = \rho'-1$, $\mu_2 = \rho\rho'(k^2+kk'+k'^2+1) - 3(k+k')^2 - 1$. This surface meets the plane in a curve, c, for which $\epsilon_0 = \rho+\rho'-2$, $\epsilon_1 = (\rho+\rho'-2)(\rho+\rho'-3)$, that is, in a curve of order $\rho+\rho'-2$ without double points, upon which the primals have double points of respective numbers $\delta = (\rho-1)^2$, $\delta' = (\rho'-1)^2$.

Ex. 2. If two primals (ρ), (ρ') have two planes in common (together forming a composite surface of characters $\mu_0 = 2$, $\mu_1 = 0$, $\mu_2 = 0$, $\nu_2 = 0$, $d = 1$), they meet further in a surface of characters

$$\mu_0 = \rho\rho' - 2, \quad \mu_1 = (\rho\rho'-4)(\rho+\rho'-2), \quad \nu_2 = (\rho\rho'-4)(\rho-1)(\rho'-1),$$

for which also, with $k = \rho-1$, $k' = \rho'-1$,

$$\mu_2 = \rho\rho'(k^2+kk'+k'^2+2) - 6(k+k')^2 + 4;$$

this surface meets the planes in a composite curve, c, for which $\epsilon_0 = 2(\rho+\rho'-2)$, $\epsilon_1 = 2(\rho+\rho'-2)(\rho+\rho'-3) - 4$, that is, it meets each plane in a curve of order $\rho+\rho'-2$, with a single double point at the point of intersection of the two planes, upon which the primals have double points, other than at the intersection of the planes, of respective numbers $\delta = 2\rho(\rho-2)$, $\delta' = 2\rho'(\rho'-2)$; or, with the double point at the intersection of the planes, they have, respectively, $(\rho-1)^2$ and $(\rho'-1)^2$ double points on each plane.

Ex. 3. If, in space $[r]$, with $r > 4$, there be put $r - k$ primals through a manifold of dimension k, M_k, where $k < \tfrac{1}{2}r$, these primals meeting again in a M_k', which has with M_k a manifold common of dimension $(k-1)$, say m_{k-1}, then the tangent $[k-1]$ of this m_{k-1} at any point of this lies in the two tangent spaces $[k]$ of M_k and M_k' at this point, and these two tangent spaces define a space $[k+1]$. Denote by X the number of such spaces $[k+1]$ which meet an arbitrary space $[r-2k]$; likewise denote by z the number of tangent $[k]$, of M_k, at points of m_{k-1}, which meet an arbitrary space $[r-2k+1]$. In particular, when $k=2$, and there is a surface M_2, through which there pass $r-2$ primals of orders $\rho_1, \ldots, \rho_{r-2}$, let s_1 denote $\Sigma(\rho_i-1)$, and s_2 denote $\Sigma(\rho_i-1)(\rho_j-1)$, for $i \neq j$; then for general $r \geqslant 4$, we also have

$$\mu_0 + \mu_0' = \rho_1 \ldots \rho_{r-2}, \quad \mu_1' - \mu_1 = (\mu_0' - \mu_0)s_1, \quad \nu_2' - \nu_2 = (\mu_0' - \mu_0)s_2,$$

$$\mu_2 + \nu_2 + z = \mu_1 s_1, \quad \mu_2' + \nu_2' + z' = \mu_1' s_1, \quad X = z + z' - \epsilon_1 = \mu_0 s_2 - \nu_2 - 2d,$$

$$\epsilon_0 = \mu_0 s_1 - \mu_1, \quad \epsilon_1 = \epsilon_0(s_1-1) - 4d,$$

where for $r > 4$, we have $d = 0$, and the primals are not supposed to have double points on the surface through which they are drawn.

Ex. 4. As an example of the formulae just given, prove that if, in [5], three quadrics be drawn through the Del Pezzo surface—that namely whose prime sections are represented by plane cubic curves with four common base points—for which $\mu_0 = 5$, $\mu_1 = 10$, $\mu_2 = 12$, $\nu_2 = 8$, then these quadrics meet again in a rational ruled cubic surface (lying actually in a space [4]) for which $\mu_0' = 3$, $\mu_1' = 4$, $\mu_2' = 3$, $\nu_2' = 2$, having common with the Del Pezzo surface an elliptic quartic curve, a prime section of this.

Ex. 5. In fact an irreducible algebraic manifold $M_k{}^n$, of order n and dimension k, is necessarily contained in a space $[r]$ for which $r < n+k$. Or, if the manifold be contained in a space $[r]$, and not in a space of less dimension, its order n is necessarily greater than the dimension $(r-k)$ complementary to that of the manifold. This fact was suggested in Clifford, "Classification of Loci", *Papers*, 1882, p. 307; a complete proof is given in Bertini, *Geom. d. Iperspazi* (Pisa, 1907), p. 193. But, summarily, we may assume that there is at least one space of complementary dimension, $r-k$, whose intersections with the manifold, which are n in number, do not lie in a space of less than $r-k$ dimensions, while the dimension of the linear space of least dimension which contains n points is certainly less than n; thus $r-k < n$.

V. Residual intersection of three primals having a surface in common.

By a combination of III and IV above, we can now prove that if, in space [4], three primals Π, Π', Π'', of respective orders ρ, ρ', ρ'', have a surface, ψ, in common, of characters μ_0, μ_1, μ_2, but without double points, and these primals meet further in a curve of order ϵ_0' and rank ϵ_1', having i intersections with ψ, then

$$\epsilon_0' = \rho\rho'\rho'' - \mu_0(s_1+1) + \mu_1, \quad \epsilon_1' = s_1\epsilon_0' - 2i, \quad i = \mu_0 s_2 - \mu_1 s_1 + \mu_2,$$

where s_1, s_2 are respectively the sum, and the sum of the three products of pairs, of the three numbers $\rho-1$, $\rho'-1$, $\rho''-1$, which we may denote by k, k', k''.

The first formula is obvious by taking an arbitrary prime section, whereby we obtain three surfaces of orders ρ, ρ', ρ'', in space [3], having in common a curve of order μ_0 and rank μ_1; these curves intersect further in ϵ_0' points (Vol. v, Chap. viii). For the second formula, we may consider the Jacobian of the three primals and of two arbitrary primes, meeting in a plane ϖ. This Jacobian is of order s_1, and meets the curve $(\epsilon_0', \epsilon_1')$ in the points whereat the tangent line meets the plane ϖ. It may also be proved to have two intersections with the common curve of the three primals at each of the points where this curve meets the surface ψ common to the primals. This second formula may also be found by the method we now employ to obtain the third formula: let the primals Π, Π' meet, beside in ψ, in the surface ψ', of characters $\mu_0{}', \mu_1{}', \ldots$, and let ψ' meet ψ in the curve c, of characters ϵ_0, ϵ_1, with a class of immersion in ψ' which we denote by z'. The primal Π'' passes through c, and meets ψ' also in another curve, say γ, which is the residual curve of intersection of Π, Π', Π''. As c has a class of immersion in the primal Π'' which is equal to the number of intersections of the first polar of this, taken in regard to an arbitrary point, with this curve, namely $\epsilon_0 k''$, we have, by the formulae of III above, $i = z' - \epsilon_1 + \epsilon_0 k''$. By the formulae in IV, however, we have

$$\epsilon_0 = \mu_0(k+k') - \mu_1, \quad z' - \epsilon_1 = \mu_0\mu_0{}' - \epsilon_0 - z, \quad z = \mu_1(k+k') - \mu_2 - \nu_2.$$

Thus

$$i = \epsilon_0 k'' + \mu_0\mu_0{}' - \epsilon_0 + \mu_2 + \nu_2 - \mu_1(k+k')$$
$$= \mu_2 + \nu_2 + (k''-1)[\mu_0(k+k') - \mu_1] - \mu_1(k+k')$$
$$\qquad + \mu_0[(k+1)(k'+1) - \mu_0]$$
$$= \nu_2 - \mu_0{}^2 + \mu_0 + \mu_1 + \mu_2 + \mu_0 s_2 - \mu_1 s_1,$$

which has the value given above, since the number

$$d, = \tfrac{1}{2}\mu_0(\mu_0-1) - \tfrac{1}{2}\mu_1 - \tfrac{1}{2}\nu_2,$$

of double points of the surface ψ is, by hypothesis, zero. The value of $\epsilon_1{}'$ is given by the formula of III above

$$\epsilon_1{}' + \epsilon_1 + 2i = \mu_0{}' k''(k''+1) + \mu_1{}'(k''+1),$$

and will be found to lead to the value stated above in terms of $\epsilon_0{}'$.

Corollary. Four primals (ρ), (ρ'), (ρ''), (ρ''') passing through a given surface (μ_0, μ_1, μ_2) have further in common a number of points given by $\rho\rho'\rho''\rho''' - [\mu_0\sigma_2 + (\mu_0 - \mu_1)(\sigma_1+1) + \mu_2]$, where σ_1, σ_2 are the sum, and the sum of the six products of pairs, of k, k', k'', k''', where $k = \rho - 1$, etc.

More generally if, in space $[r]$, there be $r-1$ primals through a surface, meeting further in a curve $(\epsilon_0{}', \epsilon_1{}')$, having i intersections with this surface, we have

$$\epsilon_0{}' = \rho_1 \ldots \rho_{r-1} - \mu_0(s_1+1) + \mu_1, \quad \epsilon_1{}' = s_1\epsilon_0{}' - 2i, \quad i = \mu_0 s_2 - \mu_1 s_1 + \mu_2,$$

where s_1, s_2 are formed as before from the numbers $\rho_1 - 1, \ldots,$ $\rho_{r-1} - 1$. And the point equivalence of the surface is

$$\mu_0\sigma_2 + (\mu_0 - \mu_1)(\sigma_1 + 1) + \mu_2,$$

where σ_1, σ_2 are formed from r primals. The result is given by Severi, *Intersezioni...*, No. 29; it is assumed that the surface has no multiple points or multiple curves. Of the other results given here, I–V, wide generalisations will be found in this paper.

Ex. 1. More generally, in space $[r]$, the point equivalence of a manifold $M_k(k < \frac{1}{2}r)$, without multiple points, for primals of orders $k_1 + 1, \ldots,$ $k_r + 1$, is the coefficient of t^k in the ascending expansion of the function $(1 + k_1 t) \ldots (1 + k_r t)/(1 - t)(1 + \mu t)$, if it be understood that $(1 + \mu t)^{-1}$ means the series $\mu_0 - \mu_1 t + \mu_2 t^2 - \ldots$.

Ex. 2. Find the point equivalence of a surface with d accidental double points in space $[4]$, for primals.

Ex. 3. If a manifold of order $\rho_1 \ldots \rho_h$, and dimension $r - h$, in space $[r]$, without multiple parts of dimension $r - h$, but possibly with multiple parts (or points) of less dimension, be the complete intersection of h primals of orders $\rho_1, \ldots, \rho_h$, the postulation of this manifold for a primal of any order R is the coefficient of x^{-1} in the ascending expansion of $(1 - x^{\rho_1})(1 - x^{\rho_2}) \ldots (1 - x^{\rho_h})/x^{R+1}(1 - x)^{r+1}$. The proof of this has already been indicated (Vol. v, p. 222); it will be found at length in Bertini, *Iperspazi*, 1907, p. 263. Applying to this rational function of x the theorem already referred to (Vol. v, p. 47), written for abbreviation as $[R(x)\,dx/dt]_t{}^{-1} = 0$, the values of x beside $x = 0$, to be considered are $x = 1$ and $x = \infty$. Thus the postulation in question, say $\chi(R)$, is found to be equal to the sum

$$(-1)^{r-h}\left\{\frac{[(1+\xi)^{\rho_1} - 1] \ldots [(1+\xi)^{\rho_h} - 1]}{\xi^{r+1}(1+\xi)^{R+1}}\right\}_{\xi^{-1}}$$
$$+ (-1)^{r-1-h}\left\{\zeta^{R - \Sigma\rho + r}\frac{(1 - \zeta^{\rho_1}) \ldots (1 - \zeta^{\rho_h})}{(1 - \zeta)^{r+1}}\right\}_{\zeta^{-1}};$$

of these the latter gives no contribution when $R \geqslant \Sigma\rho - r$, but gives $(-1)^{r-1-h}$ when $R = \Sigma\rho - r - 1$; putting $k_i = \rho_i - 1$, the former is equal to

$$(-1)^{r-h}\mu_0\{(1+\xi)^{-(R+1)}\,\Pi\,[1 + \tfrac{1}{2}k_i\xi + \tfrac{1}{6}k_i(k_i - 1)\,\xi^2 + \ldots]\}_{\xi^{r-h}},$$

where $\mu_0 = \rho_1 \ldots \rho_h$. We consider the cases $h = r - 1$, $h = r - 2$.

For $h = r - 1$, we obtain $\chi(\Sigma k - 2) = \mu_0(\frac{1}{2}\Sigma k - 1) + 1$, in fact equal to the genus, p, of the curve which is the complete intersection in question; for $R > \Sigma k - 2$, however, $\chi(R) = \mu_0 R - p + 1$. If the curve break up into two curves of orders m and m', with i intersections, of genera p and p', and we take account of the equations

$$2m + 2p - 2 + i = m\Sigma k, \quad 2m' + 2p' - 2 + i = m'\Sigma k,$$

the postulation $\chi(\Sigma k - 2)$, which is $(m + m')(\frac{1}{2}\Sigma k - 1) + 1$, is equal to $p + p' + i - 1$. For $R > \Sigma k - 2$, we have

$$\chi(R) = (m + m')R - (p + p' + i - 1) + 1,$$

which is $\quad (mR - i - p + 1) + (m'R - i - p' + 1) + i$.

For $h = r - 2$, we obtain (cf. p. 247, Ex. 3, and p. 262, Ex. 1)

$$\chi(\Sigma k - 3) = \tfrac{1}{12}\mu_0[2\Sigma k_i{}^2 + 3\Sigma k_i k_j - 8\Sigma k_i + 12] - 1$$

and, denoting this by p_n, we have, for $R > \Sigma k - 3$,

$$\chi(R) = \mu_0[\tfrac{1}{2}(R+1)(R+2) - \tfrac{1}{2}(R+1)\Sigma k_i + \tfrac{1}{6}\Sigma k_i(k_i - 1) + \tfrac{1}{4}\Sigma k_i k_j],$$

which is the same as $\frac{1}{2}\mu_0 R(R-1) - R(p-1) + p_n + 1$, where p is the genus of a prime section of the surface, equal to $\mu_0(\frac{1}{2}\Sigma k - 1) + 1$. When the surface breaks up into surfaces ψ, ψ' of orders μ_0 and μ_0', this last formula remains true if we replace μ_0, p, p_n respectively by $\mu_0 + \mu_0'$, $p + p' + \epsilon_0 - 1$

and $\chi(\Sigma k-3)$, where ϵ_0 is the order of the curve of intersection of the two surfaces.

When the surface ψ possesses a canonical system of curves, of at least one curve $(p_g > 0)$, these (together with the exceptional curves if there are such) are all given by the primals of order $\Sigma k - 3$ through the surface ψ'.

Ex. 4. Assuming Severi's theorem quoted in the footnote, p. 262, above, prove in space [4] that if two cubic primals can be put through a regular surface of order μ_0, which has no double point, the irregularity of the residual surface of intersection is $\mu_0 - p - 3$ if this is not negative, p being the genus of the prime section of the original surface. One example is when the surface is that obtained by projection of the Veronese surface, in which the residual intersection is a ruled elliptic quintic surface.

Ex. 5. The following results may also be proved (Severi, *Rend. Palermo*, XVII, 1903; also *Rend. Palermo*, XXVIII, 1909, p. 5): (*a*), An irreducible curve, without multiple points, which is the complete intersection of $r - 1$ primals in space $[r]$, is normal; (*b*), A manifold M_h, which is irreducible, and without multiple M_{h-1}, and is a complete intersection of $r - h$ primals, is normal; (*c*), A surface, without multiple points, which is the complete intersection of $r - 2$ primals, is regular $(p_g = p_n)$, and without exceptional curves (except when $r = 4$, $\rho_1 = \rho_2 = 2$). On this surface the canonical system of curves is given by the primals of order $\Sigma k - 3$, while primals of order $R > \Sigma k - 3$ determine thereon a linear system of curves which is regular and complete.

Ex. 6. The following theorems, proved by Severi, *Atti...Torino*, XLI, 1906, have connexion with the results of Ex. 3: (*a*), The general one of a linear system of manifolds M_{k-1}, of dimension $k - 1$, existing on a given manifold M_k, has no multiple points other than at the base points common to all the M_{k-1} or at multiple points of the M_k; If $F_0, ..., F_r$ be $r + 1$ homogeneous polynomials in the $r + 1$ variables $x_0, ..., x_r$, of orders $\rho_0, ..., \rho_r$, any other such *form*, of order $> \Sigma(\rho_i - 1)$, is expressible in the form $A_0 F_0 + ... + A_r F_r$, where $A_0, ..., A_r$ are suitably chosen forms; (*c*), If, for $h < r + 1$, the $h + 1$ forms Φ, $F_1, ..., F_h$ have only ∞^{r-h-1} common zeros, and there holds an equation $\Phi F = A_1 F_1 + ... + A_h F_h$, where F, $A_1, ..., A_h$ are suitable forms, then F is likewise expressible as $B_1 F_1 + ... + B_h F_h$, for suitable forms $B_1, ..., B_h$. Cf. also Lasker, "Zur Theorie der Moduln und Ideale", *Math. Ann.* LX, 1905.

Ex. 7. Recalling that to a linear system of curves $|C|$, without base points, of grade n and genus π, there exists an adjoint system $|C'|$, defined by $|C_J - 2C|$, whose curves cut canonical sets on curves of $|C|$, and further that the system $|C' - C|$ is independent of $|C|$ (see Chap. v, pp. 217, 222), it follows, if $|C''|$ be the adjoint system of $|C'|$, that $|C'' - C'| = |C' - C|$, and hence $|C'' + C| = |2C'|$. Wherefore, if n', π' be the grade and genus for $|C'|$, we have

$$2n' = (2C', C') = (C'' + C, C') = (C'', C') + (C', C), = 2\pi' - 2 + 2\pi - 2;$$

so that $n' = \pi' - 1 + \pi - 1$ (p. 225, Ex. 4).

Ex. 8. We have assumed above, in obtaining the equation

$$r = n - \pi + 1 + p_n,$$

for a linear system $|C|$, of curves of sufficiently high order, that, for this system the characteristic series is of deficiency $p_g - p_n$. It is convenient to enunciate also a theorem stated by Picard (*Crelle*, CXXIX, 1905; Picard-Simart, *Fonctions algéb.* II, 1906, p. 438. Cf. Severi, *Rend. Lincei*, XVII, 1908, p. 465), that for *any* general linear system $|C|$, the series of canonical sets cut on a curve of this system, by the curves $|C'|$ of the adjoint system, is of deficiency $p_g - p_n$. Hence, if r' be the freedom of the system $|\, C'\,|$, since p_g curves of this system contain any curve of $|\, C\,|$, we have $r' - p_g = \pi - 1 - (p_g - p_n)$, or $r' = \pi - 1 + p_n$. Hence, by the result of the preceding example, we have $r' = n' - \pi' + 1 + p_n$ (cf. p. 169).

Ex. 9. If, in space [4], two primals F, F', of orders ρ, ρ', which have in common no surface which is multiple, have as their intersection two surfaces ψ, ψ', and if ψ be regular ($p_g = p_n$), then the formula

$$\tfrac{1}{2}\mu_0' R(R+1) - R(p'-1) + p_n' + 1 - d'$$

is valid to compute the postulation of the surface ψ' for primals of order $R = \rho + \rho' - 5$ put through it; and such primals, if $p_g > 0$, after subtraction of primals of this order containing both ψ and ψ' (which are given by an equation $UF + U'F' = 0$) give the complete canonical system on ψ (Severi, *Rend. Palermo*, xvii, 1903). Shew that this interprets the first identity of Ex. 1, p. 262.

Ex. 10. For a quintic surface with two tacnodes (cf. p. 241, Ex. 6) the adjoint system $|\, C'\,|$, of the plane sections $|\, C\,|$, is given by quadric surfaces through the tacnodes (verifying the relation $r' = \pi - 1 + p_n$ of Ex. 8 with $\pi = 6$, $p_n = p_g = 2$); and the adjoint system of the canonical system is given by quadric surfaces touching the given surface at the tacnodes (verifying $r'' = \pi' - 1 + p_n$, with $\pi' = 2$).

Ex. 11. It is easily seen that the acquisition of a tacnode lessens the class of a surface by 12 (as does an ordinary triple point). Hence, accepting the result of Ex. 1, p. 236, (4), the tacnode lessens the invariant I by 10, and the invariant ω by 2. Thus, for the quintic surface of the previous example we find $I = 31$, $\omega = 2$. More simply, for a quartic surface with one tacnode, we have $I = 10$, $\omega = -1$; this last can be tested in the manner of pp. 201–6 above (which are for *ordinary* multiple points only); for the quartic surface can be transformed to a cubic surface (Vol. v, p. 110); thus we arrive at $I_0 = 9$, $\omega_0 = 0$.

Ex. 12. By applying Clifford's theorem (Vol. v, p. 80) to the characteristic series on a canonical curve of a surface, prove that $p^{(2)} > 2(p_g - 2)$. Noether's proof of this (*Math. Ann.* viii, 1875, §11) is effectively a deduction of Clifford's theorem (1877) from the theory of special series.

CHAPTER VII

ILLUSTRATIVE EXAMPLES AND PARTICULAR THEOREMS

THE present chapter contains various examples and results which may serve to illustrate foregoing theories, the earlier cases being very simple applications of the account given in the preceding chapter of the residual intersection of two primals, in space [4], passed through a given surface. In general a surface, in space [r], of order n, with prime sections of genus p, will be denoted by $^p\psi^n[r]$, or, when $r=4$, by $^p\psi^n$.

I. Particular examples of intersections.

1. The formulae for the characters of a surface $^4\psi^6$ which is the complete intersection of two primals in space [4], Chap. VI, p. 247, when applied to the intersection of a quadric and a cubic primal, in general positions, lead to

$$\mu_0=6, \quad \mu_1=18, \quad \mu_2=42, \quad \nu_2=12, \quad p_n=1, \quad I=20, \quad \omega=1, \quad t=0, \quad d=0.$$

The adjoint system of the prime sections is given by the prime sections themselves (in accordance with the formulae $r'=\pi-1+p_n$, $n'=\pi'-1+\pi-1$ of Ex. 8, p. 269, Chap. VI), so that the canonical system of the surface is a single curve of zero order, or $p_g=1$. The postulation of the surface for primals of order ρ (Chap. VI, p. 263) is $3\rho^2+2$; thus for $\rho=2$ it is 14 and there are $15-14$, or only one quadric containing the surface, as is obvious, since the surface is of order 6; and, for $\rho=3$ it is 29, so that there are $35-29$, or 6, cubic primals containing the surface, evidently given by an equation $(a_0x_0+...+a_4x_4)Q+U=0$, where $Q=0$ is the unique quadric and $U=0$ a cubic primal.

2. If we consider the surface $^2\psi^5$ which is the residual intersection of a cubic primal with a quadric primal when these have a plane in common, so that the quadric is a cone, the formulae (Chap. VI, p. 265) give

$$\mu_0=5, \quad \mu_1=12, \quad \mu_2=20, \quad \nu_2=8, \quad p_n=0, \quad I=7,$$
$$\omega=2, \quad d=0, \quad \epsilon_0=3, \quad \epsilon_1=6, \quad \delta=1, \quad \delta'=4,$$

and the canonical series on any prime section is given by primes through the fixed plane, and hence by planes through the trisecant chord of this section which lies in this plane. Thus $p_g=0$. The surface is in fact rational, its prime sections being representable in a plane by a system of quartic curves with one double and seven simple base points, say by a system $c^4(2, 1^7)$, the equations

$I-8=-1$, $\omega+8=10$ being in accordance with the equations $I_0-e=I_0{}'-e'$, etc. of Chap. v (p. 198). Starting from such a system of plane curves we may verify the definition of the surface as the partial intersection of a quadric and cubic. More generally, it has been proved by Castelnuovo (*Rend. Palermo*, IV, 1890, pp. 73–88, No. 5; cf. Enriques, *Math. Ann.* XLVI, 1895, p. 179; Scorza, for manifolds of 3 dimensions, *Ann. d. Mat.* XV, 1908, p. 217; Castelnuovo-Enriques, *Ann. d. Mat.* VI, 1901, p. 38), that any surface of which the prime sections are hyperelliptic curves is necessarily rational. In this case, if the quadric and cubic primals have equations $xt-yz=0$, $xU+yV=0$, where U, V are quadric polynomials in the coordinates x, y, z, t, u, their common plane being $x=0$, $y=0$, the cubic curve in which the surface meets this plane is given by $zU_0+tV_0=0$, where U_0 is U with $x=0$, $y=0$, etc. The cubic primal has double points (as we may verify by forming the partial derivatives) at the four points of the cubic given by $U_0=0$, $V_0=0$. The adjoint system of the prime sections of the surface are the conics given by such equations as $y=mx$, $t=mz$, $U_1+mV_1=0$, where U_1 is U for $y=mx$, $t=mz$, etc.; each of these meets the cubic curve in the plane $x=0$, $y=0$ in two points collinear with the vertex of the quadric primal. These conics of the surface correspond to the lines of the plane passing through the common double point of the plane quartic curves, each of these lines being a rational curve having two variable intersections with the quartic curves of the system. The surface also contains seven families of rational cubic curves; each corresponding to the pencil of lines through one of the seven simple base points of the quartic curves. But the curves of the surface seem best studied by regarding the surface as the projection from a space [6] of a rational surface of order 12 in space [11], whose prime sections are represented by plane quartic curves with one double base point. For this study we refer to Castelnuovo's paper. The number of primals of order ρ containing the surface is (Chap. VI, p. 263) at least $(\rho+4, 4)-\frac{5}{2}\rho(\rho+1)+\rho-1$. Thus, for $\rho=2$, the surface lies on the one quadric cone; and, for $\rho=3$, the surface lies on 7 cubic primals, of which 5 will be given by the aggregate of the quadric cone and an arbitrary prime. It will be found (cf. Chap. VI, p. 242) that the chords drawn to the surface ${}^2\psi^5$, from an arbitrary point, meet the surface in an octavic curve of genus 5, forming a quartic conical sheet of genus 1. Thus the ${}^2\psi^5$ projects, on to space [3], into a quintic surface with an elliptic quartic curve as double curve. It will appear below, under (4), that if two cubic primals be put through the surface their residual intersection is the Segre quartic surface which is the intersection of two quadric primals.

3. A simple and familiar surface is the cubic ruled surface, ${}^0\psi^3$ (4), representable by equations $x_0/x_1=x_1/x_2=x_3/x_4$. The general

formulae for a ruled surface (Chap. IV, p. 165) lead in this case to $\mu_0 = 3$, $\mu_1 = 4$, $\mu_2 = 3$, $\nu_2 = 0$, $p = 0$, $d = 0$, $I = 0$, $p_n = 0$, $\omega = 9$, in accordance with the plane representation of the prime sections of the surface, which is by conics having a common simple base point. The postulation of the surface for quadrics containing it is 12, so that the surface lies on ∞^2 quadrics; it is in fact projected by a quadric cone from any point of itself. The postulation of the surface for cubic primals containing it is 22, so that there are 13 such cubics. And in fact, with the coordinates indicated above, three quadrics containing the surface are given by

$$x_0 x_2 - x_1{}^2 = 0, \quad x_1 x_3 - x_0 x_4 = 0, \quad x_1 x_4 - x_2 x_3 = 0,$$

and, denoting these by $u = 0$, $v = 0$, $w = 0$, thirteen cubics containing the surface are in fact given by the equation $uL + vM + wN = 0$, where $L = 0$, $M = 0$, $N = 0$ represent arbitrary primes, of which the last may be taken through the line $x_2 = 0$, $x_3 = 0$, $x_4 = 0$. It will be found on examination that this equation contains 13 independent primals. Thus two cubic primals through the $^0\psi^3$ intersect further in the sextic surface given by the vanishing of the determinants of the third order in a matrix of three rows and four columns,

$$\begin{vmatrix} x_0, & x_1, & N, & N' \\ x_1, & x_2, & M, & M' \\ x_3, & x_4, & L, & L' \end{vmatrix}$$

wherein L, M, N, L', M', N' are linear in the coordinates. Thus this residual surface is the locus of intersection of four corresponding primes $\nu x_0 + \mu x_1 + \lambda x_3 = 0$, $\nu x_1 + \mu x_2 + \lambda x_4 = 0$, $\lambda L + \mu M + \nu N = 0$, $\lambda L' + \mu M' + \nu N' = 0$, in four projective systems of primes, of which each system consists of the primes through a line. This surface is evidently rational, any point having its coordinates rational in λ, μ, ν; and it is easy to see that its prime sections are represented in a plane by quartic curves $c^4(1^{10})$ with 10 simple base points. The characters of this surface $^3\psi^6$ can then be found at once (Chap. VI, p. 242); or they can be found from the formulae for the residual intersection of two primals through a given surface (Chap. VI, p. 252). They are in fact $\mu_0' = 6$, $\mu_1' = 16$, $\mu_2' = 27$, $\nu_2' = 14$, $I' = 9$, $p_n' = 0$, and the $^3\psi^6$ meets the $^0\psi^3$ in a curve of order 8 and genus 5, say $^5c^8$; upon this curve the primes of the space [4] cut the complete canonical series (Chap. VI, p. 262, footnote). This curve, it may be seen, meets each generator of the $^0\psi^3$ in three points, and thus contains a linear series $g_1{}^3$; it is not the most general curve of genus 5, of which a plane representative is a sextic curve with 5 double points; but is that which can be represented by a plane quintic curve with only one double point. It can be determined on the

${}^0\psi^3$ by a cubic primal, not containing this surface but containing one generator of it. The surface ${}^3\psi^6$ was considered by Bordiga, *Mem....Lincei*, IV, 1887, p. 182 (see also Veronese, *Math. Ann.* XIX, 1882; White, *Proc. Camb. Phil. Soc.* XXI, 1923, p. 221; Room, *Proc. Roy. Soc.* A, CXI, 1926, p. 386). The sextic curve in [3] which is obtained by a prime section has been considered by Schur, *Math. Ann.* XVIII, 1881. By the plane representation of the surface ${}^3\psi^6$ we infer that the surface contains 10 lines, corresponding to the base points of the quartic curves; also 10 elliptic plane cubics, corresponding to the plane cubics through 9 of the base points, each of these curves meeting 9 of the lines; also 10 pencils of rational cubic curves, etc.

4. For the Segre quartic surface ${}^1\psi^4[4]$, the complete intersection of two quadrics, we have

$$\mu_0 = 4, \quad \mu_1 = 8, \quad \mu_2 = 12, \quad \nu_2 = 4, \quad d = 0, \quad p = 1, \quad I = 4, \quad \omega = 5, \quad p_n = 0,$$

in accordance with the fact that the surface is rational, being represented by plane cubic curves having 5 simple base points, say by a system $c^3(1^5)$. The characters for a general Del Pezzo surface will be given below. The postulation of the surface for primals of order ρ (Chap. VI, p. 263) becomes here $\rho^2 + (\rho+1)^2$; there are therefore at least 2 quadrics and 10 cubic primals containing the surface. Thus if $U = 0$, $V = 0$ be the defining quadrics, ten cubics are given by $uU + vV = 0$, where $u = 0$, $v = 0$ are arbitrary primes. The formula (Chap. VI, p. 262) $\delta + 2d = \mu_0(\rho-1)^2 - \mu_1(\rho-1) + \nu_2$ shews that a primal of order ρ through the ${}^1\psi^4$ has $4(\rho-2)^2$ double points at ordinary points of the surface; such a primal is given by an equation $uU + vV = 0$, where u, v are of order $\rho-2$, and the double points in question are the intersections of $U = 0$, $V = 0$, $u = 0$, $v = 0$. Two cubic primals through the ${}^1\psi^4$ meet again in a quintic surface; if we use the formulae of Chap. VI, p. 261, we find for the characters of this residual surface $\mu_0' = 5$, $\mu_1' = 12$, $\mu_2' = 20$, $\nu_2' = 8$, which are the same as for the ${}^2\psi^5$ considered under (2), the residual intersection of a cubic primal and a quadric cone having common a plane. In fact two cubic primals $uU + vV = 0$, $u'U + v'V = 0$, meet, beside where $U = 0, V = 0$, also where $uv' - u'v = 0$; this represents a quadric cone having the plane $u = 0$, $v = 0$ which lies on $uU + vV = 0$. Further (Chap. VI, p. 261), the ${}^2\psi^5$ meets the ${}^1\psi^4$ in a curve ${}^5c^8$, of order 8 and genus 5, which is thus the intersection of the three quadrics $U = 0$, $V = 0$, $uv' - u'v = 0$, and is a general curve of genus 5.

4 a. There is a particular Segre quartic surface which has a double line, obtainable as the intersection of two quadric *line* cones, say $x_0 x_4 = x_2^2$, $x_1 x_4 = x_3^2$, the double line being $x_2 = x_3 = x_4 = 0$. This surface is representable, in terms of ξ, η, ζ, in the form

$x_0/\xi^2 = x_1/\eta^2 = x_2/\xi\zeta = x_3/\eta\zeta = x_4/\zeta^2$, and its prime sections project from the double line into conics meeting a given line in pairs of an involution. By equating these five ratios to a ratio $x_5/\xi\eta$ we obtain the parametric representation of a Veronese surface in space [5], and the particular Segre surface is obtainable by projecting this Veronese surface from the point common to the chords joining any two points $(\xi, \eta, 0)$, $(\xi, -\eta, 0)$ of the Veronese surface.

4*b*. We have already defined the characters μ_0, μ_1, μ_2, ν_2 for a surface in space [3] so as to correspond to characters denoted by these symbols for a surface in space [4]; we may similarly adopt definitions for higher space which, on projection into space [4], may give the characters as there defined. Thus, for space [5], μ_0 will be the order of a surface; μ_1 will be the order of the curve on the surface at which the tangent plane meets an arbitrary given plane (in a point), which is in fact the Jacobian curve of the net of curves on the surface determined by primes through the given plane; μ_2 will be the number of tangent planes of the surface which meet a given arbitrary solid, or space [3], in a line; ν_2 will be the number of tangent planes which meet an arbitrary line; d will be the number of chords of the surface which pass through an arbitrary point; t will be the number of trisecant chords of the surface which meet an arbitrary line, and so on. And similarly for space of higher dimension.

The general Del Pezzo surface in space [μ_0] may be understood to mean that surface of order μ_0 whose prime sections are represented by the cubic curves in a plane which pass through $9 - \mu_0$ simple base points. For such a surface we then have $p_n = 0$, $p = 1$, $I = 8 - \mu_0$, and hence, by the formulae of Chap. VI, p. 242, we have

$$\mu_1 = 2\mu_0, \quad \mu_2 = 12, \quad \nu_2 = 4(\mu_0 - 3), \quad d = \tfrac{1}{2}(\mu_0 - 3)(\mu_0 - 4),$$
$$t = \tfrac{1}{6}(\mu_0 - 2)(\mu_0 - 3)(\mu_0 - 4).$$

Thus, for the $^1\psi^5[4]$, obtained by projecting the Del Pezzo surface in space [5], we have $\mu_1 = 10$, $\mu_2 = 12$, $\nu_2 = 8$, $d = 1$, $t = 1$; and, by the formula (Chap. VI, p. 263) $\tfrac{1}{2}\mu_0\rho(\rho+1) - \rho(p-1) + p_n + 1 - d$, this surface lies on no quadrics, but on 5 cubics. The surface is thus obtainable as the partial intersection of two cubic primals. The remaining intersection of these will then be a quartic surface $^0\psi^4$, for which (by Chap. VI, p. 257) we have

$$\mu_0' = 4, \quad \mu_1' = 6, \quad \mu_2' = 4, \quad \nu_2' = 4, \quad p' = 0, \quad \epsilon_0 = 10, \quad \epsilon_1 = 26, \quad d' = 1,$$

while, by $\delta = \mu_0(\rho-1)^2 - \mu_1(\rho-1) + \nu_2 - 2d$, we have 6 for the number of double points, at ordinary points of $^1\psi^5$, which a cubic primal through this must possess. Comparing these formulae for μ_0', μ_1', ... with the results $\mu_1 = 2\mu_0 + 2p - 2$, $\mu_2 = \mu_0$, $\nu_2 = 2\mu_0 + 4p - 4$ holding for a ruled surface (Chap. IV, p. 165) we infer that two cubic primals through the projected Veronese surface meet again in a ruled $^0\psi^4$ (projection of the normal rational ruled quartic surface in [5]), which meets the projected Del Pezzo surface in a curve of order 10 and genus 4. In accordance with the footnote, Chap. VI, p. 262, the complete canonical series on this curve is determined by primes through its four-fold point. We shall see below (under (6)), that this $^0\psi^4$ is also the residual intersection of a quadric cone with a cubic primal containing two planes of the cone of the same system.

For the Del Pezzo surfaces in general, reference may be made to G. Timms, *Proc. Roy. Soc.* A, CXIX, 1928, p. 213. Those of these surfaces which are not representable on a plane by non-singular cubic curves are there referred to on p. 218 and in the diagram on p. 238.

5. For the quartic surface in [4] which is the projection of the Veronese surface in [5], we easily find

$$\mu_0 = 4, \quad \mu_1 = 6, \quad \mu_2 = 3, \quad \nu_2 = 6, \quad d = 0, \quad t = 1, \quad p_n = 0,$$

and the postulation, $\frac{1}{2}\mu_0\rho(\rho+1) - \rho(\rho-1) + p_n + 1$, for primals of order ρ is $(\rho+1)(2\rho+1)$; in accordance with this, the surface lies on no quadrics, and on 7 cubic primals. It can in fact be shewn that the surface can be represented on a plane (ξ, η, ζ) by taking the coordinates $x_0, \ldots, x_4$ respectively equal to $\xi^2, \xi\eta, \frac{1}{3}(\xi\zeta + 2\eta^2), \eta\zeta, \zeta^2$; and that the coefficients in the sextic covariant of the quartic form $x_0 t^4 + 4x_1 t^3 + 6x_2 t^2 + 4x_3 t + x_4$, separately equated to zero, give 7 cubic primals containing the surface. Any two of these cubic primals meet further in a surface $^1\psi^5$, for which $\mu_0' = 5$, $\mu_1' = 10$, $\mu_2' = 5$, $\nu_2' = 10$, $\epsilon_0 = 10$, $\epsilon_1 = 30$, that is in a ruled quintic surface of genus 1, which meets the projected Veronese surface in a curve $^6c^{10}$, of order 10 and genus 6. And in fact, the definite trisecant chord which can be drawn to the Veronese surface from any point, outside this surface, which lies on a cubic primal containing the surface, will meet this primal in 4 points, and will therefore be entirely upon it; thus the residual surface of intersection of two cubic primals through the Veronese surface is composed of lines (cf. Bertini, *Iperspazi*, 1907, p. 232 ff.). In accordance with this (see Chap. VI, p. 266), it can be shewn that 3 of the cubic primals have as residual common intersection a set of five lines, and that 4 of them have no residual intersection; and further (as will be remarked below in more detail) the equation of the ∞^6 system of cubic primals is expressible in terms of four polynomial functions of the five coordinates. The surface $^1\psi^5[4]$ is obtainable, as is known (Segre, *Rend. Palermo*, II, 1888, No. 7), as the locus of lines meeting 5 arbitrary planes in the space [4] (cf. Vol. IV, p. 143); it is normal in the space [4], in accordance with the rule (Segre, *Rend. Palermo*,

II, 1888, p. 45) that an elliptic ruled surface of order n is normal in space $[n-1]$. We have already (under (4)b) referred to a *rational* $^1\psi^5$, obtainable by projection of a Del Pezzo $^1\psi^5[5]$; in general, the normal space of a rational surface $^p\psi^n$ is $[n-p+1]$, as we see by considering the complete characteristic series of the representing plane curves. The cubic primals through the projected Veronese surface have a number of double points given by

$$\delta + 2d = \mu_0(\rho - 1)^2 - \mu_1(\rho - 1) + \nu_2,$$

namely, for $\rho = 3$, we have $\delta = 10$. These primals are then Segre rational cubic primals with 10 nodes (cf. Semple, *Trans. Roy. Soc.* A, CCXXVIII, 1929, p. 253), and the trisecants of the Veronese surface involved are one of the 6 systems of lines on such a primal. The ∞^6 system of primals meets an arbitrary prime space [3] in cubic surfaces of this space containing the rational quartic curve in which the prime cuts the Veronese surface, and hence in the complete system of such cubic surfaces through this curve, which is also ∞^6; and hence, as a unique trisecant of the Veronese surface passes through an external point, the primals are in (1, 1) correspondence with these cubic surfaces (or with the cubic cones which project these, and the quartic curve, from an arbitrary point).

Ex. 1. If we use a, b, c, d, e for the coordinates in the space [4], the system of cubic primals through the Veronese surface may be represented by

$$\lambda_1(a^2d - 3abc + 2b^3) + \lambda_2(a^2e + 2abd - 9ac^2 + 6b^2c)$$
$$+ 5\lambda_3(abe - 3acd + 2b^2d) + 10\lambda_4(b^2e - d^2a)$$
$$- \lambda_7(e^2b - 3cde + 2d^3) - \lambda_6(ae^2 + 2bde - 9ec^2 + 6d^2c)$$
$$- 5\lambda_5(ade - 3bce + 2bd^2) = 0,$$

where the cubic expressions are those of the sextic covariant of the binary form $at^4 + 4bt^3 + \ldots + e$; if h be the Hessian of this form f, and we recall that the sextic covariant $J_{\kappa\lambda}$ of the quartic form $\kappa f + \lambda h$ is a multiple of the covariant of f (in fact, with proper multiples of f and h, if I, J be the invariants of f, we have $J_{\kappa\lambda}/J_{10} = \kappa^3 - \frac{1}{2}I\kappa\lambda^2 - \frac{1}{3}J\lambda^3$ (Clebsch-Lindemann-Benoist, *Géométrie*, I, 1879, p. 302)), it follows that, save for a common factor, the seven cubic forms above are unaffected by writing for a, b, c, d, e respectively $\kappa a + ac - b^2$, $\kappa b + \frac{1}{2}(ad - bc)$, $\kappa c + \frac{1}{6}(ae + 2bd - 3c^2)$, $\kappa d + \frac{1}{2}(be - cd)$, $\kappa e + ce - d^2$ (cf. for instance, Salmon, *Higher Algebra*, 1885, p. 192), where κ is arbitrary. We may in particular take κ so that the middle term of these vanishes; the cubic forms are then in the ratios of the forms obtained by writing A, B, 0, D, E respectively for a, b, c, d, e, where $A = a(ae + 2bd - 3c^2) - 6c(ac - b^2)$, etc.; namely

$$A = a^2e + 2abd - 9ac^2 + 6b^2c, \quad B = abe - 3acd + 2b^2d,$$
$$E = ae^2 + 2bde - 9ec^2 + 6cd^2, \quad D = ade - 3bce + 2bd^2.$$

Thereby the ∞^6 cubic primals are represented by

$$\lambda_1(A^2D + 2B^3) + 10\lambda_4(B^2E - D^2A) - \lambda_7(E^2B + 2D^3)$$
$$+ (AE + 2BD)(\lambda_2A + 5\lambda_3B - 5\lambda_5D - \lambda_6E) = 0.$$

In the space [4], the coordinates A, B, D, E are in fact those of the inter-section of the trisecant line of the Veronese surface from the point (a, b, c, d, e) with the prime $x_2 = 0$; and the last equation may then be regarded as that of the cubic surfaces in this prime which pass through the quartic curve given by $x_0/4\lambda^4 = x_1/2\lambda^3 = -x_3/\lambda = x_4$ (which has two points of inflexion). In order that a binary sextic form $\xi_0 t^6 + 6\xi_1 t^5 + \ldots + \xi_6$ should be of the form of the sextic covariant of a binary quartic, that is, should have for roots three pairs each harmonic in regard to both the other pairs, the point $(\xi_0, \ldots, \xi_6)$, of a space [6], must lie on 5 quadrics (expressing the identical vanishing of the covariant $(ab)^4 a_x^2 b_x^2$ of a binary sextic a_x^6)

$$\xi_0 \xi_4 - 4\xi_1 \xi_3 + 3\xi_2^2 = 0, \quad \xi_2 \xi_6 - 4\xi_5 \xi_3 + 3\xi_4^2 = 0, \quad \xi_0 \xi_5 - 3\xi_1 \xi_4 + 2\xi_2 \xi_3 = 0,$$
$$\xi_1 \xi_6 - 3\xi_2 \xi_5 + 2\xi_3 \xi_4 = 0, \quad \xi_0 \xi_6 - 9\xi_2 \xi_4 + 8\xi_3^2 = 0$$

(Clebsch, *Binäre Formen*, pp. 440, 447), and the point $(\xi_0, \ldots, \xi_6)$ then lies on a certain manifold of order 5 and dimension 3, M_3^5, in the space [6]. By what we have seen, the prime sections of this manifold are represented by the cubic surfaces through a rational quartic curve in space [3]. Precisely, with $\Omega = AE + 2BD$, the coordinates $(\xi_0, \ldots, \xi_6)$ in [6] are respectively

$$\xi_0 = A^2D + 2B^3, \quad \xi_3 = \tfrac{1}{2}(B^2E - D^2A), \quad \xi_6 = -(E^2B + 2D^3), \quad \xi_1 = \tfrac{1}{3}\Omega A,$$
$$\xi_2 = \tfrac{1}{3}\Omega B, \quad \xi_4 = -\tfrac{1}{3}\Omega D, \quad \xi_5 = -\tfrac{1}{3}\Omega E;$$

and the cubic surfaces in the space [3] are obtained by projection of the prime sections of the M_3^5 from the plane $\xi_1 = \xi_2 = \xi_4 = \xi_5 = 0$. As the so-called octahedral relation of the six roots of the binary sextic form is unaffected by a linear transformation $t' = (mt + n)/(pt + q)$ of the variable t of the form, it follows that the M_3^5, and the original space [3], are capable of a continuous group of birational transformations involving 3 arbitrary parameters, which is linear in $\xi_0, \ldots, \xi_6$, and interchanges the cubic surfaces of the space [3] passing through a rational quartic curve among themselves. See Enriques-Fano ("Sui gruppi continui di transformazioni cremoniane dello spazio"), *Ann. d. Mat.* xxvi, 1897, p. 92, and Scorza, *Ann. d. Math.* xv, 1908, where M_3^5 is studied as a manifold M_k^n whose curve sections are elliptic. If we represent the lines of space [4] by points in a space where the 10 coordinates are the line coordinates, we obtain a $M_6^5[9]$. The M_3^6 is the section of this by a general space [6].

Ex. 2. Fano has proved (*Atti...Torino*, xxxix, 1904; see below, III, p. 295) that on a general cubic primal in space [4] there exists no surface (meeting all the lines of the primal in the same number of points) which is not the complete intersection of the cubic primal with another primal. The projected Veronese surface considered here (of order 4) is not a complete intersection of the 10-nodal Segre cubic primal on which it lies. It is, however, when taken with a degenerate quintic surface, the complete intersection of the cubic primal with a cubic primal (a cone). It can be shewn that through a node of the Segre primal there pass six planes lying thereon, consisting of three planes, say ξ, η, ζ, which meet only at the node, and three planes, say α, β, γ, also meeting only at the node, but all meeting each of ξ, η, ζ in lines through the node (these being the lines joining the node to the other 9 nodes of the primal). Further that ∞^3 cubic cones, each a primal, can be described with vertex at the given node, to have the plane ξ as a double plane and to contain the planes α, β, γ. The residual intersection of such a cubic cone with the Segre primal is a projected Veronese surface. We may consider, for the Veronese surface, the plane ξ traced by the trisecants of the surface drawn from the point of it which is the node of the Segre primal. Through any one of these trisecants pass two planes meeting the Veronese surface in conics. The locus of these pairs of planes constitutes the cubic primal cone spoken of. The result may be verified in detail by considering the Segre primal as represented by the quadric surfaces in [3] which have 5 common points; the Veronese surface is represented in this space by a plane.

Ex. 3. We have seen that two cubic primals through the Veronese surface in space [4] intersect further in a $^1\psi^5$ meeting the Veronese surface in a curve $^6c^{10}$. Prove that the cubic primals through this curve meet the Veronese surface again in the ∞^2 conics lying thereon.

Ex. 4. Through any point of $^1\psi^5$ pass two planes each meeting $^1\psi^5$ in a cubic curve. The ∞^1 planes so obtainable generate a $M_3{}^5$ which is the construct dual to the $^1\psi^5$ (cf. Segre, *Rend. Palermo*, II, 1888, p. 45).

Ex. 5. Through the $^1\psi^5$ there pass 5 cubic primals, all Segre cubics with 10 nodes, as the formulae shew. And four of these primals (by Chap. VI, p. 267) meet otherwise in $81 - [24\mu_0{}' + 9(\mu_0{}' - \mu_1{}') + \mu_2{}']$, that is in 1 point. By such a homaloidal system of primals, therefore, we can define a (1, 1) transformation to another space [4]. It can be shewn that the primals of the reverse system are quadrics through an elliptic quintic curve (cf. Semple, *loc. cit.* p. 333).

6. We have, under 4*b*, obtained a ruled surface $^0\psi^4$, the projection of the normal ruled quartic surface in [5], as the residual intersection of two cubic primals passing through a projected Del Pezzo $^1\psi^5[4]$. Now consider the residual intersection of a quadric cone with a cubic primal which contains two planes of the cone, of the same system. Regarding this pair of planes as a degenerate quadric for which $\mu_0 = 2$, $\mu_1 = 0$, $\mu_2 = 0$, $\nu_2 = 0$, $d = 1$, the cone and the cubic meet further in a quartic surface for which (Chap. VI, p. 265) $\mu_0 = 4$, $\mu_1 = 6$, $\mu_2 = 4$, $\nu_2 = 4$; this is the projection of the ruled $^0\psi^4[5]$, whose double point in this case is at the intersection of the two planes; it meets each of the planes in a cubic curve, having a double point at the intersection of the two planes. The cubic primal has 3 nodes on each of these cubic curves, beside a node at the common point of the two planes. The $^0\psi^4$ may be represented, in terms of θ, ϕ, by

$$x/\theta^2\phi = y/\theta\phi = z/\theta(\theta - \phi - 1) = t/(\theta - \phi - 1) = u/1,$$

lying on the quadric $xt - yz = 0$ and the cubic

$$uz(x - y) - yt(y + z) = 0,$$

the two common planes being $x = 0 = y$, $z = 0 = t$. The postulation of the $^0\psi^4$ for primals of order ρ, given by

$$\tfrac{1}{2}\mu_0\rho(\rho + 1) - \rho(p - 1) + p_n + 1 - d,$$

is 14 for quadrics and 27 for cubics. The surface thus lies on one quadric, say $Q = 0$, and on 8 cubic primals given by an equation $(ax + by + cz + dt + eu)Q + \lambda U + \mu V + \nu W = 0$; among these will be the ∞^2 cubic cones obtained by projecting the surface from a general point of itself.

7. Consider a surface Ψ_m for which

$$\mu_0 = \tfrac{1}{2}m(m+1), \quad \mu_1 = \tfrac{2}{3}m(m^2-1), \quad \mu_2 = \tfrac{1}{8}m(m^2-1)(5m-6),$$
$$\nu_2 = \tfrac{1}{12}m(m^2-1)(3m-2).$$

For this surface it can then be deduced that the number of accidental double points, the genus of a prime section, p, and the invariants I, p_n, are given by

$$d=0, \quad p = \tfrac{1}{6}(m-1)(m-2)(2m+3),$$
$$I+4 = \tfrac{1}{24}m(m+1)(15m^2-65m+86),$$
$$12(p_n+1) = \tfrac{1}{2}m(m+1)(3m^2-17m+26),$$

an alternative form for p_n being $\tfrac{1}{24}(m-1)(m-2)(m-3)(3m+4)$, and further that there are $4m+5$ primals F^{m+1} of order $m+1$ passing through this surface, each having $\tfrac{1}{12}m(m+1)^2(m+2)$ nodes at ordinary points of the surface.

Shew that two such F^{m+1} through Ψ_m intersect further in a Ψ_{m+1}, which meets Ψ_m in a curve of order ϵ_0 given by

$$\epsilon_0 = \tfrac{1}{3}m(m+1)(m+2).$$

Shew that Ψ_m is the surface represented by the vanishing of all minors of order m in a matrix of m rows and $m+1$ columns in which each element is linear in the coordinates.

The surface Ψ_1 is a plane, the surface Ψ_2 is the ruled cubic, the surface Ψ_3 is the Bordiga surface $^3\psi^6$, and so on. The *quartic* primals through Ψ_4 are ∞^4 and give a Cremona transformation of the space [4], expressible by four lineo-linear relations connecting the coordinates of the corresponding points. In fact these quartic primals determine, on the Bordiga surface which is the residual intersection of two of them, the net of rational quartic curves corresponding to the lines of the plane on which the Bordiga surface may be represented. In general 4 primals of order m through Ψ_m meet further in $\tfrac{1}{24}m(m-1)(m-2)(m-3)$ points.

8. The surface which is the residual intersection of two cubic primals having a plane in common, has, by the results of Chapter VI, p. 265, for characters

$$\mu_0=8, \quad \mu_1=28, \quad \mu_2=68, \quad \nu_2=28, \quad \omega-1=0, \quad I+4=36, \quad p_n+1=3, \quad d=0;$$

it is a surface $^7\psi^8$, meeting the plane in a general quartic curve, on which curves of the canonical system are given by primes through the plane.

9. If, in space [4], two cubic primals be taken through a quadric surface (in ordinary space, lying in the [4]), for which

$$\mu_0=2, \quad \mu_1=2, \quad \mu_2=2, \quad \nu_2=0,$$

the formulae of the last chapter give, for the residual $^5\psi^7$,

$$\mu_0'=7, \quad \mu_1'=22, \quad \mu_2'=48, \quad \nu_2'=20, \quad p_n'=1, \quad \omega'=0, \quad I'=21,$$

and this meets the quadric surface in a curve $^4c^6$. Each of the two cubic primals contains, in the space of the quadric surface, also a plane; the line of intersection of these planes lies on the $^5\psi^7$. Canonical curves of this surface are given by the solid, or space [3], containing the quadric surface, which meets the $^5\psi^7$ (beside in the $^4c^6$) only on the line of intersection of these two planes. If the equations of the two cubic primals be written $u^3 + u^2P_1 + uP_2 + Q_1\Omega = 0$ and $u^3 + u^2P_1' + uP_2' + Q_1'\Omega = 0$, the surface is the projection of the surface $^5\psi^8$ in space [5], which is the intersection of the three quadrics $u^2 + uP_1 + P_2 + Q_1v = 0$, $u^2 + uP_1' + P_2' + Q_1'v = 0$, $\Omega - uv = 0$ (the coordinates in the [5] being x, y, z, t, u, v). For this surface $p_g = p_n = 1$, $\omega = 1$, $I = 20$; the exceptional line of $^5\psi^7$ lies in the tangent plane of the new surface at the point of projection $(0, 0, 0, 0, 0, 1)$, and is the transformation of this point.

10. The surface which is the residual intersection of two cubic primals which have two planes in common is, by the formulae of Chap. VI, p. 265, of characters

$$\mu_0 = 7, \quad \mu_1 = 20, \quad \mu_2 = 34, \quad \nu_2 = 20, \quad d = 1, \quad p_n = 0, \quad \omega = -2, \quad I = 11,$$

so that the surface is a $^4\psi^7$; it meets each plane in a quartic curve with a double point (at the intersection of the two planes); each of the cubic primals has also 3 double points at ordinary points of each plane quartic curve. The characters are those of the rational surface whose prime sections are represented by plane quintic curves with 2 double and 10 simple base points, $c^5(2^2, 1^{10})$. The formula for the postulation gives two cubic primals passing through the $^4\psi^7$.

The $^4\psi^7[4]$ is evidently the projection of a rational $^4\psi^8[5]$ represented by quintic curves $c^6(2^2, 1^9)$. When the 11 base points of this system of plane curves lie on a cubic curve, the surface $^4\psi^8[5]$ has a double line (Babbage, referred to, p. 246, Ex. 2, above). The reader may prove that the surface represented by the system of plane septimics $c^7(3, 2^8, 1)$, with one triple point, eight double points, and one simple base point, is a $^4\psi^7[4]$ which has a double line (cf. Ex. 2, p. 246).

11. Consider the residual intersection of two cubic primals passing through the aggregate of a quadric surface (in space [3]) and a plane not lying in the space of the quadric surface. For the plane we have $\mu_0 = 1$, $\mu_1 = 0$, $\mu_2 = 0$, $\nu_2 = 0$, and for the quadric surface we have $\mu_0 = 2$, $\mu_1 = 2$, $\mu_2 = 2$, $\nu_2 = 0$; thus as neither of these surfaces has a double point (though they intersect in two points, each therefore an accidental double point of the composite surface) we may take for the composite surface (cf. Ex. 1, p. 262), $\mu_0 = 3$, $\mu_1 = 2$, $\mu_2 = 2$, $\nu_2 = 0$ and $p_n = 1$. Whence we compute for the residual intersection of two cubic primals through the composite surface

$\mu_0' = 6$, $\mu_1' = 14$, $\mu_2' = 20$, $\nu_2' = 12$, $\epsilon_0 = 10$, $\epsilon_1 = 22$, $d' = 2$, $p' = 2$, $p_n' = 0$, $I' = 6$, $\omega' = 3$, which are the characters of the rational surface ${}^2\psi^6$ whose prime sections are represented by the system of plane quartic curves $c^4(2, 1^6)$, with one double point and six simple base points. Using the formula $\delta + 2d = \mu_0(\rho - 1)^2 - \mu_1(\rho - 1) + \nu_2$, we find that a cubic primal through the composite surface has 4 double points not at the intersections of the quadric surface with the plane; and detailed algebra shews that two of these are on the quadric surface and two on the plane. Similarly the formula

$$\tfrac{1}{2}\rho(\rho + 3)\mu_0 - \tfrac{1}{2}\rho\mu_1 + p_n + 1 - 2$$

applied to the composite surface leads to 11 as the number of cubic primals containing this composite surface, which likewise may be directly verified. (Cf. Segre, *Mem. Torino*, xxxix, 1888.)

Many of the simple cases considered here for intersections of cubic primals occur in the papers of J. G. Semple, *Trans. Roy. Soc.* A, ccxxviii, 1929, p. 331; *Proc. Camb. Phil. Soc.* xxv, 1929, p. 145; *Proc. Lond. Math. Soc.* xxxii, 1931, p. 369. For the characters of composite surfaces, a paper by L. Roth, *Proc. Camb. Phil. Soc.* xxix, 1933, p. 88, and a paper by J. G. Semple, *Proc. Royal Irish Academy*, 1933, may be consulted. (Also, Segre, *Atti Torino*, xxii, 1887.)

II. **The surface representing pairs of points of one or two curves.** As an example of an irregular surface, for which the geometric genus p_g exceeds the arithmetic genus p_n, we have already considered (Chap. v, p. 203), after Castelnuovo, the surface of order $2n$ having a n-ple node at each of the 8 associated intersections of three quadric surfaces. We consider now another wide class of irregular surfaces, those namely of which the general point represents a pair of points one on each of two given curves. A particular case is that in which the curves coincide in one curve, and we consider the *ordered* pairs of points thereon. But we may also consider the surface representing the pairs of points of a single curve, without regard to the order of the points of a pair; each point of this surface corresponds then to two points of the surface representing the *ordered* pairs of points. We give an account first of a method followed by De Franchis (*Rend. Palermo*, xvii, 1903; see also Maroni, *Atti...Torino*, xxxviii, 1903); and then, for the surface representing the pairs of points of a single curve without regard to order, of a descriptive method followed by Severi (*Atti...Torino*, xxxviii, 1903).

Consider two curves f, ϕ, of respective genera p, π. We suppose that, to every pair of points, one on each curve, corresponds a definite point of a surface (with coordinates rational in the coordinates of the two points of the curves), and, conversely, to any point of this surface corresponds a unique point of each curve (with coordinates rational in the coordinates of the point of the surface). Such a surface is obtainable by supposing the curves f, ϕ, to be in perfectly general positions, in space of sufficiently high order; then that every point of either curve is joined by a line to every point of the other curve; and then, that a section is taken by a prime of the space, which will meet every joining line in one point. The locus of these points is the surface in question. More analytically, if we suppose the two curves to be plane curves, in general positions in different planes of a space of 4 dimensions, with equations respectively $f(x, y) = 0$ and $\phi(\xi, \eta) = 0$, the surface in question may be that of which a point has (non-homogeneous) coordinates x, y, ξ, η. If A be any definite point of the curve f, and Q a point of the curve ϕ, the point (A, Q) of the surface considered, as Q describes the curve ϕ, will describe a curve on the surface, of genus π; and, for different positions of A on the curve f, we thus obtain a system of curves on the surface, with the property that one such curve passes through an arbitrary point (A, B) of the surface. These form then what has been previously defined as an irrational pencil of curves on the surface (Chap. v, p. 190), the genus of this pencil being that of the curve f, of which every point A determines one such curve; namely p. In particular when $p = 0$, the pencil is rational. In the same way we have another irrational pencil, of curves each of genus p, the genus of this pencil being π, of which every curve is the locus of points (P, B) of the surface, wherein B is a definite point of the curve ϕ, as P describes the curve f. Through any point (A, B) of the surface there passes one curve of each pencil, and conversely any curve of either pencil has just one intersection with any curve of the other pencil. If we consider any (algebraical) curve on the surface, the curve of the first pencil spoken of which passes through an arbitrary point U of this curve, will cut thereon a set of points determined by U, or equally by any other point of the same set, that is an involution; and the genus of this involution will be that of the first pencil, namely p. On the same curve there will likewise be another involution of genus π, determined by curves of the second pencil. Unless both p and π be zero, one of these involutions is irrational; and no irrational involution exists on a rational curve (Vol. ii, p. 136). Thus, unless both p and π be zero, the surface under consideration contains no rational curves. In particular, therefore, the surface contains no exceptional curves of the kind which can arise by transformation of simple points of another surface. In what follows we do not consider cases in which either p or π is zero, for a surface which contains a pencil of rational curves is either rational ($p = \pi = 0$), or reducible to a ruled surface of genus equal to that of the pencil, as we have remarked (Chap. iii, p. 147).

We can at once make statements, based on the representation of the surface as the locus of points (x, y, ξ, η) in space of four dimensions, where $f(x, y) = 0$ and $\phi(\xi, \eta) = 0$, in regard to the everywhere finite simple integrals belonging to the surface (Picard et Simart, *Fonctions algébriques*, I, p. 130), and in regard to the everywhere finite double integrals (*loc. cit.* p. 197). The number of the simple integrals is in fact $p + \pi$, and the number of the double integrals is $p\pi$. In regard to the former, if $\int l_i(x, y)\, dx/f'(y)$ and $\int \lambda_j(\xi, \eta)\, d\xi/\phi'(\eta)$ be finite integrals for the curves f, ϕ respectively $(i = 1, \ldots, p;$ $j = 1, \ldots, \pi)$, it is clear that the sum of these is a finite simple integral for the surface. Conversely, every such integral is necessarily of the form $\int P(x, y, \xi, \eta)\, dx + \Pi(x, y, \xi, \eta)\, d\xi$, where P and Π are rational functions; for this to be everywhere finite, the integral $\int P(x, y, \xi, \eta)\, dx$, for every definite (ξ, η) satisfying the equation $\phi(\xi, \eta) = 0$, must be everywhere finite on $f(x, y) = 0$; this integral must then be expressible as a sum of the integrals $\int l_i(x, y)\, dx/f'(y)$, each multiplied by a number independent of (x, y), and hence, here, depending on (ξ, η), namely by a rational function of (ξ, η). Any infinity of such a multiplier, regarded as a function of (ξ, η), cannot be compensated by infinities arising in the part $\int \Pi(x, y, \xi, \eta)\, d\xi$, since this must be finite everywhere in regard to (ξ, η). All the multipliers in question must then be absolute constants. In the same way the part $\int \Pi(x, y, \xi, \eta)\, d\xi$ must be a sum of constant multipliers of the everywhere finite integrals for the curve $\phi(\xi, \eta)$. We thus obtain $p + \pi$ everywhere finite simple integrals for the surface, and no more. A very similar argument shews that the general everywhere finite double integral for the surface is a sum of constant multipliers of the $p\pi$ integrals

$$\iint l_i(x, y)\, \lambda_j(\xi, \eta)\, dx\, d\xi / f'(y)\, \phi'(\eta).$$

From this it follows, defining the geometric genus p_g of the surface, in the manner originally proposed by Clebsch, as the number of linearly independent everywhere finite double integrals, that we have $p_g = p\pi$; and, defining the canonical curves of the surface as those for which the integrand of the double integral vanishes (that is, by the number of adjoint surfaces of appropriate order), we see that such a canonical curve is here equivalent on the surface to the composite curve consisting of $2p - 2$ curves of the irrational pencil of which each curve arises from a particular point of the curve f, taken with $2\pi - 2$ curves of the irrational pencil of which each curve arises from a particular point of the curve ϕ. From this last fact we can obtain the grade of the canonical system, and the genus of a canonical curve. For, if C, Γ denote respectively two such aggregates of respectively $2p - 2$ and $2\pi - 2$ curves from these two pencils, and C', Γ' two other such aggregates, since no two curves of the same pencil have an intersection, and any curve of

either pencil has one intersection with any curve of the other pencil, this grade $p^{(2)}$ of the canonical system, $(C+\Gamma, C'+\Gamma')$, is equal to $(C, \Gamma') + (C', \Gamma)$, namely to $2(2p-2)(2\pi-2)$. For the genus, $p^{(1)}$, of a curve of the canonical system, if we use the formula $p_1 + p_2 + i - 1$ for the genus of a curve consisting of two curves of genera p_1 and p_2 with i intersections, we have, first, for a composite curve consisting of $2p-2$ curves of the first pencil, the genus $(2p-2)\pi - (2p-3)$, and, similarly, for the aggregate of $2\pi - 2$ curves of the second pencil, the genus $(2\pi-2)p - (2\pi-3)$; also, the total number of intersections of the former composite curve with the latter is $(2p-2)(2\pi-2)$. Whence, the genus of a canonical curve is given by
$$p^{(1)} = (2p-2)\pi - (2p-3) + (2\pi-2)p - (2\pi-3) + (2p-2)(2\pi-2) - 1,$$
which is $2(2p-2)(2\pi-2)+1$. This verifies the relation previously (Chap. v, p. 223) remarked, $p^{(2)} = p^{(1)} - 1$. From the same point of view we can obtain the invariant I of the surface, if we recall the formula previously obtained (Chap. v, p. 190), $I + 4 = \Delta + (2p-2)(2p'-2)$, for a surface on which there exists an irrational pencil, of genus p', of curves of genus p, among which there are Δ curves with actual double points. Applying this to either of the pencils on the surface under consideration, with the assumption that any point of this surface corresponds to only one point of both curves, so that $\Delta = 0$, we have $I + 4 = (2p-2)(2\pi-2)$. The invariant ω of the surface, as there are no exceptional curves obtainable by transformation from simple points of another surface, is (see Chap. v, pp. 184, 228) given by $\omega = p^{(1)}$. Whence, Noether's relation $I + 4 + \omega - 1 = 12(p_n + 1)$ (*ibid.* p. 185), gives here
$$p_n + 1 = (p-1)(\pi-1),$$
for the arithmetic genus.

And we thus prove that the number of everywhere finite simple integrals belonging to the surface is equal to $p_g - p_n$. This is a very noticeable result, the proof that this is always the case, for any surface, having been completed only in 1905, and with the help of results not employed in this earlier theory. Summarising the results obtained we may say that
$$\tfrac{1}{8}(\omega-1) = \tfrac{1}{4}(I+4) = p_n + 1 = (p-1)(\pi-1), \quad p^{(1)} = \omega, \quad p_g = p\pi.$$
It has been assumed, as stated, that $p > 0$ and $\pi > 0$; if either p or π be 1, the results remain, as may be proved directly.

We pass now to consider the surface of which any point represents a pair of points of a given curve, without regard to the order of these points. We study this surface, ψ, in association with the surface, ψ_0, which represents the pairs of points of the same curve when the order of the points of a pair is regarded. Thus ψ_0 is the surface which we have just discussed in the case when the two curves f, ϕ are replicas of one another (or identical). Any general point of the surface ψ thus corresponds to two points of the surface ψ_0, or these surfaces (ψ, ψ_0) are in (1, 2) birational correspondence. But there is a curve c on the surface ψ which represents the pairs of coincident points of the fundamental curve, f, and there is a curve c_0 on the surface ψ_0 likewise representing the pairs of coincident points of the curve f; and the curves c, c_0 are birationally equivalent with one another, and with the curve f. Upon the surface ψ there will be a curve representing the couples of f which consist of a general point of f taken with all other points of f; this curve, γ, is birationally equivalent with f; and such curves γ will constitute a system, of which two pass through any general point of ψ (or a system of *index* 2), while any two of the curves have one point of intersection. In another point of view, thinking of f as a curve which does not lie in a plane, any curve γ, of this system on ψ, corresponds to the cone of chords projecting f from any point of itself; any chord of this cone belongs to two such cones, and any two such cones have one generator in common. The common generator of two such cones, with vertices at O and O', becomes the tangent line of the curve f at O when O' moves to coincidence with O. Thus two consecutive curves of the system (γ) intersect on the coincidence curve c. In other words all the curves γ touch c. To every curve γ on ψ corresponds two curves on the surface ψ_0, intersecting one another on the curve c_0, at the point which corresponds to the point of contact of γ with c.

To any linear system $|C|$ on the surface ψ, whereon the co-ordinates of a point are rational functions of the coordinates of either of the two corresponding points of the surface ψ_0, there corresponds a linear system $|C_0|$ on the surface Ψ_0, there being two points of a curve C_0 arising from a single point of C. And, if π, π_0 be the genera respectively of C and C_0, and so the number of coincidences on C_0 of a pair of points arising from the same point of C, it follows, from the Zeuthen correspondence formula (Chap. I, p. 19) that $2(2\pi-2)=2\pi_0-2-s_0$; the number s_0 is the number of intersections of the curve C_0 with the coincidence curve c_0 on ψ_0, and therefore also the number of intersections of C, on ψ, with the coincidence curve c on ψ; thus we write s for s_0. If ν, ν_0 be the respective grades of the systems $|C|$, $|C_0|$, we have $\nu_0=2\nu$; and if d, d_0 be the respective numbers of double points of linear pencils of curves contained in $|C|$ and $|C_0|$, the invariants I, I_0 of ψ and ψ_0 are such that $I=d-\nu-4\pi$, $I_0=d_0-\nu_0-4\pi_0$. To a double point of a curve of $|C|$ which does not lie on the coincidence curve c, correspond two double points of the corresponding curve of $|C_0|$, not lying on c_0; but to a double point of a curve C_0 which lies on c_0, corresponds a contact of the corresponding curve C with the curve c; and conversely. Thus we can express d_0 in terms of d; for a linear pencil of curves in the system $|C|$ determines, on the coincidence curve c of ψ, a linear series $g_1{}^s$, of sets of s points, of freedom 1; and there are $2(s+p-1)$ coincidences on c of pairs of a set of this linear series, where p is the genus of c, and therefore of the fundamental curve f; each of these coincidences determines, in the simplest case, a double point of a curve of the corresponding linear pencil of curves contained in $|C_0|$. Wherefore we have $d_0 = 2(d+s+p-1)$. From this then we deduce

$$2(I+4)-(I_0+4) = 2d - d_0 - 2\nu + \nu_0 - 2[2(2\pi-2)-(2\pi_0-2)]$$

$$= -2(s+p-1)+2s = -2(p-1);$$

we have proved, however, above, that $I_0 + 4 = 4(p-1)^2$. Wherefore,

$$2(I+4) = 4(p-1)^2 - 2(p-1), \quad \text{or} \quad I+4 = (p-1)(2p-3).$$

We consider now the adjoint systems of curves, on ψ and ψ_0, of the systems $|C|$ and $|C_0|$, denoting these adjoint systems respectively by $|C'|$ and $|C_0'|$. We have seen previously, in connexion with the proof of the Zeuthen correspondence formula quoted above, that a canonical set on any curve C_0 contains, beside the set directly corresponding to a canonical set on C, also the set of coincidences on C_0, where two points of C_0 corresponding to a point of C coincide with one another. Whence, if $t|C'|$ denote the linear series on ψ_0 arising by direct transformation of the system $|C'|$ on ψ, we have $|C_0'| = t|C'| + c_0$, it being assumed, to save explanations, that there are no exceptional curves on ψ or ψ_0. Whence, the canonical systems on ψ, and ψ_0, which we may define by $K = |C'-C|$ and $K_0 = |C_0'-C_0|$, are also such that $|K_0| = t|K| + c_0$, where $t|K|$ denotes the direct transformation of $|K|$. From this equation we can infer an equivalence for the curves of $|K|$ which gives us much information in regard thereto. For, first, there are curves γ on ψ, each representing couples of f of which one point is fixed, each of which corresponds to two of the curves of ψ_0 which have a similar definition; and we have seen that a degenerate curve of K_0 consists of $2(2p-2)$ curves of this latter kind; this is then the direct transformation of $2p-2$ curves γ on ψ. And second, the direct transformation of the curve c on ψ is twice the curve c_0 on ψ_0. We may then regard a curve of $|K_0|-c_0$ as equivalent to the direct transformation of $(2p-2)$ curves γ on ψ less half the curve c on ψ. Hence we may regard any curve of $|K|$ as equivalent to $(2p-2)$ curves γ, less half the curve c, and write $K \equiv (2p-2)\gamma - \frac{1}{2}c$.

Hence, if ξ denote the virtual grade of the curve c, and s' the number of intersections of c with a curve of the adjoint system $|C'|$, recalling that the curves γ all touch c, we infer that

$$s'-s = 2(2p-2)-\tfrac{1}{2}\xi.$$

This, however, being the number of intersections of c with a curve of the canonical system of ψ, is equal to $2p-2-\xi$ (Chap. v, p. 222), provided it be assumed that the curve c has no multiple points, so that its complete genus is the same as its actual genus. Whence we infer that $\xi = -2(2p-2)$, and $s'-s=3(2p-2)$.

In passing we remark that the virtual grade ξ_0, of c_0, on ψ_0, is given by the fact that the virtual grade of $t(c)=2\xi$, that is the virtual grade of $2c_0=2\xi$, so that $4\xi_0=2\xi$, leading to $\xi_0 = -(2p-2)$. We can also find this directly on ψ_0, since $2(2p-2)=$ number of intersections of a canonical curve on ψ_0 with c_0, $=2p-2-\xi_0$.

From the equivalence for the canonical curve on ψ, we can find the invariant ω for the surface, assuming that $\omega-1$ is the grade of the canonical curves. For such curve being $\equiv (2p-2)\gamma - \frac{1}{2}c$, and because $(\gamma, \gamma)=1$, $(\gamma, c)=2$, $(c, c)=\xi$, we have

$$\omega-1=(2p-2)^2-2(2p-2)-\tfrac{1}{4}\xi=(p-1)(4p-9).$$

We have found however that $I+4=(p-1)(2p-3)$. Hence $p_n+1=\frac{1}{12}[I+4+\omega-1]=\frac{1}{2}(p-1)(p-2)$. A somewhat different method of obtaining $\omega-1$ is to remark that this is the difference, $\zeta'-\zeta$, of the canonical numbers, ζ' and ζ, of the adjoint system $|C'|$ and the system $|C|$, as is clear because the canonical number of a curve is the number of its intersections with a canonical curve. If, as before, C_0 be the curve on ψ_0 obtained by direct transformation of the curve C on ψ, and ν, ν_0 be the grades of $|C|$ and $|C_0|$, and ζ, ζ_0 their canonical numbers, we have

$$\zeta = 2\pi - 2 - \nu, \quad \zeta_0 = 2\pi_0 - 2 - \nu_0, \quad 2(2\pi - 2) = 2\pi_0 - 2 - s, \quad \nu_0 = 2\nu;$$

these lead to $2\zeta = \zeta_0 - s$. If this result be applied to the adjoint system $|C'|$, we have $2\zeta' = \zeta_1 - s'$, where ζ_1 is the canonical number of the curve obtained by direct transformation of C'; this direct transformation, we have remarked, is $C_0' - c_0$, and we have found that the canonical number of c_0 is $(2p-2-\xi_0$ or) $2(2p-2)$. Whence $2\zeta' = \zeta_0' - s' - 2(2p-2)$. But also $\zeta_0' - \zeta_0$, or $\omega_0 - 1$, was proved to be $8(p-1)^2$, and also $s'-s=3(2p-2)$. Wherefore

$$2(\omega-1)=2(\zeta'-\zeta)=\zeta_0'-\zeta_0-(s'-s)-2(2p-2)$$
$$=8(p-1)^2-5(2p-2),$$

giving the equation $\omega-1=(p-1)(4p-9)$ found above.

This argument may be regarded as an introduction to the general investigation of the relation between the invariants of two surfaces in multiple correspondence (see Note, p. 294, below).

To obtain the invariant p_g of the surface ψ, we investigate the everywhere finite double integrals belonging to the surface. If the fundamental curve be regarded as a plane curve, given by $f(x,y)=0$, of which the general everywhere finite integral is u_i^x, of the form

$\int \phi_i(x) dx/f'(y)$, the general everywhere finite double integral of the surface, as in the case of the surface ψ_0, is $\iint \Sigma A_{ij} du_i{}^x du_j{}^\xi$, for $i, j = 1, \dots, p$, namely is $\Sigma A_{ij} \iint F_{ij} dx d\xi / f'(y) f'(\eta)$, where, subject to $f(\xi, \eta) = 0$, we have $F_{ij} = \phi_i(x) \phi_j(\xi) - \phi_i(\xi) \phi_j(x)$. But as, in this case, the points (x, y), (ξ, η) may be taken in either order, we must have $A_{ji} = -A_{ij}$, and in particular $A_{ii} = 0$. Hence $p_g = \frac{1}{2} p(p-1)$. And, with the value found above for p_n, this leads to

$$p_g - p_n = \tfrac{1}{2} p(p-1) - \tfrac{1}{2}(p-1)(p-2) + 1,$$

which is p. This again is the same as the number of everywhere finite simple integrals belonging to the surface, for it may be proved as before that these integrals are

$$\int \phi_i(x) dx/f'(y) + \int \phi_i(\xi) d\xi/f'(\eta).$$

We may interpret the function F_{ij} as follows: In every set of the pencil of canonical sets given by the equation $\phi_i(x) + \lambda \phi_j(x) = 0$, there are $(p-1)(2p-3)$ couples. The locus of the points of the surface ψ representing these couples is a canonical curve on ψ. For from $\phi_i(x) + \lambda \phi_j(x) = 0$ and $\phi_i(\xi) + \lambda \phi_j(\xi) = 0$, we have

$$\phi_i(x) \phi_j(\xi) - \phi_j(x) \phi_i(\xi) = 0.$$

For instance, if the curve f be the intersection of a quadric and a cubic surface, a pencil of canonical sets is given on the curve by planes through an arbitrary line; and the points of the surface ψ representing the 15 couples of such a set describe such a curve. Thus, more generally, the canonical system of curves on ψ, in this case, represents the chords of the sextic curve which belong to a linear complex.

The values found above for a general value of p become, for $p = 1$, respectively $I = -4$, $\omega = 1$, $p_n = -1$, $p_g = 0$, which (since, for a ruled surface of genus p, we have $I = -4p$, $\omega - 1 = -8(p-1)$) are the values appropriate to an elliptic ruled surface. For $p = 1$ the surface ψ must contain an elliptic pencil of rational curves (there is an involution on a plane cubic curve determined by lines through an arbitrary point of the curve), and is hence (cf. Chap. III, p. 147) reducible to an elliptic ruled surface. The general formulae have been found on the assumption $p > 0$; but, in fact, for $p = 0$ they lead to the values $I = -1$, $\omega = 10$, $p_n = 0$, $p_g = 0$, appropriate to a rational surface, in accordance with the fact that a rational symmetric function of the parameters θ, θ' of any two points of a rational curve, is a rational function of the sum and product, $\theta + \theta'$ and $\theta \theta'$, of these. In general, however, the preceding theory requires further examination when the fundamental curve is hyperelliptic with $p > 1$, since there exists, then, a rational series $g_1{}^2$ of pairs of points on the curve, giving a rational curve on the surface ψ; while we have assumed that the surface has no (rational) exceptional curves.

Ex. Examine the canonical curves for the surface, in space (x, y, z), expressed by $x = t_1 + t_2$, $y = t_1 t_2$, $z = s_1 + s_2$, where $s_1{}^2 = F(t_1)$, $s_2{}^2 = F(t_2)$, in which F is a polynomial of order 6. Shew that for this surface $p_g = 1$, $p_n = -1$. Also shew that for the surface expressed by $x = t_1 + t_2$, $y = t_1 t_2$, $z = s_1 s_2$, we have $p_g = 1$, $p_n = 1$.

We pass now to a more geometrical investigation (due to Severi, as has been said). Consider the canonical curve ${}^p c^{2p-2}[p-1]$, of order $2p-2$ and genus p, in space $[p-1]$. The ∞^2 chords of this curve are representable by a surface, lying in a space of dimension $\frac{1}{2}p(p-1)-1$, wherein any point has for coordinates the line co-ordinates of a line of the space $[p-1]$. The general problem is to find the order of this surface, which we denote by $p^{(2)}$ (with a subsequent justification of the notation), and the genus of its prime section, which we denote by $p^{(1)}$.

The character of the work will be sufficiently represented by taking the case when $p=4$, the canonical curve being the complete sextic intersection of a quadric and a cubic surface in ordinary space. The lines of this space are then represented by the points of a quadric Ω in space $[5]$; the surface to be examined is the surface lying on Ω which represents the chords of this sextic curve. The order of this surface is the number of its points lying on two primes of $[5]$; hence it is the number of chords of the curve ${}^4 c^6 [3]$ which belong to two arbitrary linear complexes; hence also it is the number of these chords which meet two arbitrary lines; and thus finally is the sum of the number of chords of the sextic curve which lie in an arbitrary plane and the number of chords which pass through an arbitrary point. For a curve of order n and genus p this would give $\frac{1}{2}n(n-1)+\frac{1}{2}(n-1)(n-2)-p$, or $(n-1)^2-p$; for $n=2p-2$ this is $(p-1)(4p-9)$, which for $p=4$ is 21. A prime section of the surface consists of points representing the chords of the sextic curve which belong to a linear complex; in particular chords meeting an arbitrary line; these form a ruled surface. These chords can be arranged as sets of a "linear series", g_1^{15}, of which one set consists of the chords in a plane through the arbitrary line. For a curve of order n and genus p one of these sets would consist of $\frac{1}{2}n(n-1)$ chords; a coincidence of two chords of such a set can arise only in a plane through the given line which contains a tangent line of the curve, and then there are $n-2$ such coincidences, the lines which join the point of contact, of the plane with the curve, to its remaining $n-2$ intersections. The total number of coincidences of pairs of lines in the same set of the linear series $g_1^{\frac{1}{2}n(n-1)}$ is thus $(n-2)(2n+2p-2)$. Thus, if q be the genus of the ruled surface formed by the chords of the curve which meet the given line, we have $2[\frac{1}{2}n(n-1)]+2q-2=(n-2)(2n+2p-2)$, leading to $q=\frac{1}{2}(n-2)(n-1+2p-2)$, which is $(p-2)(4p-5)$ when $n=2p-2$; this exceeds by 1 the number $(p-1)(4p-9)$ found for the order of the surface; and for $p=4$ it is equal to 22.

Ex. The chords of the sextic curve that meet a line may also be arranged as sets of 6, all passing through a point of the line, forming thus a linear series g_1^6. Determine the number of coincidences of lines of the same set, and hence find the genus q.

Let the curve under consideration, of which we have taken the sextic intersection of a quadric and a cubic surface as typical, be denoted by σ; and the ruled surface formed by the chords of the curve σ which meet a given arbitrary line, l, be denoted by ρ. The lines of ρ may be represented, as we have said, by points, in another space, forming a curve on the surface which represents all the chords of the given curve σ. It is therefore clear what is meant if we speak of a canonical set, K, of lines of the ruled surface ρ. We may now consider a multiple correspondence $(2, n-1)$ between the points of the curve σ, and the lines of the ruled surface ρ, that namely in which, to each point P of σ correspond the $n-1$ (or 5) lines of ρ which pass through P (that is, join P to the $n-1$ other points in which the plane Pl meets σ), and to each generator, G, of ρ, correspond the two points in which G meets σ. There is, on σ, a linear series g_1^n (or g_1^6), of sets of n points all lying in a plane through l; of this series the Jacobian set is formed by the points of σ whereat the tangent line meets l. This Jacobian set is then equivalent, on σ, to two sets of g_1^n together with a canonical set of σ, or, as σ is a canonical curve, to three times a canonical set of σ. Let a set of generators of ρ which lie in a plane (through l) be denoted by A; to a canonical set on σ, namely a set of coplanar points, corresponds in virtue of the correspondence spoken of, a set of $n(n-1)$ lines, forming a set $2A$ on the ruled surface ρ. Thus the points of σ whereat the tangent lines meet l give rise, by this correspondence, to a set on ρ equivalent to $6A$. The lines on ρ to which they give rise consist however of the $(2n+2p-2$, or 18) tangents (say B) of σ which meet l, together with the $18(n-2)$ chords of σ through a point of contact of such a tangent which lie in a plane through l; these latter chords, however, are the double elements of the various sets A, and are thus equivalent, on ρ, to a set $2A+K$. Thus we infer that on ρ we have the equivalence $6A \equiv B+2A+K$, or $K \equiv 4A-B$, where B denotes the lines of ρ which touch σ.

If, however, C denote a set of (6) chords of σ through any point of l, we can shew that, on ρ, $B \equiv 2A-2C$. For the lines of ρ which meet an arbitrary line form a set equivalent to the aggregate $A+C$ (of a set A lying in a plane through l, and a set C passing through a point of l); thus the lines of ρ which touch the quadric surface containing the curve σ (all such tangents forming a quadratic complex), being equivalent on ρ to twice a set belonging to a linear complex, are equivalent to a set $2A+2C$. On the other hand, the lines of ρ which touch this quadric surface consist of the set B touching the curve σ, and the generators of this quadric surface through the two points where it is met by l; the two generators through one of these points constitute the whole of a set C of chords of σ through this point; but, beyond touching the quadric surface,

they lie upon it, and are, therefore, we assume, equivalent to a set
2*C*. On the whole, therefore, we reach the conclusion

$$2A + 2C \equiv B + 4C,$$

as was stated.

Hence a canonical set of lines on ρ satisfies the equivalence
$K \equiv 2A + 2C$. Thus, on the surface in space [5] which represents the
chords of the curve σ, whereof a prime section represents the
generators of a ruled surface such as ρ (or more generally the chords
of σ which belong to a linear complex), the canonical series on a
prime section consists of sets, $2A + 2C$, each the double of a set,
$A + C$, of the characteristic series of the prime section. The canonical
system of curves of the surface thus cuts a prime section in sets
equivalent with a characteristic set. Wherefore the canonical
system of the surface contains the prime sections, there being no
exceptional curves on the surface.

That the canonical system is not more ample than, and is therefore equivalent with, the system of prime sections, follows from the
analytical theory given above; for it was shewn, from consideration
of the everywhere finite double integrals of the surface, that the
canonical system of the surface is linearly dependent on curves of
which one is the locus of points representing the pairs of points, on
the original curve, which can be chosen from a set of a series
g_1^{2p-2} of freedom 1, chosen from the canonical series (in the notation
employed, curves formed from functions $\phi_i(x)\,\phi_j(\xi) - \phi_i(\xi)\,\phi_j(x)$).
In our case, such curve of the canonical system represents the pairs
of points of the curve σ whose joining line meets an arbitrary
general line; thus the canonical system consists of curves representing pairs of points, on chords of σ, which belong to a general
linear complex (whose equation is linearly dependent from that
of six special linear complexes); and such linear complex gives a
prime section of the representative surface.

The geometric proof is however instructive. Denote by γ a curve
on the surface which represents the generators of a cone projecting
the curve σ from a point of itself, and by c the curve which, similarly, represents the tangents of σ. This curve is touched by all the
curves γ, in accordance with the fact that the osculating plane at
any point of the curve σ is tangent to the cone projecting σ from
this point. It can be shewn that the curves γ are plane quintic
curves, of genus 4, each with two double points, the joining line of
which contains the point where the curve γ touches c; and further
that the surface ψ contains two conics (corresponding to the trisecants of the curve σ), while through any point of either conic pass
three curves γ, each having a double point thereat. Two curves γ
have a point in common, and thus lie in a prime. We have seen that
any prime section of the surface ψ is a canonical curve; we can

estimate the freedom of canonical curves which contain two of the curves γ, by their intersections with a third curve γ, and hence infer that the only such canonical curve is the prime section through the two curves γ. From this it can be proved that the canonical curves are all prime sections.

We can also prove that a canonical curve on the surface ψ is equivalent to $(2p-2$ or$)$ 6 curves γ, less a curve equivalent to half the curve c (with resulting order $6.5-8$ or 22). For consider a set of 6 cones projecting the curve σ from the 6 points of a special coplanar set of points thereon. The vertices of these cones project from any other point of σ into a canonical set of generators of the cone whose vertex is this other point. Hence, on the surface ψ, 6 curves γ which form a special set cut, on an arbitrary curve γ, a set of 6 points equivalent with (the 6 intersections of this γ with a conic in its plane through its two double points, and hence with) the set constituted by 5 collinear points of this γ, taken with the point where this γ touches the curve c; that is equivalent to the intersection of this γ with the aggregate of a prime section and a curve equivalent to half the curve c. We may assume however that two curves (of the same order), which cut equivalent sets on every curve of an algebraic system of curves, are equivalent, or are equivalent save for curves which are fundamental for the algebraic system, that is have no variable intersections with these*.

Particular consequences are, (a), for the virtual grade of the curve c, we have $2\pi-2-\xi$ equal to the number of its intersections with a prime, equal to the rank of the curve σ, which is $3(2p-2)$, so that $\xi=-2(2p-2)$, as was previously found; and (b), the order of the surface ψ is $p^{(2)}=(2p-2)^2-2(2p-2)-p-1$ or $(p-1)(4p-9)$, equally found before on the supposition that ψ has no exceptional curves. For $p=4$ this gives $p^{(2)}=p^{(1)}-1=21$.

For the invariant I we may proceed directly on the surface ψ, considering a pencil of curves obtained by primes through a fixed solid, or space [3], which meets the quadric Ω in two planes. It can be shewn that there are 96 prime sections with a double point, beside 6 prime sections which degenerate into a curve γ and a residual curve which meets this in 4 points, so leading to 24 further double points. Hence we infer

$$I=96+24-p^{(2)}-4p^{(1)}=120-21-88, =11;$$

and, from Noether's equation, $I+p^{(1)}=12p_n+9$, we can infer $p_n=2$. For, the prime sections, of the surface ψ, spoken of, are those

---

* Severi, "Alcune relazioni di equivalenza", *Atti...Veneto*, LXX, 1911, p. 380, enunciates the general result: If on a manifold of k dimensions, M_k, two manifolds M_{k-1}, of $k-1$ dimensions, cut equivalent manifolds on the M_h of an algebraic system Σ, which is ∞^{k-h}, where $h \leqslant k-1$, and is of index $\nu \geqslant 1$, then they are equivalent or differ by manifolds fundamental for Σ.

representing the ruled surfaces ρ formed by chords of the curve σ which meet a variable line of a flat pencil of lines, say (O, ϖ), formed of lines in the plane ϖ which pass through a point O of this plane. Now, through O there pass 96 bitangent planes of the curve σ (the number of bitangents of the projection of σ on to a plane, which is $d + \frac{1}{2}(n' - n)(n' + n - 9)$, where $d = 6$, $n = 6$, $n' = 30 - 2d = 18$, Vol. v, Chap. viii), each giving a double generator in the particular ruled surface ρ associated with the line, of the pencil, in which the bitangent plane meets ϖ. Again, the ruled surface associated with one of the six lines OP, where P is a point of the curve σ lying on ϖ, breaks up into the cone γ projecting σ from P, and other chords of σ meeting OP, of which 4 belonging to this cone γ are obtained by the plane joining O to the tangent line of σ at P.

The results obtained by the geometrical reasoning thus agree with those which were found more analytically above.

Remark. Another irregular surface extremely worth study, is the surface $^{46}\psi^{45}[9]$, of order 45, with prime sections of genus 46, in space [9], whose points are those representing the lines lying on a general cubic primal in space [4]. On this also, the canonical curves are those given by the primes of its space. For it $p_g = 10$, $p_n = 5$, $I = 23$, $\omega = 46$. Of these numbers, p_g and p_n are the same as for the surface representing the chords of a curve of genus 5, for which, however, $I = 24$, $\omega = 45$. For the system of lines of a cubic primal in space [4] which has $5 - p$ independent nodes, with $0 \leqslant p < 5$, all the numbers $p_g = \frac{1}{2}p(p-1)$, $p_n = \frac{1}{2}p(p-3)$, $I = (p-1)(2p-3) - 4$, $\omega = (p-1)(4p-9) + 1$, are the same as for the system of chords of a curve of genus p. See Fano, *Atti...Torino*, xxxix, 1904, p. 778; *Ann. d. Mat.* x, 1904, p. 251; *Rend....Lombardo*, xxxvii, 1904, p. 554.

Note on the multiple correspondence of two surfaces. We have had occasion to consider the (1, 2) correspondence between the surfaces ψ, ψ_0, in the preceding. We quote now the general results, first for an $(1, \alpha')$ correspondence between two surfaces ψ, ψ', and then for an (α, α') correspondence (Severi, "Sulle relazioni...di due superficie in corrispondenza algebrica", *Rend....Lombardo*, xxxvi, 1903). In the case of an (α, α') correspondence, we denote *the branch curve* on ψ, locus of points for which two of the α' corresponding points of ψ' coincide with one another, by b; and the coincidence curve on ψ', which corresponds to this, by c', with a similar notation, b' for the branch curve on ψ', c for the corresponding curve of coincidence on ψ. Unless the contrary is stated we suppose that there is no point of either surface transforming, by the relations which express the correspondence, into a curve of the other surface, which is stated by saying that, unless remarked, there is no *fundamental point of the correspondence* on either surface. We denote the *unreduced* canonical system on ψ by K; this is of the form $K_0 + \epsilon$, if ϵ denote the exceptional curves on ψ and K_0 the reduced canonical system; the n-canonical system will then be $K_{n,0} + n\epsilon$, where $K_{n,0}$ denotes the reduced n-canonical system; this we denote by K_n. Similarly for canonical systems on ψ'. We recall the theorems previously (Chap. I, p. 19) considered for two curves C, C' in $(1, \alpha')$ correspondence: (a), on C', a canonical set is equivalent with the set obtained by direct transformation of a canonical set on C, augmented by the set of coincidence points on C'; (b), on C, the aggregate of α' canonical sets, augmented by the set of branch points on C, corresponds to a canonical set on C'. We have similar results in the $(1, \alpha')$ correspondence of the surfaces ψ, ψ', for the canonical systems of curves; namely, using the symbol t for transformation, (a), on ψ', $K' \equiv t(K) + c'$; (b), on ψ, $K_{\alpha'} + b \equiv t(K')$; hence in case of an (α, α') correspondence, we have, *on* ψ', $K_{\alpha'} + b' \equiv t(K) + c'$. This latter is true if there be, on ψ', fundamental curves of the correspondence, say ϕ', when we put $c' + \phi'$ in place of c'. Neglect this possibility, as before, and denote by ϖ, θ, κ respectively the genus, canonical number, and number of cusps of the branch curve b, putting $\varpi - 1$, θ and κ zero when there is no branch curve, with similar notation for the branch curve b' on ψ'. Then we have, for the invariants ω, I and the arithmetic genus p_n,

$$\alpha(\omega' - 1) + \varpi' - 1 + \tfrac{3}{2}\theta' = \alpha'(\omega - 1) + \varpi - 1 + \tfrac{3}{2}\theta,$$

$$\alpha(I' + 4) + 2(\varpi' - 1) - \kappa' = \alpha'(I + 4) + 2(\varpi - 1) - \kappa,$$

$$12\alpha(p_n' + 1) + 3(\varpi' - 1) + \tfrac{3}{2}\theta' - \kappa' = 12\alpha'(p_n + 1) + 3(\varpi - 1) + \tfrac{3}{2}\theta - \kappa,$$

and, in particular, for a $(1, \alpha')$ correspondence, putting $\varpi' - 1$, θ' and κ' zero, we have

$$\omega' - 1 = \alpha'(\omega - 1) + \varpi - 1 + \tfrac{3}{2}\theta, \quad I' + 4 = \alpha'(I + 4) + 2(\varpi - 1) - \kappa,$$

which can be modified, by addition of terms, to include the case when there are fundamental points for the correspondence.

III. **On complete sections of a non–singular primal.** It was suggested as probable by Cayley in 1859 (*Papers*, IV, p. 455) that on a surface in space of three dimensions, represented by the vanishing of a perfectly general polynomial in the coordinates, of order > 3, there exist no algebraical curves other than complete intersections of this surface each with one other surface. For consideration of this result, see Noether, "Zur Grundlegung, u.s.w.", *Berlin. Abh.* 1882, §§ 11, 12). It was then proved by Klein, in 1872 (*Ges. Abh.* I, p. 153) that upon a quadric primal of sufficient generality, in space $[r]$, $(r > 3)$, any algebraic manifold of one less dimension, that is, of dimension $r - 2$, is the complete intersection of the quadric with another primal; the necessary generality of the quadric, which may be a cone, may be briefly stated by saying that its equation must involve 5 (homogeneous) coordinates at least. It was next

proved by Fano, in 1904 (*Atti...Torino*, XXXIX, 1904, p. 613), that
a general cubic primal in space [4] contains only surfaces each a
complete intersection with another primal, the surfaces considered
in the first instance being supposed to cut every one of the ∞^2 lines
lying on the cubic primal in the same number of points. (The
theorem proved however is more general than here stated; for
instance, the ruled surface of order 15, formed by the lines of the
primal which meet any one of the lines, is a complete intersection.
And the primal may have nodes. Cf. Ex. 2, p. 278, above.) It was then
proved by Severi, in 1906 (*Rend. Lincei*, XV, p. 691), that *on any
primal, which is without multiple points, in space* [*r*], *if* $r \geqslant 4$, *every
algebraic locus of dimension* $r-2$ *is a complete intersection of the
primal with one other primal*. This was extended in 1909 by Fano
(*Atti...Torino*, XLIV, p. 633) to apply, *in general*, to a (non-singular)
manifold M_k, of k dimensions, which is the complete intersection of
$r-k$ primals in space [*r*], with $k \geqslant 2$ (excluding *rational* surfaces if
$k=2$), any M_{k-1} thereon being the complete intersection with one
further primal. And, in 1915, it was proved by Severi (*Ann. d.
Mat.* XXIV) that on a Grassmannian (the locus of points, in space
of appropriate dimension, each representing a linear space [*k*] of
space [*r*]), of dimension *t*, where $t=(k+1)(r-k)$, any algebraic
locus of dimension $t-1$ is the complete intersection with one primal.
In particular, when $r=3$, $k=1$, this last gives Klein's theorem for a
general quadric in [5].

In the present place we limit ourselves to giving a summary
account of Severi's proof of the theorem for surfaces, without
double points, which can exist upon a non-singular primal in space
[4]; it is to be shewn that such a surface is the complete intersection
of the primal with another primal.

Denote the primal by V, and the surface in which this is met by
a prime ϖ by (ϖ, V). On V there is a surface, denoted by ψ. We
limit ourselves to the case when ψ has no double points. We shew,
first, as a consequence of the fact that V is without double points,
that the curve (ϖ, ψ), in which ψ is met by a prime ϖ, is the complete intersection of the surface (ϖ, V), on which it necessarily lies,
with a surface, lying in the prime ϖ. The proof depends on two
results due to Noether (*Berlin. Abh.* 1882; see Vol. V, Chap. VIII):
(*a*), that the genus, *p*, of a curve, lying on a surface of order ρ, in
ordinary space [3], when the order of the curve is $\rho m - \nu$ with $\nu < \rho$,
is such that $p \leqslant \frac{1}{2}(\nu-1)(\nu-2)+\frac{1}{2}(\rho m-2\nu)(\rho+m-4)$ (*loc. cit.*
p. 25); (*b*), that if the order and genus of the curve are respectively
ρm and $1+\frac{1}{2}\rho m(\rho+m-4)$, then the curve is the complete intersection of the surface on which it lies with a surface of order m
(*loc. cit.* p. 42). Let the surface ψ, given on V, have the characters
μ_0, μ_1; it can have no accidental double points, since these would be

nodes of V, and V cannot have nodes at ordinary points of ψ. Whence, by the formula $\delta + 4d = \rho\,(\rho-2)\,\mu_0 + \mu_0{}^2 - \rho\mu_1$, of the preceding chapter, where ρ is the order of V, we have

$$\rho\mu_1 = \mu_0[\mu_0 + \rho(\rho-2)].$$

From μ_0 and ρ we can obtain uniquely two integers m, ν by putting $\mu_0 = m\rho - \nu$, with $0 \leqslant \nu < \rho$ and $m \geqslant 1$; whence, considering the curve (ϖ, ψ) as lying on the surface (ϖ, V), which is of order ρ, we have, for its genus p, the inequality above. As the curve is (ϖ, ψ) we also have $\mu_1 = 2\mu_0 + 2p - 2$, which is $\mu_1 = 2(m\rho - \nu) + 2p - 2$. Utilising $\mu_0 = m\rho - \nu$, $\mu_1 = \mu_0[\mu_0 + \rho(\rho-2)]/\rho$, and the inequality for p, we find on reduction the simple result $\nu(\rho-1)(\rho-\nu) \leqslant 0$. But $\rho - \nu > 0$, and, if $\nu > 0$, this gives also $\rho - 1 > 0$. Thus we infer $\nu = 0$, and $\mu_0 = m\rho$; thence $2p - 2$, equal to $\mu_1 - 2\mu_0$, is found equal to $m\rho(m + \rho - 4)$. Wherefore, by (b) above, the curve (ϖ, ψ) is the complete intersection of the surface (ϖ, V) with another surface, in the prime ϖ, of order m.

Now, the surface ψ, regarded as having no prescribed base loci, determines on V a complete system of surfaces, which we denote by $|\psi|$; just as a curve on a surface determines a complete series of curves thereon, or a set of points on a curve determines a complete series. This we assume; we also assume that primals of order m determine on V a complete system of surfaces (each the complete intersection with such a surface); this latter is in consequence of the fact that V is a primal and without multiple points; the same is true of a manifold of dimension k, in space $[r]$, which is the complete intersection of $r - k$ primals and without multiple points (Severi, *Rend. Palermo*, XVII, 1903, and XXVIII, 1909, already referred to in the last chapter). This assumption is equivalent to the statement that if, through the intersection of V with a primal of order m, there be drawn a primal of order m', $> m$, having a residual intersection, of order $(m'-m)\rho$, with V, then any other primal of order m', through this residual intersection, meets V again in a surface which is the complete intersection of V with a primal of order m. From these two assumptions it follows that, to prove ψ to be the complete intersection of V with a primal of order m, it is sufficient to shew that the linear system $|\psi|$ is identical on V with that determined by primals of order m. This last system we denote by $|m\varpi|$. The identity of these linear systems of surfaces on V follows then if we prove the identity of two systems $|H+\psi|$, $|H+m\varpi|$, where $|H|$ is any other system of surfaces on V. In virtue of another theorem (p. 293, above; Severi, *Atti...Veneto*, LXV, 1906, LXX, 1911) these systems of surfaces will be identical if they cut equivalent (not necessarily complete) systems of curves on any prime section, ϖ, of V; this assumes, what is evidently true, that there is no surface on V, common to all its prime sections, which is part of all the surfaces of either of the two systems. For the system $|H|$ we employ the system of surfaces cut on V by primals of order $\rho - 5$ (thus assuming $\rho \geqslant 5$); thus we are to prove that the systems of surfaces on V expressed by $|(\rho-5)\varpi+\psi|$ and $|(\rho+m-5)\varpi|$ cut equivalent systems of curves on the surface of intersection of V with an arbitrary prime. We shall not consider the removal of the imitation $\rho \geqslant 5$ which is convenient for the statement of the proof.

We have shewn that the curve (ϖ, ψ) is the complete intersection of the surface (ϖ, V), of order ρ, with a surface of order m, in the prime ϖ. Thus (Vol. v, Chap. viii), the canonical series on this curve is determined by its intersections with surfaces of order $\rho + m - 4$ lying in the prime ϖ. From this we see, recalling the definition of the canonical system of curves on a surface (pp. 217, 222, above), that primals of order $\rho + m - 5$ cut canonical curves on *the surface* ψ, though not necessarily, in default of proof, the complete canonical system of *curves*. On the other hand, the complete system of canonical *surfaces*, on the primal V, is known (after Noether, *Math. Annal.* ii, 1870, § 3, reasoning with a multiple integral) to be cut thereon by primals of order $\rho - 5$; and we assume from this (arguing in a manner analogous to that employed for surfaces) that canonical *curves*, on the surface ψ, though not necessarily the complete system, are cut thereon by curves on ψ, given by $|\psi+(\rho-5)\varpi|$. Whence, on the *surface* ψ, which is without multiple points, the two complete systems of curves expressed by the last, and by $|(\rho+m-5)\varpi|$, are the same. Hence the two series, of sets of points, on the *curve* (ϖ', ψ), section of ψ by any prime ϖ', cut by the surfaces

$$|\psi+(\rho-5)\varpi|, \quad |(\rho+m-5)\varpi|$$

are equivalent. This curve (ϖ', ψ) however, we have seen, is the complete intersection of the surface (ϖ', V) with a surface of order m of the prime ϖ', and is thus, regarded as belonging to the surface (ϖ', V), one of the complete system cut thereon by surfaces of order m. The fact that the surfaces $|\psi+(\rho-5)\varpi|$, $|(\rho+m-5)\varpi|$ cut equivalent series on this curve, whatever ϖ' may be, suffices (by the theorem of equivalence to which reference has been made) to shew that these surfaces cut equivalent systems of curves on the surface (ϖ', V). And, by what has been said, this is sufficient for our purpose.

IV. Miscellaneous theorems and examples.

Ex. 1. (Chap. i, pp. 13, 14.) In regard to the double tangents of a plane curve, it was remarked by Dersch, *Math. Ann.* vii, 1873, p. 497, that if the tangent at a point (x) of a plane curve $a_x{}^n = 0$ meet the curve again in $(n-2)$ points (y), then there exists the relation

$$(abc)^2 \sum_{\lambda+\mu+\nu=n-2} (a_y b_x c_x)^\lambda (b_y c_x a_x)^\mu (c_y a_x b_x)^\nu = 0;$$

this generalises Salmon's theorem for a plane cubic curve, Vol. v, p. 22. The equation is of order $n-2$ in (y), and of order $2n-4$ in (x). Thus (x) will be one of the points of contact of a bitangent if this curve, in (y), be touched by the tangent line $a_x{}^{n-1}a_y=0$.

The condition that a curve of order m be touched by a line $u_y=0$ is of degree $2(m-1)$ in the coefficients of the curve, and of order $m(m-1)$ in the line coefficients (u) (Vol. v, p. 86). Hence, in the case in question, the condition resulting is, in (x), of order

$$2(n-3) \cdot 2(n-2) + (n-2)(n-3)(n-1), \text{ or } (n-2)(n^2-9);$$

the curve has therefore $\frac{1}{2}n(n-2)(n^2-9)$ bitangents, whose points of contact (x) lie on the curve resulting from the condition of contact. Cf. Cayley, *Papers*, iv, pp. 186, 347; Salmon, *Higher Plane Curves*, 1879, pp. 342–57. Klein, *Math. Ann.* xxxvi, 1889, p. 20, remarks that the left side of Dersch's relation is, in virtue of $a_x{}^n=0$, the same as

$$6a_x{}^{n-1}a_y \cdot N \cdot (hxy)^{-2},$$

where

$$N = \sum_{\kappa=1}^{n} (a_y b_x)^{\kappa-1} (b_y a_x)^{n-\kappa} a_h b_h - \sum_{\kappa=1}^{n-1} a_y{}^{\kappa-1} a_x{}^{n-\kappa-1} a_h{}^2 b_y{}^{n-\kappa} b_x{}^{\kappa},$$

and that N is characterised by being covariant, of degree 2 in the coefficients of $a_x{}^n$, and of order $n-1$ in both x and y, and such that $N \cdot (hxy)^{-2}$ is independent of (h). See Pick, *Math. Ann.* xxix, 1887, p. 261.

Ex. 2. (See p. 85, above.) The Grassmann manifold whose points have for coordinates the coordinates of a general linear space $[k]$ containing $k+1$ points of a rational normal curve of order n, in space $[n]$, may be regarded as representing the primals of order $n-k$ of space $[k+1]$ (Brown, *Journ. Lond. Math. Soc.* v, 1930, p. 168). On a curve of genus p, of general moduli, the number of special series $g_r{}^n$, when $p=(r+1)(r'+1)$, in which $r=n-p+r'+1$, is finite (Vol. v, p. 91); this number is equal to the order of the Grassmann manifold which represents all the linear spaces $[r]$ existing in $[r+r'+1]$, namely, equal to the number of spaces $[r]$, in $[r+r'+1]$, which meet $(r+1)(r'+1)$ spaces $[r']$ (Castelnuovo, *Rend. Lincei*, v, 1889, p. 130). See p. 85 above.

Ex. 3. (See p. 121, above.) According to Wiman (*Math. Ann.* xlviii, 1897, p. 195) every finite group of birational transformations of a plane is one of nine types: (1), a group of collineations; (2), groups with one, or two, flat pencils invariant; (3), a group in which a system of cubic curves with 3, ..., 7 fixed points, is invariant; (4), a group in which a sextic curve with 8 fixed double points, and the adjoint pencil of cubic curves, is invariant.

The collineation groups contain: (a), groups which have three points, or a point and a line invariant; (b), the Hessian group of order 216, in which the nine inflexions of a plane cubic curve are invariant; (c), the group of order 168, changing the curve $y^3z+z^3x+x^3y=0$ into itself; (d), the Valentiner group of order 360 leaving the sextic curve

$$10x^3y^3 + 9(x^5+y^5)z - 45x^2y^2z^2 - 135xyz^4 + 27z^6 = 0$$

invariant. This last is of the same structure as the even permutation group of six elements.

Ex. 4. (See p. 272, above.) In connexion with Castelnuovo's paper, prove, if u_i denote a homogeneous polynomial of order i, in x and y, with variable coefficients, and ω_k be such a polynomial of order k with fixed coefficients, that the equation

$$\omega_k{}^2 u_{p-k}z^2 + \omega_k u_{p+1}z + u_{p+k+2} = 0, \qquad (k=0, 1, ..., p),$$

represents, for any k, a system of hyperelliptic curves of genus p, of grade $4p+4$, and freedom $3p+5$. The multiplicity at $x=0, y=0$ is described in Noether's phraseology as a $(p+k)$-ple point, with k double points infinitely near, in its first neighbourhood, in the directions given by $\omega_k = 0$. A simplifying birational transformation is given by

$$\xi/xy^k = \eta/y^{k+1} = \zeta/z\omega.$$

Also, another system of hyperelliptic curves, of the same genus, grade and freedom, is given by

$$u_{p+1}z^2 + v_{p+1}zy + w_{p+1}y^2 = 0,$$

where v_{p+1}, w_{p+1} are homogeneous in x, y, of order $p+1$, with variable coefficients. (See also Castelnuovo, *Atti...Torino*, xxv, 1890, and *Ann. d. Mat.* xviii, 1890, p. 119; Scorza, *Ann. d. Mat.* xvi, 1909; and dal Re, *Rend. Napoli*, xxx, 1924, p. 80.)

Ex. 5. (See pp. 140, 222.) Prove the following theorems:

(*a*), A surface containing a pencil of irreducible rational curves can be transformed to a ruled surface (Noether, *Math. Ann.* iii, 1870; Enriques, *Math. Ann.* lii, 1899. Cf. Picard-Simart, *Fonct. algéb.* i, p. 194).

(*b*), A surface containing a linear pencil of irreducible elliptic curves, with a simple or double base point, is rational, or transformable to an elliptic ruled surface. A surface containing a linear pencil of irreducible curves of genus 2, with one or more base points of multiplicities $s_1, s_2, ...$, where $\Sigma s > 2$, is rational, or transformable to a ruled surface, of genus 1 or 2 (Castelnuovo-Enriques, *Rend. Palermo*, xiv, 1900; Castelnuovo, *Atti...Torino*, 1901).

(*c*), A surface containing a linear pencil of irreducible curves of genus p, >2, of freedom $r \geqslant 3p-5$, is rational, or reducible to a ruled surface (Castelnuovo-Enriques, *Ann. d. Mat.* vi, 1901, p. 38).

Ex. 6. (See p. 282.) Elaborate the theory of the surface representing the pairs of points on a curve of genus 3. (Cf. Severi's paper, referred to, No. 7; Humbert, *Compt. rend.* 1895, pp. 365, 425.)

INDEX

The references are to the pages of the volume

Accidental double point of a surface in four-fold space, 157, 170

Adjoint system of a linear system of curves, 219, 222, 246; relation connecting grade and genus with genus of original system, 225

Albanese, elimination of multiple points of a surface, 158

Alexander, on Castelnuovo's reduction of Cremona transformations, 117; identification of the Zeuthen-Segre and the Poincaré topological invariants, 214

Babbage, transformation of surfaces considered by Noether to nonsingular surfaces, 232; deduction of values for invariants when a surface has nodes, 235; normal surfaces in five-fold space with a double line, 246, 281

Benoist, *see* Clebsch

Bertini, representation of the pairs of a de Jonquières involution, 119; theorem for certain hyperelliptic plane curves, 120; reduction of plane involutions of two points to four types, 121; Bertini involution in a plane, 124; on reduction of plane involutions and rationality of double planes, 131; theorem for the variable multiple points of the curves of a linear system, 222; Clifford's theorem for least order of a manifold, 266; on the postulation of a complete intersection, 268; primals whose intersection consists of spaces, 276; *Geometria proiettiva degli iperspazi*, 1907 (second edition, 1923), referred to, 120, 268, 276

Berzolari, in Pascal's *Repertorium der höheren Mathematik*, referred to, 31, 34, 38, 188; on contacts of curves of two plane pencils, 240

Bitangents of a plane curve, 13, 298

Bordiga's surface, 273, 274

Brill, on quadrisecants of curves, 34; on multiple contacts, 42; Brill-Noether, Report to German Mathematical Society on algebraic functions, 1894, referred to, 117, 130

Brown, the Grassmann representation of the secant spaces of a rational curve, 299

Burnside, a particular group of Cremona transformations, 121

Calculus of conditions, 69

Campedelli, *see* Enriques-Campedelli

Canonical series on a curve, transformation in a correspondence of two curves, 18; canonical number of curve, 140, 224; canonical system of curves on a surface, 215, 217, 262; canonical curves as defined by Noether, 220, 227; canonical system of a general double plane, 237; undesigned base points of canonical system, 241

Caporali, a theorem for plane curves, 120; class of an involution, 140; number of double points of a plane pencil, 240; on intersections in fourfold space, 247

Castelnuovo, condition for an involution on a curve to belong to a linear series, 37; enumerative formulae for curve, 45; reduction of Cremona transformation to quadratic transformations, 117; surface whose prime sections are hyperelliptic, 120, 272; rationality of plane involutions, 139; theorem used by Picard, 169; on fundamental and exceptional curves, 195, 246; and Enriques, the Zeuthen-Segre invariant deduced from an irrational pencil, 191; surface with multiple points at eight associated points, 203, 227; surface having an irrational pencil of curves is irregular, 205; example in regard to transformation of exceptional curve to a simple point, 220; and Enriques, undesigned base points of the canonical system, 241; theorem for deficiency of characteristic series, 263; number of special series on a curve as the order of a Grassmannian, 299; hyperelliptic plane curves representing sections of a rational surface, 299; and Enriques, section conditions for a surface to be rational or ruled, 140, 300

Cayley, on correspondence, 12; on quadrisecants of a curve, 34; on ruled surfaces, 38; on multiple contacts of curves, 42; reduction of plane Cremona transformations to quadratic transformations, 117; pinch points of a double curve of a surface, 158; use of name *cotangent*, 177; example of pinch point, 179; direct calculation of numerical genus of a surface, 200; quartic surface with nodes at eight associated points, 203; on a particular quintic surface, 205; the numerical genus of a ruled surface, 227; birational relation of Kummer and Weddle surfaces, 237; curves on a general surface are complete intersections, 250; on double tangents of a plane curve, 299

Characteristic series of a linear system of curves, 124; deficiency of, 263

Chisini, resolution of multiple points of a surface, 158; *see also* Enriques-Chisini

Chord, common to two curves, 29; chord-curve and trisecant-curve of a surface in four-fold space, 242

Class of an involution (Caporali), 140; of immersion and canonical number, 222, 225, 251

Clebsch, on multiple tangents of a surface, 90; in lectures, on Cremona transformation, 117; everywhere finite double integral of a surface, 183, 184; plane representation of double curve of a rational surface, 246; Hessian of a pencil of binary quartics, 277; conditions for roots of binary sextic to be three pairs mutually harmonic, 278

Clifford, reduction of plane Cremona transformation to quadratic transformations, 117; least order of a

manifold belonging to a given space, 266

Coble, Cremona transformations, 117

Coincidences in a correspondence with valency, 8; of $r+1$ points in a linear series of freedom r, 10; in a general correspondence, 54, 142

Completed grade and genus of a curve of a linear system, 139, 140, 221; complete system of sections of a non-singular primal, 295

Composite curve, genus and grade in terms of components, 223

Correspondence, of points on a curve, elementary methods, 1; transcendental methods, 46; connexion with theory of degenerate integrals, 59; correspondence of points of two manifolds, 92; of linear spaces, 105; enumerative formulae for, 108; multiple, between two curves, analytic treatment of, 20; of two surfaces, relations of invariants of, 294

Cotangent of a surface at a pinch point (Cayley), 177

Coxeter, on lines of a cubic surface, 15

Cremona, transformations in a plane, 112; on a certain curve on a cubic surface, 130; the Zeuthen-Segre invariant in plane transformations, 240; the Zeuthen-Segre invariant for pencils of surfaces, 241; transformations, groups of, 121, 278, 299

Curve, fundamental, 194, 246

Cyclical sets in a correspondence, 143

Darboux, birational relation of Kummer and Weddle surfaces, 237

Dedekind, on an algebraical reversion formula, 144

Defective integrals, in the correspondence of two curves, 22, 59; regular system, 62

Deficiency, of the characteristic series on a surface, 263; of series on a curve by curve of adjoint system (Picard's theorem), 269

Dersch, on bitangent curve, 13, 298

Double, curve of a surface, general property for, 169; points of a surface, in four-fold space, accidental, 171, 259; double plane, canonical system of general, 237

Edge, on a certain curve on a cubic surface, 130; treatise on ruled surfaces, 166
Eliminant, rule for degree of, in parameters which enter, 186
Enriques, and Chisini, *Teoria geometrica delle equazioni*, referred to, 34, 117, 232; on cubic primal in four dimensions, 137; on rationality of surface representing an involution, 139; on surface with an irrational pencil of rational curves, 147; on ruled surfaces, 168; use of Jacobian of net of curves to obtain canonical system of surface, 183; and Campedelli, *Superficie algebriche*, 1932, referred to, 184, 232, 234, 236, 237, 241; researches on theory of surfaces of 1893, referred to, 184; exceptional curves in a plane, 113; exceptional curves on a surface, 184, 195, 246; and Castelnuovo, the Zeuthen-Segre invariant from an irrational pencil, 191; surface whose canonical system consists of elliptic curves, 205; phraseology for multiple base points of a linear system of curves, 215; on transformation of exceptional curve to simple point, 220; relation of exceptional curve to transformed adjoint system, 236; on surfaces whose prime sections are hyperelliptic, 272; and Fano, groups of Cremona transformations in threefold space, 278; and Castelnuovo, section conditions for a surface to be rational or ruled, 300
Enumerative formulae for manifolds defined by matrices, 109
Equivalence of two manifolds, Severi's theorem for, 293, 297
Exceptional curves, in plane transformations, 113; on a surface, 184, 195, 246; arising in birational transformation, 201, 202, 203, 210; as considered by Noether, 220, 227

Fano, on the rationality of a three-fold manifold, 137; and Enriques, groups of Cremona transformations in ordinary space, 278; theorem for

surfaces lying on a cubic primal in four-fold space, a particular case, 278; lines lying on cubic primal in four-fold space, 294; manifold which is complete intersection contains only complete intersections, 296
Fouret, common normals of two primals, 104; tangent planes to a surface from a multiple line, 188
Franchis, De, on surface representing points of two curves, 227, 282
Fundamental curve of a linear system of curves, 194, 210, 246

Geiser involution in a plane, 122
Genus of a curve, completed, 139, 221
Giambelli, number of spaces meeting a curve in higher space, 44; manifolds defined by matrices, 111
Godeaux, tangent planes to a surface from a multiple line, 188
Grade of a curve of a linear system, completed, 139, 221
Grassmannian, of linear spaces, order of, 85; giving number of special series on a curve, 299
Groups of Cremona transformations in a plane, 299; in space, 278
Guccia, number of double points of plane pencil, 240; Zeuthen-Segre invariant for pencil of surfaces, 241

Halphen, common lines of two congruences, 28; results anticipatory of Schubert's ideas, 69
Hessian of a ruled surface, Salmon on, 27
Hilton, on a certain curve on a cubic surface, 130
Hodge, canonical system of a double plane, 237
Hudson, H. P., treatise on Cremona transformations, referred to, 117, 124, 128, 200
Hudson, R. W. H. T., treatise on Kummer's surface, referred to, 237
Humbert, on a particular curve, 24; on hyperelliptic surfaces, 227
Hurwitz, transcendental methods for theory of correspondence, 46, 54, 56
Hyperelliptic curves representing sections of a rational surface, 299

Immersion, class of, and canonical number, 222, 225 (and 251)

Improper double points of a surface in four-fold space, 157, 170

Inflexional lines of a surface, 148

Intersections, residual, of three primals through a curve, 248; of two surfaces, 251; of a primal and a surface, 255; of two primals, 257; of three primals through a surface, 266

Invariants, the Zeuthen-Segre, 185, 186, 189; the Segre-Severi, 191, 214, 221; modified for isolated nodes, 199, 200; proof of invariance, 206, 210, 221; of a surface which is determined by a matrix, 280; of two surfaces in multiple correspondence, relations of, 294; of an involution, 135

Involution, not rational, on a curve, 17; number of sets belonging to a linear series, 35; in space of three dimensions, leading to the Geiser involution, 124; rationality of, 132; surface representing, 133; in ordinary space, presumably not rational, 137

Irrational pencil of curves, not existing on a surface representing an involution, 138; used for defining Zeuthen-Segre invariant, 190; existence involves irregularity of surface, 205

Irregularity of surface, 227; of surface with nodes at eight associated points, 204; of surface representing points of two curves, 285

Jacobian, of a net of curves on a surface, 152, 192

Join of two spaces, 70

Jonquières, de, on multiple contacts of curves, 39, 42; manifolds defined by matrices, 111; transformation, 115; involution, 118

Jung, the Zeuthen-Segre invariant in plane transformations, 240

Kantor, on periodic Cremona transformations, 131, 145

Klein, manifold on quadric in higher space, 250, 295; on double tangents of a plane curve, 299

Kohn, manifolds defined by matrices, 111

Kummer surface, transformation to a Weddle surface, 237

Lasker, on the theory of Moduls and Ideals, 269

Lateral, direct, correspondence, 11

Lefschetz, on correspondence between two curves, 59

Levi, Beppo, resolution of multiple points of a surface, 158

Lindemann, *see* Clebsch

Line, formula for product of two line conditions, 81

Lüroth, rationality of involutions and surfaces, 131, 132

Maroni, surface representing points of two curves, 282

Matrices, manifolds defined by, enumerative formulae for, 109; invariants of, 280

Maxwell, surface with nine exceptional lines, 232; a general formula for invariants, 235

Meet of two spaces, 70

Milinowsky, on the Geiser involution, 122

Moebius, an algebraic reversion formula, 144

Moore, and Slaught, a particular group of Cremona transformations, 121

Nanson, manifolds defined by matrices, 111

Net of curves on a surface, 152

Netto, on a theorem for symmetrical functions, 136

Noether, reduction of Cremona transformations to quadratic transformations, 117; on a certain curve of genus three, 130; on the double planes which are rational, 130; on cubic primal in four-fold space, 137; on surfaces with a rational pencil of rational curves, 146; determination of canonical series on a curve in space, 161; on multiple curves of a surface, 163; formulae for the class, stationary points, genus and curve-genus of a surface, 180; procedure by enveloping cone for invariants of a surface, 183; number of conditions for a surface imposed by a given multiple line, 200; surface whose canonical curves are aggregates of elliptic curves, 205; on a particular sextic surface, 205; on conditions for an adjoint surface at a tacnode, 220; fifteen examples of the invariants of surfaces, 226; on condition of adjointness at an isolated multiple point, 233; incidental proof of Clifford's theorem for special series on a curve, 270; curves on a general surface are complete intersections, 295; particular theorems for a curve in space, 296; canonical surfaces on a primal in four-fold space, 298

Painlevé, surface with a finite group of birational transformations, 147
Palatini, manifolds determined by matrices, 111
Pannelli, Zeuthen-Segre invariant for pencil of surfaces, 241
Pascal, *Repertorium of higher mathematics, geometry*, 1922, referred to, 199, etc.
Pencil, irrational, of curves on a surface, 138, 189
Pezzo, Del, surfaces whose order is one less than the dimension of the space to which they belong, 120; Del Pezzo surfaces, characters, 275
Picard, on deficiency of series by adjoint system, 169, 269; influence of tacnode on adjunction, 220, 234; on hyperelliptic surfaces, 227
Pick, on a covariant arising for bitangents of a plane curve, 299
Pieri, formula for multiple correspondence, 99; applications, 102, 104; coincidences of corresponding lines, 106; manifolds defined by matrices, 111; Zeuthen-Segre invariant for pencils of surfaces, 241
Pinch points of a double curve of a surface, 158, 176
Poincaré, theorem of complementary systems of degenerate integrals, 68; topological invariant generalising Euler's, 214
Poisson, on a theorem for symmetrical functions, 136
Postulation, of a surface for primals, 263; of a manifold which is complete intersection, 268
Primals, intersection of three, with a common curve, 248; point equivalence of common curve of four, 249; and surface, residual intersection of, 255; intersection of two with a common curve, 257; intersection of three with a common surface, 266
Product, and sum, of two correspondences, 6; product of two curves, 57

Quadratic transformation, reduction of Cremona transformations to, 117
Quadrisecants of a curve, number of, 33

Rank and grade for a complete intersection, 75
Rationality, of an involution, 132, 137; of surface representing a plane involution, 139; rational surfaces, their plane representation, 245
Re, dal, theorem for surfaces, 300
Reduced canonical system on a surface, 227
Richmond, on cubic primal in four-fold space, 137
Roberts, manifolds defined by matrices, 111
Rohn, on a plane sextic curve with six nodes, 156
Room, on generalisation of the Bordiga figure, 274
Rosanes, on reduction of Cremona transformations to quadratic transformations, 117
Rosati, correspondence and degenerate integrals, 68
Roth, multiple tangents, 91; a numerical property of the characters of a surface, 164; inflexional lines for a surface in four-fold space, 175; a surface with a double line, in five-fold space, 246; on composite surfaces, 282
Ruled surface, by joins of corresponding points of two curves, 17; genus of a curve thereon, 25; number of torsal generators, 26; condition for a surface to be transformable to, 140; characters of 164, 167

Salmon, bitangent curve, 13, 299; Hessian of a ruled surface, 27, 156; quadrisecants of a curve, 34; multiple tangents of a surface, 90, 91; multiple correspondence, 92;

manifolds defined by matrices, 111; general formulae for a surface, 159, 169; generalisation to four-fold space, 174; use of four-fold space to solve a problem of ordinary space, 251; treatise on Higher Algebra, 277

Schottky, on a certain curve, 130

Schubert, torsal generators of a ruled surface, 26, 102, 166; refers to Halphen as precursor, 28; enumerative formulae for contacts of spaces with curves, 46; reference to his treatise, 28, 69, 96, 102, 103, 104; composite condition for a linear space, 71; on multiple tangents of a primal, 90, 91; notation for calculus of conditions, 98; on umbilici, 102; coincidences of corresponding spaces, 106; manifolds defined by matrices, 111; on quadric surfaces satisfying nine conditions, 251

Schur, on sextic curve, 274

Scorza, correspondence and degenerate integrals, 68; on three-folds with curve sections of given genus, 272; on a manifold of order five and dimension three in six-fold space, 278

Segre, genus of a curve on a ruled surface, 25; sets common to an involution and a linear series, 37; Report on higher space in *Encyklopädie der Mathematischen Wissenschaften*, III, C 7, 1912, referred to (under *Enzyk. Math. Wiss.*), 44, 110, 111, 174; enumerative formulae for manifolds defined by matrices, 110, 111; reduction of Cremona transformations to quadratic transformations, 117; on resolution of multiple points of a surface, 158; inflexional lines of a ruled surface in four dimensions, 174; use of pencil of curves on a surface to obtain invariant, 183; the Zeuthen-Segre invariant, 185; from irrational pencil, 189; Segre-Severi invariant, 191; as grade of canonical system, 197; class of a surface with isolated nodes, 199; influence of nodes on invariants, 199, 200; proof of the invariance, 206, 210; generalisation of

the Zeuthen-Segre invariant, 214; Segre-Severi invariant defined by grade of canonical system, 224; modification in birational transformation, 226; Segre quartic surface with double line by projection of Veronese surface, 274; surface formed by lines meeting five planes in four-fold space, 276; elliptic ruled surface normal in space of dimension one less than its order, 277; duality of an elliptic ruled surface and a manifold of order five, 279

Semple, on a birational transformation in four and five dimensions, 250; cubic primals through a projected Veronese surface are rational cubic primals, 277; on a birational transformation in four-fold space, 279; papers on transformation, and composite surfaces, 282

Series, characteristic for a linear system of curves, 124, 263; on double curve of a surface by surfaces through the triple points, 169

Severi, *Trattato di geometria algebrica*, 1926, referred to, 37; enumerative formulae for contacts of spaces with curves, 45; correspondence between two curves, by considering their product, 59; formula for coincidences of corresponding spaces, 105; number of cyclical sets in a correspondence, 145; on improper double points of a surface in four-fold space, 157, 171; elimination of multiple points of a surface, 158; on residual intersection of a surface with a surface through its double curve, 168; use of net of curves on a surface to obtain invariant, 183; on virtual curves, 217; class of immersion and canonical number, 222; on surface representing pairs of points of a curve, 227, 282; modification of Zeuthen-Segre invariant in case of isolated nodes, 235; relation of exceptional curves to transformed canonical system, 236; enumerative formulae for a net of curves on a surface, 241; memoir on intersection of manifolds, 247; application of theory of intersections, 250; canonical curves on part intersection of two primals, 262;

postulation of surface residual intersection of two primals, 265; postulation of composite surface, 265; point equivalence of a surface, 267; theorems for a complete intersection, 269, 270; extension of a theorem of Bertini, 269; on Picard's theorem for deficiency of series cut by adjoint system, 269; on modular expression of a manifold, 269; fundamental theorem of equivalence of two manifolds, 293, 297; relations of invariants of two surfaces in multiple correspondence, 294; theorem that non-singular primal contains only complete intersections of appropriate dimension, 296; for Grassmannian, 296; theorem of completeness of system of surfaces on a primal, 297

Simart, *see* Picard

Slaught, *see* Moore and Slaught

Spaces, linear, numbers satisfying given conditions, fundamental theorems, 70

Stationary contact, of a tangent plane with a surface, 149

Steiner, locus of points of contact of curves of two plane pencils, 240

Sturm, normals of a surface, 104

Sum, and product, of two correspondences, 6; sum, and difference, of linear systems of curves, 197

Surface, representing the sets of an involution, 133; containing a pencil of rational curves, 145; in ordinary space, preliminary properties of, 148; Salmon's formulae for, 159, 164; with only double curve and triple points, 157; in space of four dimensions, 169; in four dimensions which is complete intersection, characters of, 175, 247; residual intersection of two with a common curve, 251; a characteristic equation proved from theory of intersections, 254; and primal, residual intersection of, 255; in four-fold space, number of accidental double points, 259; two, forming complete intersection of two primals, sum of invariants, 262; whose prime sections are hyperelliptic is rational 272; representing points of two curves, 282; representing pairs of

points of one curve, 285, 289; representing pairs of points of a curve of genus two, 289; representing lines of a cubic primal in four-fold space, 294

Tacnode, effect on invariants of surface, 220, 234, 270

Tangent, multiple, of a manifold, 86; tangent planes to a surface from a multiple line, 187

Timms, on the general Del Pezzo surface, 276

Torelli, on de Jonquières formula for multiple contacts of curves, 42

Torsal chords of a curve, 17; torsal generators of a ruled surface, 26

Transformations, Cremona, in a plane, 112; groups of, 278, 299

Trisecants, ruled surface by, 32; to a surface in four-fold space, 172; trisecant curve of a surface in four-fold space, 242

Ursell, on quadric surfaces satisfying nine conditions, 251

Vahlen, manifolds defined by matrices, 111

Val, Du, on bitangent curve for a plane quartic curve, 14

Valency, of a correspondence, 3, 54

Veronese, relations of two curves forming a complete intersection, 251; on a rational surface, 274

Voss, umbilici of a general surface, 102

Weddle surface, transformation to a Kummer surface, 237

Welchman, on contact primes of the canonical curve, 43

White, example of a (2, 1) correspondence between two curves, 24; on the Bordiga surface, 274

Wiman, groups of Cremona transformations in a plane, 121, 299

Zeuthen, formula for correspondence between two curves, 19; on torsal generators of a ruled surface, 27; on ruled surface by trisecants of a curve, 32; on quadrisecants, 34; *Lehrbuch der abzählenden Methoden der Geometrie*, 1914, referred to, 19, 27, 34, 42, 69, 91, 181, 183, 235;

number of triple points of the nodal curve of a given curve, 38; on de Jonquières' formula for multiple contacts of curves, 42; on multiple tangents, 91; on multiple correspondence, fundamental results, 92; on pinch points and close points, memoir, 178; procedure by enveloping cone for invariants of a surface, 183; Zeuthen-Segre invariant, 185, 189, 199, 206, 210, 233, 247; for pencil of surfaces, 214, 241; enumerative formulae for a net of curves on a surface, 241

CAMBRIDGE: PRINTED BY WALTER LEWIS, M.A., AT THE UNIVERSITY PRESS

Printed in the United States
By Bookmasters